Springer Proceedings in Mathematics & Statistics

Volume 232

Springer Proceedings in Mathematics & Statistics

This book series features volumes composed of selected contributions from workshops and conferences in all areas of current research in mathematics and statistics, including operation research and optimization. In addition to an overall evaluation of the interest, scientific quality, and timeliness of each proposal at the hands of the publisher, individual contributions are all refereed to the high quality standards of leading journals in the field. Thus, this series provides the research community with well-edited, authoritative reports on developments in the most exciting areas of mathematical and statistical research today.

More information about this series at http://www.springer.com/series/10533

Luis L. Bonilla · Efthimios Kaxiras
Roderick Melnik
Editors

Coupled Mathematical Models for Physical and Biological Nanoscale Systems and Their Applications

Banff International Research Station, Banff, Canada, 28 August–2 September 2016

Springer

Editors
Luis L. Bonilla
Department of Materials Science and Engineering and G. Millan Institute
Universidad Carlos III de Madrid
Leganés
Spain

Roderick Melnik
The MS2 Discovery Interdisciplinary Research Institute
Wilfrid Laurier University
Waterloo, ON
Canada

Efthimios Kaxiras
Lyman Laboratory of Physics, Department of Physics
Harvard University
Cambridge, MA
USA

ISSN 2194-1009 ISSN 2194-1017 (electronic)
Springer Proceedings in Mathematics & Statistics
ISBN 978-3-030-09530-7 ISBN 978-3-319-76599-0 (eBook)
https://doi.org/10.1007/978-3-319-76599-0

Mathematics Subject Classification (2010): 78A70, 82C26, 82C31, 82D37, 82D80, 92C05, 92C17, 92B05, 97M10, 97M50, 97M60

Printed on acid-free paper

This Springer imprint is published by the registered company Springer International Publishing AG part of Springer Nature
The registered company address is: Gewerbestrasse 11, 6330 Cham, Switzerland

Preface

Due to the vastness, novelty, and complexity of the interface between mathematical modeling and nanoscience and nanotechnology, many important areas in these disciplines remain barely explored. In progressing further, multidisciplinary research communities have come to a clear understanding that, along with experimental techniques, mathematical modeling and analysis have become crucial in the study, development, and applications of systems at the nanoscale. This volume puts together selected contributions from the participants in the Banff International Research Station (BIRS) workshop *Coupled Mathematical Models for Physical and Biological Nanoscale Systems and Their Applications*. The contributors are experts working on different aspects of the analysis, modeling, and applications of nanoscale systems, with particular focus on low-dimensional nanostructures and coupled mathematical models for their description. The development of such models requires concerted efforts from mathematicians, physicists (both theoreticians and experimentalists), and computational scientists, including those working on biological nanostructures. The contributions may serve as up-to-date introductions to different topics in nano- and biosystems, identification of important challenges, evaluation of current methodologies, and exploration of promising approaches.

Topics treated in the book belong to three categories:

Part I: Charge and Spin Transport in Low-Dimensional Structures.
Part II: Modeling Biological Phenomena from Nano- to Macro-scales.
Part III: Mathematics for 2D Materials and Properties of Confined Nanostructures.

In Part I and Part II, Birnir describes different models of nonlinear transport in semiconductor quantum wells and superlattice structures. This chapter reviews the obtention of envelope function and Hartree approximation for quantum wells from the Schrödinger equation, and a dynamical system and bifurcation study for the corresponding time-dependent Liouville–von Neumann matrix density under a laser field. Possible applications are listed. On a different vein, Birnir describes time-dependent electron transport using the sequential tunneling model of a weakly coupled superlattice, i.e., a certain number of identical quantum wells separated by thick barriers. A weakly coupled semiconductor superlattice is a nonlinear dynamical system whose nonlinearity is due to sequential resonant tunneling between adjacent quantum wells and to the Poisson equation between electron density and potential at the wells. For different *dc* voltage and parameter values, there are stable stationary states, time-periodic solutions, and period doubling routes to chaos. Birnir also discusses the effects of noise and imperfections on the attractors with applications to true random number generators and other devices.

H. T. Grahn et al. review experiments on weakly coupled semiconductor superlattices including recent observations of spontaneous chaotic and quasi-periodic self-oscillations in doped GaAs/AlGaAs superlattices at room temperature. Based on them, an all-electronic true random number generator has been demonstrated at room temperature with achievable bit rates of up to 80 Gbit/s. This is about two orders of magnitude larger than typical bit rates for currently available all-electronic true random number generators. The paper also reports synchronization of chaos, which is useful for secure communications and possible applications to advanced secure multiuser communication methods using large networks of coupled superlattices.

R. Sanchez discusses models of electron or heat transport in mesoscopic conductors when detailed balance is locally broken and indicates several possible realizations. Kaupusz and Melnik discuss non-perturbative approaches in nanoscience with possible applications to charge transport and optical phenomena. They focus on finite-size effects in spin systems near the critical point, based on Monte Carlo (MC) method and some analytical arguments. They report Monte Carlo simulations of the three-dimensional Ising model for small and large linear lattice sizes, providing numerical evidence that the asymptotic decay of corrections to finite-size scaling is remarkably slower than expected before. M. Willatzen discusses a continuum model for coupled acousto-optical phonons in piezoelectric materials.

In Part II, Bonilla et al. review stochastic models of tumor-induced growth of blood vessels (angiogenesis). In this complex multiscale process, diffusing vessel endothelial growth factors induce sprouting of blood vessels that carry oxygen and nutrients to hypoxic tissue. They study a model of stochastic differential equations for blood vessels coupled to reaction–diffusion equations for biochemical field densities and derive coupled integro-differential equations for the mean densities. The density of active blood vessel tips solves these equations and it approaches a soliton-like wave before the vessel tips arrive at the tumor. The authors also include a review of other approaches ranking from reinforced random walks to cellular Potts models.

Carpio et al. study formation of bacterial biofilms both inside flows and on solid surfaces. To analyze this multiscale system, they use elastic rod and plate models that incorporate information from the biomass production and differentiation process at the microscale, such as growth rates, growth tensors or inner stresses, as well as constraints imposed by the interaction with the environment.

Plata and Prados analyze the unfolding pathway of biomolecules comprising several independent modules in pulling experiments with atomic force microscopes. The unfolding pathway is the order in which modules of a biomolecule fixed to platform and tip of the microscope open when pulled with a specific velocity. The authors study a simple mechanical model of the biomolecule under pulling and predict a critical pulling velocity: Below it, the weakest module opens first, whereas, above it, the module at the pulled end opens first. They also discuss the robustness of critical velocity and its dependence on the model parameters and propose an experiment to test the theoretical predictions.

Neu et al. present a heuristic derivation of a geometric minimum action method that can be used to determine most-probable transition paths in noise-driven dynamical systems. The latter are ubiquitous in applications to physics and biology. In traditional descriptions that employ Fokker–Planck equations for the probability density, detailed balance ensures that there is a globally stable equilibrium density. The authors pay particular attention to systems that violate detailed balance and emphasize the role of the stochastic vorticity tensor in ascertaining fulfillment or violation thereof. The authors explore the general method through a detailed study of a two-dimensional quadratic shear flow which exhibits bifurcating most-probable transition pathways.

In Part III, Yatsyhin and collaborators describe classical density functional theory (DFT) and apply it to simple equilibrium models. Classical DFT is an ab initio theoretical–computational framework with a firm foundation in statistical physics. It systematically accounts for fluid spatial inhomogeneity, as well as for the non-localities of intermolecular fluid–fluid and fluid–substrate interactions. The theory expresses the grand canonical free energy of a system as a functional of its one-body density, thus generating a hierarchy of N-body correlation functions. Unconstrained minimization of a properly approximated free-energy functional with respect to the one-body density then yields the basic DFT equation. The authors review recent progress in the understanding of planar prewetting and interface unbending on planar substrates and compute substrate–fluid interfaces and wetting isotherms. Guo et al. explain how to model metastability in CdTe solar cells due to Cu migration.

Urata and Li discuss how to couple multiscale molecular dynamics with the finite element method.

Carr et al. present a general method for the electronic characterization of aperiodic 2D materials using ab-initio tight-binding models. They study the subclass of twisted, stacked heterostructures, but their formalism can be implemented for any 2D system without long-range interactions. Their method provides a multiscale approach for dealing with the ab-initio calculation of electronic transport properties in stacked nanomaterials, allowing for fast and efficient simulation of multilayered stacks in the presence of twist angles, magnetic field, and defects. Tuszynski et al. discuss molecular dynamics and related computational methods with applications to drug discovery. Hoiles and Krishnamurthy describe mesoscopic and macroscopic models for the bioelectronic interface of engineered artificial membranes.

This volume will be useful to researchers interested in coupled mathematical models, and their analysis for physical and biological nanoscale systems impacts applications in biotechnology and medicine, quantum information processing, and optoelectronics. Potential audience of the book includes researchers in applied mathematics, physicists, and biologists.

Lastly, we would like to express our gratitude to the Banff International Research Station for making the workshop possible through their hospitality and financial and logistic support. We thank Ruth Allewelt, from Springer, for help, patience, and competence in completing this book.

Leganés, Spain — Luis L. Bonilla
Cambridge, USA — Efthimios Kaxiras
Waterloo, Canada — Roderick Melnik
December 2017

Contents

Part I
Charge and Spin Transport in Low-Dimensional Structures

Nonlinear Quantum Mechanics

Björn Birnir

Abstract We show that the nonlinear bifurcations found by simulations in single quantum wells in the terahertz regime [5, 6, 25] also occur in semiconductor superlattices (SSLs) in the gigahertz range [2, 3, 11]. The only exception is the second Hopf bifurcation to quasi-periodic orbits on a torus. The advantage of experiments on SSLs in the gigahertz range is that the experiments can be conducted at room temperature and a chaotic oscillator due to the random dressing of a period two-orbit has already been measured [31, 32]. We determine [42] that the route to chaos for SSLs in the sequential tunneling regime is the period doubling cascade. Shorter (10-period) superlattices are observed to exhibit faster oscillations compared with longer (50-period) ones. Two plateaus are observed as functions of the voltage bias, and intrinsically chaotic dynamics on the second plateau are possible only for shorter SSLs, while the dynamics in the first plateau contain intrinsic chaos only for longer (N > 50) SSLs [21].

Keywords Semiconductor · Superlattice · Nonlinear quantum systems · Quantum strange attractors · Quantum bifurcations

1 Introduction

Quantum wells are fabricated (or 'grown') from semiconductors by depositing a thin layer of one material, such as gallium arsenide $GaAs$, onto a substrate of a different material, such as aluminum gallium arsenide $AlGaAs$, followed by another layer of substrate material, so that a type of 'sandwich' geometry called a heterostructure is formed. The defining property of quantum wells is that the middle layer has a

B. Birnir (✉)
Department of Mathematics and CNLS, University of California, Santa Barbara, CA 93106, USA
e-mail: birnir@math.ucsb.edu

B. Birnir
University of Iceland, 107, Reykjavík, Iceland

L. L. Bonilla et al. (eds.), *Coupled Mathematical Models for Physical and Biological Nanoscale Systems and Their Applications*, Springer Proceedings in Mathematics & Statistics 232, https://doi.org/10.1007/978-3-319-76599-0_1

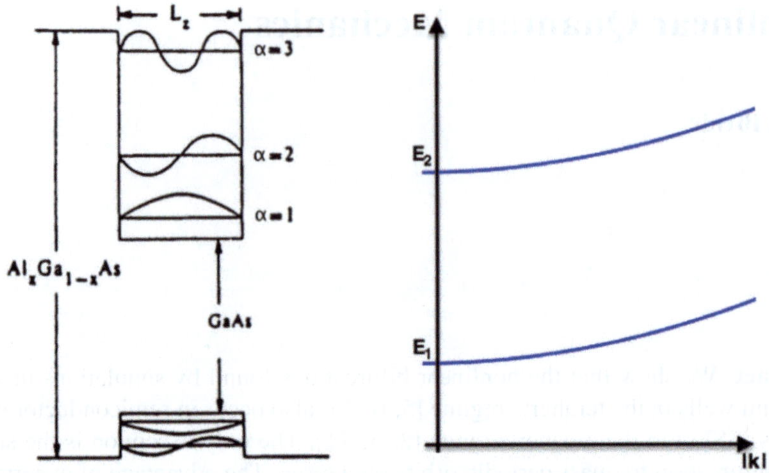

Fig. 1 Left: the conduction and valance band for a semiconductor heterostructure, showing several 'bare' electron and hole subbands (states). Right: the parabolic subbands of the energies $E_{k_x k_y \alpha}$ of the envelope wavefunctions ξ_α

significantly smaller band gap than the substrate layers, and has thickness of the same order as the de Broglie wavelength of the electron. This causes an electron occupying the conduction band of the middle layer to be confined to move freely in only two dimensions, while motion in the third dimension is only possible via transitions between quantized energy levels, called 'subbands', see Fig. 1. These wells can be populated by a density of electrons by a process called doping and this makes them ideal quantum systems for the study of nonlinear effects.

The success of nonlinear dynamical systems theory in the late 20th century, see Guckenheimer and Holmes [28], and its application in the sciences and engineering, see for example Birnir [9], lead to the conjecture that similar phenomena could be found in quantum systems. In semi-classical systems nonlinear dynamics and bifurcations of coherent solutions (solitons) have been shown to exist [7, 8, 27, 35], so it is not unreasonable to expect nonlinearity in some quantum systems far from the semi-classical limit. In the late 1990s and early 2000s, Galdrikian, Batista, Birnir et al. [5, 6, 25] studied intersubband transitions of doped quantum wells. They developed computational methods for determining the coherent states of the electron gas in an oscillating external electric field. It was found that the density matrix equations of motion were nonlinear due to the interactions of the electron gas and that these nonlinearities could be enhanced by fabricating quantum wells with certain asymmetries, so that the lowest-lying subband levels were close to one another in energy. For sufficiently nonlinear wells, it was predicted that the wells would exhibit a bistable response as the terahertz power of the electric field (laser) was ramped up and then down. For strong enough terahertz fields, period-doubling bifurcations leading to a period-doubling cascade were predicted. Galdrikian, Batista, Birnir et al.

[5, 6, 25] developed this nonlinear theory of semiconductor quantum wells, typically made out of $GaAs$ and $Al_zGa_{1-z}As$ and populated by the technique of doping, where material providing electrons is deposited close to the well structure, see Heyman et al. [29, 30]. The nonlinearity was introduced through the Hartree and Hartree-Fock local density approximation where a system of n-interacting electrons is replaced by a system of n-noninteracting electrons in a different (Hartree) potential. This latter view leads to the quantum mechanical system of coherent electron states. These states satisfy a Schrödinger equation where the potential depends on several parameters and when these parameters change the coherent electron states can bifurcate.

2 The Local Density Approximation

We now briefly describe the steps involved in obtaining the nonlinear quantum system describing the coherent electron states and their bifurcations. It is possible to add donor-type dopants at the interfaces of the middle layer with the substrate layers, so that a two-dimensional electron gas will occupy the well. In order to describe such a system theoretically, we take advantage of the fact that all the materials involved exhibit a periodic crystal lattice structure, which constrains the allowed electronic states and leads to a set of semiconductor Bloch equations. Assuming that the electron gas is not too dense, we make an effective mass approximation and work with a simplified model that will be discussed below. The starting point is the Heisenberg equation for the electron operator

$$i\hbar\frac{\partial\psi}{\partial t} = [\psi, H].$$

Let the x- and z-coordinates parameterize the lateral and growth directions of the heterostructure, respectively. Here x is a two-dimensional coordinate parameterizing both the direction of lateral and the transverse direction. Since the wavelength of the laser drive is much longer than the width of the quantum well, the vertical field will be coupled to the electrons in the active region with the dipole approximation. In the effective mass approximation, the mean field Hamiltonian including the vertical field Fz (which falls off rapidly outside the active region) is

$$H(t) = \int \psi^\dagger(x, z, t)[\frac{\hbar^2}{2m}\nabla^2 + v(x, z) + w(x, z, t) - ezF(x, t)]\psi(x, z, t)\, d^2x dz,$$

where v and w are the time-independent and time-dependent parts of the electric potential, respectively, e is the electron charge and m is the effective mass. The electric potential is coupled to the electron density n by Poisson's equation

$$\nabla^2[v(x, z) + w(x, z, t)] = -\frac{e}{\epsilon}n(x, z, t)$$

where

$$n(x,z,t) = \langle \psi^\dagger(x,z,t)\psi(x,z,t)\rangle,$$

is the electron density. The electron operator is expressed as

$$\psi(x,z,t) = \sum_\alpha \int e^{ik\cdot x}\xi_\alpha(z)a_{k\alpha}(t)\,\frac{d^2k}{2\pi} \tag{1}$$

where the envelope wavefunctions $\xi_\alpha(z)$ form a complete orthonormal basis. If the active region is filled and the bias voltage is zero (i.e. the electron density is uniform), then the self-consistent envelope wavefunctions may be calculated in the same manner as in Galdrikian, Batista and Birnir [5, 6, 25].

3 Homogeneous Quantum Wells

The theory of homogenous quantum wells with the local density approximation consists of the following steps. We discretize the integral in Eq. (1) and write the electron operator as

$$\psi(x,y,z) = A^{-1/2}\sum_{k_x,k_y,\alpha} a_{k_xk_y\alpha}\, e^{ik\cdot x}\, \xi_\alpha(z),$$

where A is a constant. The envelope wavefunctions obey the time-independent Schrodinger equation

$$\left\{-\frac{\hbar^2}{2m}\frac{\partial^2}{\partial z^2}+v(z)\right\}\xi_\alpha(z) = E_\alpha\xi_\alpha(z), \tag{2}$$

with the energies lying in parabolic subbands

$$E_{k_xk_y\alpha} = \frac{\hbar^2}{2m}(k_x^2+k_y^2)+E_\alpha, \tag{3}$$

see Fig. 1. The envelope wave functions satisfy vanishing boundary conditions at the sides of the well, which is a good proxy for a vanishing boundary conditions at $z=\pm\infty$, see [5, 25]. The self-consistent potential is determined by Poisson equation

$$\frac{\partial^2}{\partial z^2}v(z) = -\frac{e^2}{\varepsilon}n(z). \tag{4}$$

The local density (or Hartree) approximation is implemented by use of the partition function that yields the relationship between the chemical potential μ, sheet density N_s and subband energies E_α, giving the thermal weights w_α

$$N_s = \frac{m}{\pi\hbar^2\beta}\sum_\alpha \log\left\{1 + e^{-\beta(E_\alpha-\mu)}\right\} \equiv \frac{m}{\pi\hbar^2\beta}\sum_\alpha w_\alpha,$$

where $\beta = 1/T$, T being temperature. Then the electron density can be expressed as

$$n(z) = \sum_\alpha w_\alpha |\xi_\alpha(z)|^2. \tag{5}$$

4 The Hartree Iteration

The potential v is determined by a Hartree iteration:

1. Solve Schrödinger equation (2) for $\{\xi_\alpha(z), E_\alpha\}$.
2. Determine $\{\mu, w_\alpha\}$ from $\{N_s, E_\alpha\}$ and update $n(z)$ in (5).
3. Solve Poisson's equation and update $v(z)$ in (4).
4. Repeat until the iteration has converged.

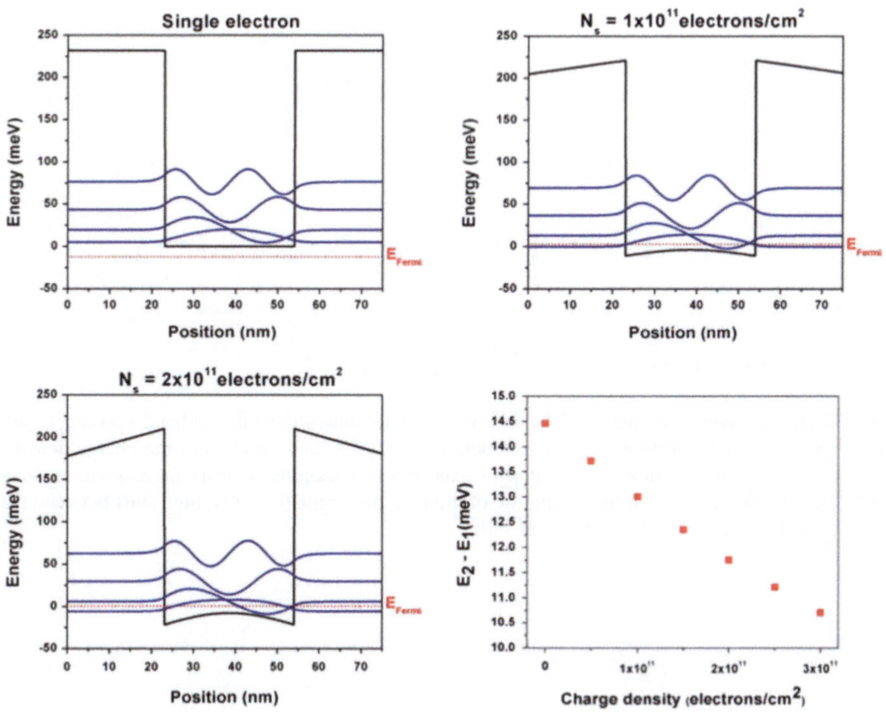

Fig. 2 The first few energy states, the shape of the potential and the Fermi energy for different doping levels and the energy difference $E_2 - E_1$ between the first and second state as a function of the charge density

Figure 2, shows the first few energy states, the shape of the potential and the Fermi energy for different doping levels and the energy difference between the first and second state as a function of the charge density.

5 Intersubband Absorption

The nonlinear effects due to the electron density in the quantum wells have been understood and measured since the mid 1990s. The intersubband absorption when the quantum well is illuminated by an auxiliary laser, can be simulated and measured. The simulations were developed by Załużny [49, 50] and the experiments were done by Craig et al. [19]. The incoming radiation first builds up a charge in the quantum well, this is a Stark effect and corresponds to a redshift of the absorption frequency. However, as the incoming radiation increases the electrons in the well shield against it, this is called the depolarization shift and is a blue shift. Eventually, the depolarization shift dominates, see Fig. 3. One can also measure the relaxation times for the densities in the quantum well, namely Γ_1; the depopulation rate and Γ_2; the depolarization rate. This was also done by Craig et al. [19].

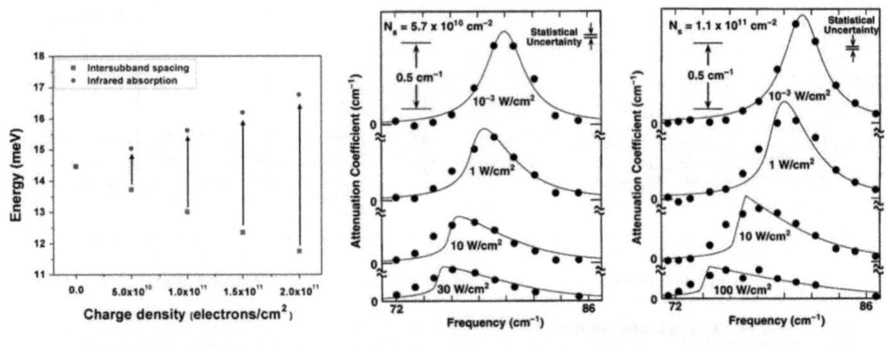

Fig. 3 The absorption frequency. The leftmost picture shows the intersubband spacing (Stark shift) and infrared absorption due to the depolarization shift, as a function of the charge density. The second and third picture show that with fixed electron (doping) density the absorption peak shifts and changes form with the amplitude of the incoming radiation. The blue shift between the two leftmost figures is the depolarization shift

6 Time-Dependent Local Density Approximation

The driving of the quantum wells with time-periodic laser fields requires the development of the time dependent version of above theory. The laser field is added to the Hamiltonian in the electric dipole approximation, this works well since the wavelength of the laser is large compared to width of the well. The intersubband absorption now occurs by collective oscillations of electrons occupying the well and the resonance is broadened and shifted away from intersubband spacing (This is the depolarization shift discussed above). Using the time-dependent Hartree (local density) approximation, we calculate the self-consistent fluctuations due to the time-periodic driving term. The time-dependent electric field $zF(t)$ now induces self-consistent fluctuations in potential $\delta v(z,t)$ and electron density $\delta n(z,t)$

$$\tilde{H}(t) = H - ezF(t) + \delta v(z,t); \qquad \frac{\partial^2}{\partial z^2}\delta v(z,t) = -\frac{e^2}{\varepsilon}\delta n(z,t).$$

These satisfy a Poisson equation with a self-consistency condition: $v(z,t) = v(z) + \delta v(z,t), n(z,t) = n(z) + \delta n(z,t)$. Then we can find a Liouville-von Neumann equation

$$\frac{\partial \rho(z,z',t)}{\partial t} = -\frac{i}{\hbar}[\tilde{H}(t), \rho(z,z',t)] - R[\rho(z,z',t)],$$

for the density matrix ρ, where R is the relaxation operator. In the final quantum mechanical dynamical system, in 4 complex dimensions, R is determined by the experimentally measured depolarization rate Γ_1 and decorrelation rate Γ_2, discussed above.

The time-dependent potential is determined by a time-dependent Hartree iteration:

1. Evolve $\rho(z,z',t)$ until a periodic response is reached.
2. Compute the electron density by the formula $n(z,t) = \rho(z,z',t)\big|_{z'=z}$.
3. Solve Poisson's equation and update $\delta v(z,t)$.
4. Repeat until converged.

7 Nonlinear Phenomena in Asymmetric Quantum Wells

Nonlinear bifurcation of the time-periodic coherent electron states were explored by the theory above. In 1996 Galdrikian and Birnir [25] found the period-doubling bifurcation of these states. In 2003 Batista et al. [6] found the Hopf bifurcation. These bifurcations were found in simulations of the above model and simulations showed the period-doubling cascade to the (Feigenbaum) strange attractor in first

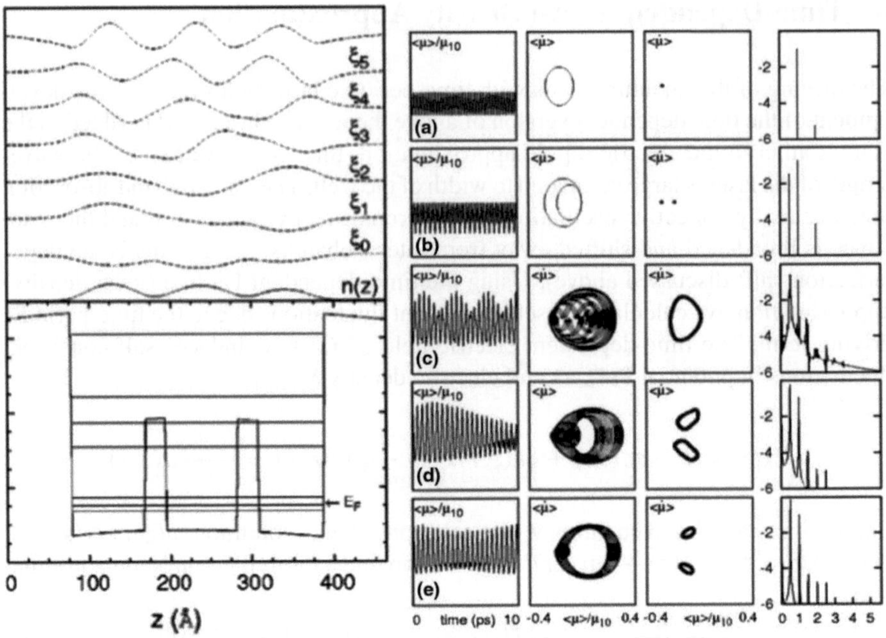

Fig. 4 Left: The asymmetric quantum well with two barriers; bottom, the corresponding eigenstates, top. Right: Time series, phase portrait, Poincaré map and power spectrum

case [25] and the quasi-periodic cascade to another strange attractor in the second case [5]. We illustrate both of these in Fig. 4, taken from Batista et al. [6]. We sample the normalized dipole momentum $\langle\mu\rangle/\mu_{10}$, $\mu = zF$ above, onto the two pertinent complex valued states, in Fig. 4, μ_{10} is a normalization. The first column in Fig. 4 is the time series of this sample. It shows if there is a simple oscillation present or if there are more oscillations superimposed. The pattern in the third row is called beating. The second column is the phase portrait, the basic mode is plotted against its derivative. Both must be sampled at the same (spatial) point. The third column is the Poincaré map, again the mode and its derivative are sampled, but now only at each period T. This is also called the stroboscopic map. A circle on the second column turns into a point in the Poincaré map. Finally the fourth column is the power spectrum or the absolute value of the Fourier transform of the solution. The fundamental frequency $\omega = 2\pi/T$ shows up as the biggest peak. Superharmonics are smaller peaks at integer valued multiples $n\omega$ of the fundamental frequency. In the first row, we see a periodic orbit, with the fundamental frequency and one superharmonic on the power spectrum. The second row shows a period doubling, we see a periodic orbit with twice the period $2T$, this is now a simple periodic orbit (two dots) on the Poincaré map in the third column, and the power spectrum now has a peak at half the frequency $\omega/2$. On the third row, we see a Hopf bifurcation in the Poincaré map, that means that we now have a torus in the phase space. The Poincaré map is a cross section (cut) of this torus. The orbits are quasi-periodic and fill the surface of the torus. On the power spectrum we now see a new incommensurate frequency, smaller than the fundamental frequency. Then on the fourth row we see the torus period double. On the fifth row the doubled torus deforms.

8 Experimental Search by Morris and Sherwin (2011)

A thorough experimental search for the period doubling bifurcation was done by Sherwin and Morris [38], the experimental configuration is illustrated in Fig. 5. They found superharmonics that are signs of the nonlinearity but no subharmonic of the fundamental frequency. The latter is the signature of the period doubling bifurcation. Optical bistability that frequently precedes a period doubling bifurcation was not observed either. It is possible that the subharmonic signal was too weak to be observed but other possible issues include: 1. Excessive heating in the quantum well. It is possible that the laser drive ionizes the electrons by excessive heating before the subharmonic is able to form. 2. Domain formation, it is possible that domains with difference charges form in the lateral direction perpendicular to the confinement direction of the well. If this is the case these may add up and cancel or minimize the effect of the nonlinearity.

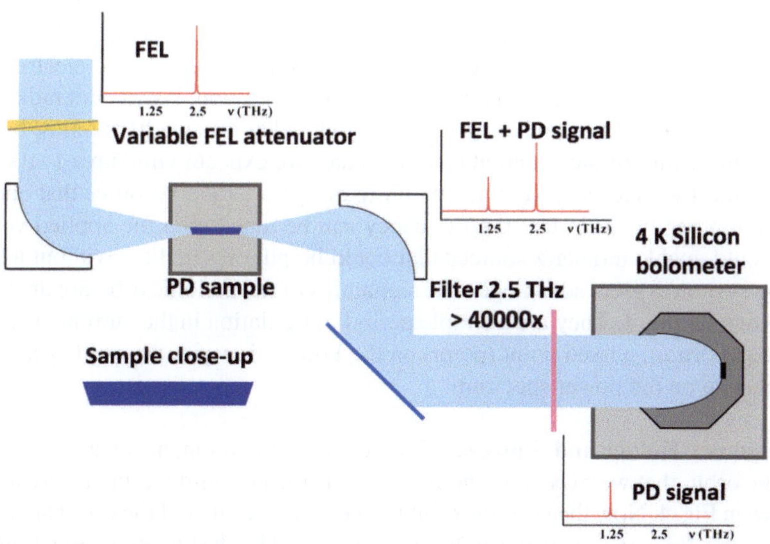

Fig. 5 The experimental setup of Morris and Sherwin (2011). FEL denotes the free-energy laser (with frequency 2.5 THz). PD denotes the period doubling. The fundamental FEL frequency is filtered out to observe the subharmonic (1.25 THz) signal

9 The Anticipated Results: The Quantum Oscillator

A natural way to strengthening the signal emitted by coherent electron states in a quantum well is to string together several quantum wells into a heterostructure called a superlattice (SL). An example is shown in Fig. 6, taken from Alvaro et al. [1].

In the proposed research we will design and simulate SLs that can be used as sources, period halvers and squeezers, random sequence generators and frequency mixers, at room temperature. Periodic, quasi-periodic and chaotic oscillations have been observed in superlattices [34, 47], at very low temperatures that require cooling with liquid nitrogen. We propose to design and simulate SLs where all these phenomena can be simulated and understood dynamically at room temperatures. The design of such SLs requires heterostructures where the background temperature is surppressed and below we will discuss the structure of such lattices. First we list the devices and dynamical phenomena that we propose to find and study:

A Source: The difference between the above analysis of the oscillations in the quantum well and the oscillations in the SL is that we drove the quantum well with an ac laser field, whereas the electron states in the SL will be driven by a dc voltage. This means that whereas we always start with a time-periodic orbit in the quantum well, in the SL we start with a steady current (stationary state). The first bifurcation we will look for is the generic Hopf bifurcation, from the stationary state, in such an SL. Here generic means that it is the bifurcation we expect to find in the SL, as the voltage bias increases, assuming it has no other symmetries. After this bifurcation we expect to find a circle in the phase portrait of the SL system (6–11), as in the first row of the second column on Fig. 4. This electron state is a *source*, namely the electron state consists of a charge, accelerated over the periodic oscillation. It will emit radiation at the frequency $\omega = 2\pi/T$, where T is the period and at its superharmonics, because of the nonlinearity of the coherent electron state. We expect to find up a 1–100 GHz source, and if we succeed, we will next try to design a terahertz source that operates at room temperatures. In fact the frequency can be tuned with the applied voltage. This would enable terahertz sources that could be put on satellites without a heavy cooling system; a great advantage. The signatures of the periodic orbit are the figures in row one on Fig. 4. They are a simple periodic oscillation in the current, a circle in the phase portrait, a fixed point (point) on the Poincaré map and a single peak at the frequency ω on the power spectrum.

A Frequency Halver and Squeezer: The second bifurcation, now starting from the periodic orbit, that we expect to find in the SL, is the period-doubling bifurcation on row two in Fig. 4. Now the periodic orbit has period twice that of the original periodic orbit, so the frequency gets cut in half: $\omega \to \omega/2$. This has many applications, the signal can be read at lower frequency where the noise may be reduced, it can be used to make squeezed states [26] for noise reduction etc. The periodic orbit makes two circles to close up again (figure two in row two) in Fig. 4 (on the right). The Poincaré map has two points in each period T and there is a new peak at $\omega/2$ in the power spectrum (the rightmost figure in row two) in Fig. 4. The current on the right has new oscillations with half the original frequency and all their superharmonics. This subharmonic signal can then be used in various ways: To produce a signal of half the original frequency. To compress the signal into a desirable frequency range or to squeeze out of it undesirable noise. The signatures of this bifurcation are: An additional oscillation in the current. An extra loop in the periodic orbit on phase portrait. Two points replacing one in the Poincaré map. Half the fundamental frequency and its superharmonics in the power spectrum.

The Chaotic Attractor and Random Sequences: The period doubling of the periodic orbit continues into a period-doubling cascade to a strange attractor. We will use this attractor to generate ultrafast random sequences at room temperatures. This is very important to do for secure data storage and transmission [23, 40, 45], stochastic modeling [4], and Monte Carlo simulations [10]. So far the fastest available technology for this task have been semiconductor lasers. We plan to find the parameter regions where the chaotic attractors exist and where their detection can be optimized. This parameter region my be enlarged by the addition of noise, as in [1], and we will explore different types noise and its effect on the simulations. The signatures of the strange attractors are: A wide range of oscillations in the current. A strange attractor in phase portrait. A complex Poincaré map. Broadband noise in the power spectrum.

The Torus (Mixer) and a Distinct Chaotic Attractor: It is still not clear whether a second Hopf bifurcation producing a torus in the phase space is possible in the superlattices that have been designed so far [31]. This bifurcation occurs on row three in Fig. 4, for the quantum well. The bifurcation is generic but may be prohibited by the symmetries of the superlattice. We will try to design a superlattice for which the (even) symmetry is broken and the second Hopf bifurcation of the periodic orbit to a torus takes place. The orbits on the torus are quasi-periodic and are composed of two frequencies. This makes the torus a candidate for a frequency mixer with two tunable frequencies. Then we will explore the cascade of quasi-periodic orbits on this torus to a strange attractor, as the voltage bias increases. This strange attractor has different properties than the strange attractor created by the period-doubling cascade above and gives an alternative method to generate ultrafast random sequences at room temperatures. We plan to compare the sequences generated by the two types of attractors and see if one has some advantage over the other. We also will find the parameter regions where these latter types of chaotic attractors exist and where their detection can be optimized. This parameter region my be enlarged by the addition of noise as above, and we will explore different types noise and its effect on the simulations.

However, we also see that there are only two occupies states below the Fermi levels (hill) on Fig. 6. This is analogous to the situation of Galdrikian and Birnir [25] and distinct from that of Batista et al. [6]. In the latter case there are three occupied states below the Fermi level. In fact, three states are necessary for the second Hopf bifurcation above and the cascade of quasi-periodic orbit the strange attractor. Thus

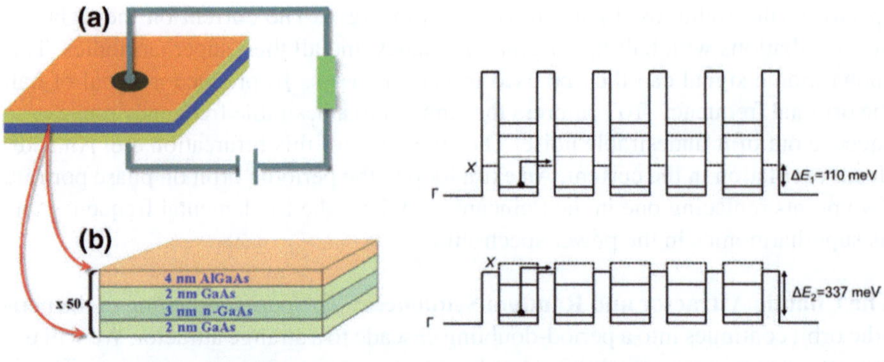

Fig. 6 Left: schematic diagram of a dc voltage-biased SSL, from Bonilla et al. [12]. Right: The band diagram of a GaAs/$Al_{0.7}Ga_{0.3}As$ SSL (**a**) and GaAs/$Al_{0.45}Ga_{0.55}As$ SSL (**b**), from Huang et al. [37]. The conduction band and the bound states of the wells are indicated by the solid horizontal lines. The bottom of the X-valley is indicated by the dashed lines

we need to redesign the SL to enable the bifurcation to the torus above and the resulting quasi-periodic orbits. We also have to make sure that the even symmetry of the SL is broken, so the signal is stronger and easier to observe.

10 Semiconductor Superlattices

The observation of Bloch oscillations in semiconductor superlattices (SSLs) [36] has led to many proposed applications of these heterostructures as sources and detectors at gigahertz and terahertz frequencies. More recently, nonlinear Gunn oscillations and chaotic dynamics have been observed in SSLs in the sequential tunneling regime. These nonlinear phenomena present the opportunity for development of new applications of SSLs, such as true random number generators and frequency mixers. Furthermore, recent advances in the design of SSLs have opened the possibility of realizing these applications at room temperature. In support of the development of applications of SSLs in the nonlinear regime, we theoretically characterize the nonlinear dynamical phenomena of the sequential resonant tunneling (SRT) model of weakly-coupled SSLs in this paper. We show the effects on the bifurcation diagram from variations of the number of periods making up superlattice, the sensitivity of the SRT model to time-dependent stochastic fluctuations in the bias voltage and local tunneling rates, and the effect of time-independent random perturbations of the widths of the wells and barriers. This and the next three sections follow [21].

Spontaneous oscillations, quasiperiodic orbits, and chaos have already been observed experimentally at very low temperatures [34, 36, 47] and at room temperature [31–33, 37, 41] in 50-period SSLs with noisy voltage sources. Simulations by Alvaro et al. [1] exhibited a strong chaotic signal. Experiments by Huang et al. [31, 32] show that heating suppresses the nonlinear phenomena in SSLs, and we suspect this also occurs in single QWs. Huang et al. also describe a way of suppressing the effect of heating, enhancing the current oscillations: It was hypothesized that at warm temperatures, phonon-assisted transport though the X-valley of AlAs allowed a thermal distribution of carriers to diffuse through the SSL, overwhelming the nonlinear quantum dynamics. This effect was suppressed by choosing the Aluminum concentration of GaAs/AlGaAs wells in order to maximize the lowest bandgap energy, i.e. make the X and Γ band gaps equal to one another (Fig. 6). As a result, current oscillations were observed in SSLs at room temperature for the first time, see [31, 32].

We consider the SRT theory of Bonilla et al., which describes the electronic dynamics of SSLs in the weakly-coupled, self-consistent regime [15]. Two different time scales are taken into account in this description of SSLs. The inter-site tunneling and inter-subband relaxation processes occur on much shorter timescales than the dielectric relaxation processes [12]. Therefore, the long timescale dynamics of semiconductor lasers [44] and superlattices [1, 14] are typically modeled using semiclassical equations, while the short timescale processes are treated separately as noise. In the case of the SRT model, the short-timescale processes are included through the addition of stochastic terms to the dynamical equations. Nonlinearities enter the model via the inter-site Coulomb interaction, which bends the conduction band of the SSL, modifying the inter-band tunneling rates by casting the energy levels of adjacent wells into or out of resonance [12]. The dynamical equations are discussed in detail in Sect. 11. The bias voltage, V_{bias}, is treated as an external parameter of the model. Gunn-like oscillations in $J(t)$ are found to occur in the SRT model over several intervals of V_{bias} [16, 20]. These oscillations undergo a series of bifurcations, which may cascade into chaotic behavior [1]. In [42], we characterized the route to chaos via analysis of the Poincaré map and power spectrum, and distinguished the effects of the noise terms from the deterministic chaos.

For sufficiently small bias voltages, the total current $J(t)$ through the SSL responds linearly to changes in the bias voltage. At higher bias voltages, $J(t)$ suddenly transitions to a time-dependent, oscillatory function, which passes through a further series of transitions leading to chaotic behavior. We summarize the behavior of $J(t)$ below:
Fixed-point: Over certain voltage intervals, $J(t)$ is attracted toward a stationary value.
Bistability: The first signal of the nonlinear dynamics is a bistable response of $J(t)$ to slow variations in V_{bias}. This behavior is observable only at sufficiently low temperatures [16, 20]. Generically, bistable behavior is found at voltages near the Hopf bifurcation described below.
Supercritical Hopf Bifurcation: As the bias voltage is increased, we next observe a supercritical Hopf bifurcation. The fixed point becomes unstable, and $J(t)$ evolves to a periodic orbit. The periodic orbit is topologically equivalent to a circle in phase space, which corresponds to a one-cycle of the Poincaré map. In this regime, the SSL acts as a GHz oscillator with a discrete power spectrum involving the frequencies $f_n = n/T$, $n = 1, 2, 3, \ldots$, where T is the period of the lowest-frequency oscillation present. The superharmonics $n > 1$ arise due to the nonlinearities present in the SRT model. In this dynamical phase, the fundamental period T varies smoothly with the bias voltage, therefore the oscillator is also *tunable*. By filtering out all but the desired

harmonic and fine-tuning it via the bias voltage, a wide range of frequences may be selected.

Period Doubling Bifurcation: In this regime, one-cycles of the Poincaré map transition to two-cycles. The fundamental period of the oscillator is doubled, $T \to 2T$, and the fundamental frequency is cut in half: $\omega \to \omega/2$. A new frequency peak will appear in the spectrum at half the fundamental frequency, and the number of superharmonics will double. Following a period doubling bifurcation, the reverse (period-halving) bifurcation may occur. We refer to the regions between these bifurcations as *period doubling bubbles*. An application of period doubling, due to the subharmonic peak, is that a signal may be read at lower frequency, where the noise may be reduced, and it can be used to make squeezed states [26].

Period Doubling Cascade: Period doublings may occur in succession over certain voltage intervals, and an infinite number of doublings is possible in a finite voltage range. The invariant phase space structures transition from compact manifolds (periodic orbits of high periods) to chaotic attractors. The Poincaré map takes on a fractal structure. Our simulations show that the SRT model does not support true quasiperiodic orbits, hence the chaotic attractor is the most complex structure in the bifurcation diagram. It is the result of a cascade of a period doubling sequence of periodic orbits. An application of the chaotic dynamics in this regime is ultrafast generation of random number sequences [37]. This has many applications in areas such as secure communication and data storage, stochastic modeling, and Monte Carlo simulations, see [4, 23, 40, 45]. Previously the generation of ultrafast random number sequences has been accomplished by fast semiconductor lasers but these require a mixture of optical and electronic components. SSLs on the other hand are entirely submicron devices that can be readily integrated into complex circuits.

In previous theoretical studies of optically-driven quantum wells, it was discovered that the introduction of one or more off-center "steps" in the confinement potential, see Fig. 7, had a profound impact on the character of the nonlinear phenomena. In the presence of a single step, a period-doubling bifurcation in the electronic response was predicted to occur near the intersubband resonant frequency at high doping densities and strong driving fields. The presence of a second step unfolded the period-doubling bifurcation into a supercritical Hopf bifurcation which gener-

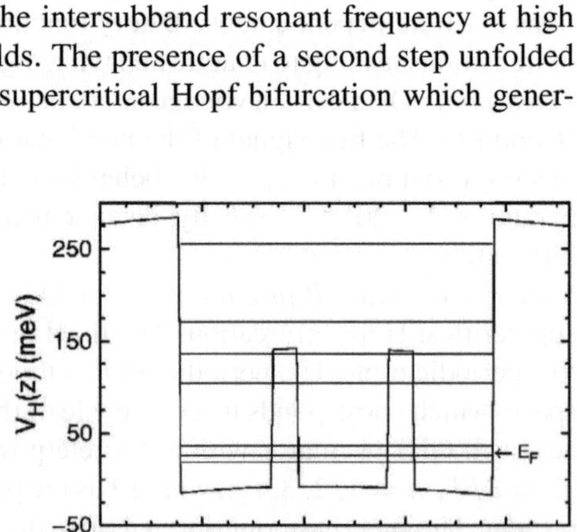

Fig. 7 The stationary self-consistent potential resulting from the band structure and Coulomb potential of the asymetric GaAs/AlGaAs quantum well, taken from Batista et al. [5]. The energy levels of the bound states are indicated by the horizontal lines

ated quasiperiodic behavior. Both single- and double-stepped assymetric quantum wells also exhibited period-doubling cascades to chaos. In analogy with these results, we consider the possibility of unfolding the period doubling bifurcation of the SRT model into a Hopf bifurcation, by breaking the periodic symmetry of the SSL, which we term "disordered superlattice."

11 Model

Many phenomena are captured by means of a quasi-one-dimensional resonant sequential tunneling model of nonlinear charge transport in SSLs [13–15]. Consider a weakly coupled superlattice having N identical periods of length l and total length $L = Nl$ subject to a dc bias voltage V_{bias}. The evolution of F_i, the average electric field at the SSL period i, and the total current density, $J(t)$, is described by Ampere's law

$$J(t) = \epsilon \frac{dF_i}{dt} + J_{i\to i+1}, \tag{6}$$

and the voltage bias condition

$$\sum_{i=1}^{N} F_i = \frac{V_{\text{bias}}}{l}. \tag{7}$$

Fluctuations of F_i away from its average value $F_{\text{avg}} = eV_{\text{bias}}/L$ arise from the inter-site tunneling current $J_{i\to i+1}$, which appears in Eq. (6). A microscopic derivation of $J_{i\to i+1}$ produces the result [12, 13]

$$J_{i\to i+1} = \frac{en_i}{l} v^{(f)}(F_i) - J^{-}_{i\to i+1}(F_i, n_{i+1}, T), \tag{8}$$

in which n_i is the electron sheet density at site i, $-e < 0$ is the electron charge and T is the lattice temperature. Here the forward velocity, $v^{(f)}(F_i)$, modeled as a Lorentzian distribution, is peaked at resonant values of F_i for which one or more energy levels at site i are aligned with the levels at site $i+1$. The backward tunneling current is given by

$$J^{-}_{i\to i+1}(F_i, n_{i+1}, T) = \frac{em^* k_B T}{\pi \hbar^2 l} v^{(f)}(F_i) \ln\left[1 + e^{-\frac{eF_i l}{k_B T}} \left(e^{\frac{\pi \hbar^2 n_{i+1}}{m^* k_B T}} - 1\right)\right], \tag{9}$$

where the reference value of the effective electron mass in $Al_xGa_{1-x}As$ is $m^* = (0.063 + 0.083x)m_e$, and k_B is the Boltzmann constant. The n_i are determined self-consistently from the discrete Poisson equation,

Table 1 (Top) The design parameters of the superlattice. (Bottom) Values of the potential barrier and energy levels for GaAs/$Al_{0.7}Ga_{0.3}As$ and GaAs/$Al_{0.45}Ga_{0.55}As$ superlattices, first and second row, respectively

T (K)	N_D (cm^{-2})	d (nm)	w (nm)	s (μm)
295	6×10^{10}	4	7	60
V_{barr} (meV)	E_1 (meV)	E_2 (meV)	E_3 (meV)	
600	53	207	440	
388	45	173	346	

$$n_i = N_D + \frac{\epsilon}{e}(F_i - F_{i-1}), \tag{10}$$

where N_D is the doping sheet density and ϵ is the average permittivity. The field variables F_i are constrained by boundary conditions at $i = 0$ and $i = N$ that represent Ohmic contacts with the electrical leads

$$J_{0\to 1} = \sigma_0 F_0, \quad J_{N\to N+1} = \sigma_0 \frac{n_N}{N_D} F_N, \tag{11}$$

where σ_0 is the contact conductivity. Shot and thermal noise can be added as indicated in [1, 17].

Table 1 gives the numerical values of the parameters used in the simulations. The GaAs/$Al_{0.45}Ga_{0.55}As$ configuration corresponds with the configuration used in experiments [31, 37, 48]. The rest of the parameters are as follows: $A = s^2$ is the transversal area of the superlattice, d and w are the barrier and well lengths, and $l = d + w$ is the SSL period. The contact conductivity is a linear approximation of the behavior of $J_{0\to 1}$, which depends on the structure of the emitter; the value has been taken to reproduce the experimental results produced by Huang et al. with $N = 50$: $\sigma_0 = 0.783$ A/Vm for $V_{barr} = 388$ meV ($x = 0.45$) and $\sigma_0 = 0.06$ A/Vm for $V_{barr} = 600$ meV ($x = 0.7$), where V_{barr} is the height of the barrier [1, 31].

11.1 Noise

To model the unavoidable fluctuations in the bias voltage, as well as the short-timescale processes in the electronic dynamics, stochastic terms are introduced into evolution equations [1]. To account for the noise in the bias voltage, Eq. (7) is modified to

$$\sum_{i=1}^{N} F_i = \frac{V_{\text{bias}} + \eta(t)}{l}, \tag{12}$$

where $\eta(t)$ is taken to be a Gaussian random variable with standard deviation σ_η. To account for the short-timescale processes at each site of the SSL, Eq. (6) is modified to include fluctuations in the tunneling current

$$\varepsilon\frac{dF_i}{dt} + J_{i\to i+1}(F_i) + \xi_i(t) = J(t), \tag{13}$$

where

$$\langle \xi_i(t)\xi_j(t')\rangle = \frac{e}{A}\left[\frac{ev^{(f)}(F_i)}{l}n_i + J^-_{i\to i+1}(F_i, n_{i+1}, T) + 2J^-_{i\to i+1}(F_i, n_i, T)\right]\delta_{ij}\delta(t-t'). \tag{14}$$

We may observe that $\eta(t)$ is independent of i, while $\xi_i(t)$ are independent Gaussian random variables associated with each site of the SSL. The strength of the fluctuations in the bias voltage may be tuned via the empirical parameter σ_η, while the strength of the fluctuations in the local tunneling current is completely determined by the other parameters, see Table 1.

11.2 Disorder

We also consider time-independent perturbations that break the periodic symmetry of the superlattice. We introduce variations in the widths of the wells and barriers via the scaling parameters β_i and ζ_i. The perturbed well and barrier lengths are

$$w_i = \beta_i w \tag{15}$$

$$d_i = \zeta_i d. \tag{16}$$

Both β_i and ζ_i are taken to be uncorrelated, normally-distributed random numbers with standard deviation σ. We constrain the total length of the SL to preserve the validity of the bias equation, (7), by forcing the condition that $\sum_i \beta_i = \sum_i \zeta_i = N$. To lowest order, the energy levels scale with β_i according to

$$\varepsilon_i^{C,m} = \frac{\varepsilon^{C,m}}{\beta_i^2}. \tag{17}$$

These modifications imply that the effective dielectric constant becomes dependent on i,

$$\varepsilon_i = l_i/(w_i/\varepsilon_w + d_i/\varepsilon_d). \tag{18}$$

Equation (7) becomes

$$V_{\text{bias}} = \sum_{i=1}^{N} F_i l_i. \tag{19}$$

Following Bonilla et al. [16], Eqs. (8) and (9) are modified to account for the effects of disorder on $v^{(f)}(F_i)$ and τ_i:

$$v^{(f)}(F_i) = \frac{\hbar^3}{2m^{*2}} \sum_{m=1}^{3} \frac{l_i\big(\gamma_{C,1} + \gamma_{C,m}\big)\tau_i(\varepsilon^i_{C,m})}{\big(\varepsilon_i^{C,1} - \varepsilon_{i+1}^{C,m} + eF_i\big(d_i + \frac{w_i + w_{i+1}}{2}\big)\big)^2 + \big(\gamma_{C,1} + \gamma_{C,m}\big)^2} \tag{20}$$

$$\tau_i = \frac{16 k_i^2 k_{i+1}^2 \alpha_i^2}{(k_i^2 + \alpha_i^2)(k_{i+1}^2 + \alpha_i^2)\big(w_i + \frac{1}{\alpha_{i-1}} + \frac{1}{\alpha_i}\big)\big(w_i + \frac{1}{\alpha_{i+1}} + \frac{1}{\alpha_i}\big) e^{2\alpha_i d_i}} \tag{21}$$

where

$$\hbar k_i^m = \sqrt{2m^* \varepsilon_i^{C,m}} \tag{22}$$

$$\hbar k_{i+1}^m = \sqrt{2m^*\left(\varepsilon_i^{C,m} + e\left[d_i + \frac{1}{2}(w_i + w_{i+1})\right] F_i\right)} \tag{23}$$

and

$$\hbar \alpha_{i-1}^m = \sqrt{2m^*\left(eV_b + e\left[d_{i-1} + \frac{w_i}{2}\right] F_i - \varepsilon_i^{C,m}\right)} \tag{24}$$

$$\hbar \alpha_i^m = \sqrt{2m^*\left(eV_b - e\left[\frac{1}{2} w_i\right] F_i - \varepsilon_i^{C,m}\right)} \tag{25}$$

$$\hbar \alpha_{i+1}^m = \sqrt{2m^*\left(eV_b - e\left[d_i + \frac{1}{2} w_i + w_{i+1}\right] F_i - \varepsilon_i^{C,m}\right)} \tag{26}$$

12 Numerical Methods

The evolution Eqs. (6)–(11) are evolved in time using the forward Euler method and the trajectory $(F_i(t), n_i(t), J(t))$ through the $(2N + 1)$-dimensional phase space is stored. When applicable, the stochastic terms ($\xi_i(t)$ and $\eta(t)$) are included using the Euler-Maruyama method. We utilize the standard methodology of dynamical systems theory (analysis of the Poincaré map) in order to characterize the dynamical phenomena of the SRT model: The first step is to construct the phase portrait, i.e. to project the evolution onto a two-dimensional surface in phase space. We choose the surface spanned by the the coordinates (F_i, F_j) for some values of i and j near the anode and cathode respectively of the SSL.

The next step is to compute the Poincaré map of the phase portrait. After sufficient time has elapsed, and regardless of the initial conditions, the phase space trajectory settles onto one of the following time-invariant structures: (a) Fixed point, (b) periodic orbit, (c) strange attractor. The Poincaré map is used in order to easily distinguish between these structures. It is computed according to the following procedure: First, the transient behavior associated with the initial conditions is excised from the trajectory and only the remaining data is considered in what follows. When $F_i(t)$ passes through its central value, the time t^* and the field $F_j(t^*)$ are stored. We also compute the quantity $\dot{F}_i(t^*)$ from Eq. (6). We then discard all of the values of t^* for which $\dot{F}_i(t^*) > 0$, in order to prevent sampling the same orbit more than once per cycle. The Poincaré map transforms the essentially continuous trajectory through phase space into a discrete map from the one-dimensional interval onto itself [18]. We represent it visually in one of two ways: Either (a) phase portraits, plotting $F_j(t^*)$ against $\dot{F}_j(t^*)$, c.f. Fig. 9, or (b) bifurcation diagrams, plotting $F_j(t^*)$ against some external parameter such as the bias voltage, c.f. Fig. 8.

Both fixed points and periodic orbits appear as a single point in the visualization of the Poincaré map. However, fixed points are easily distinguished from periodic orbits (one-cycles) by computing the power spectrum associated with the current $J(t)$:

$$P[J](f) = \left| \int_{t_i}^{t_f} dt\, e^{-i2\pi f t} J(t) \right|^2 , \qquad (27)$$

where f is the frequency. A period-doubling bifurcation is identified when one-cycles transition to two-cycles, producing two points in the Poincaré map. Chaotic regions are identified where a proliferation of period-doubling bifurcations occur, and the number of points in the Poincaré map increases without bound, yielding a fractal structure in the bifurcation diagram.

Dynamical structures revealed by the Poincaré map are associated with various power spectra as follows: (a) periodic orbits correspond to a series of peaks with widths of the same order as the frequency bin size, falling at integer multiples of the fundamental frequency, (b) period doubling bifurcations are recognized when the number of peaks in the spectrum changes by a factor of two, and a new peak appears in the power spectrum at half the fundamental frequency, (c) chaotic attractors exhibit power spectra containing both sharp peaks and broadband noise.

The boundary conditions are more subtle that indicated in Sect. 3 in the THz regime or Sect. 11 in the GHz regime. The leads should be treated as a semiclassical system that we are coupling with (highly) quantum system, namely the superlattice. This coupling involves a very large number of states and is best treated numerically. For an effective treatment using perfectly matched layers (PML), see Chaps. 2 and 3 in [22].

13 Results

We simulate superlattices and characterize the dynamical instabilities that may be applied to create sources, period halvers and squeezers, random sequence generators and frequency mixers, even at room temperature. The dynamical equations (6)–(11)

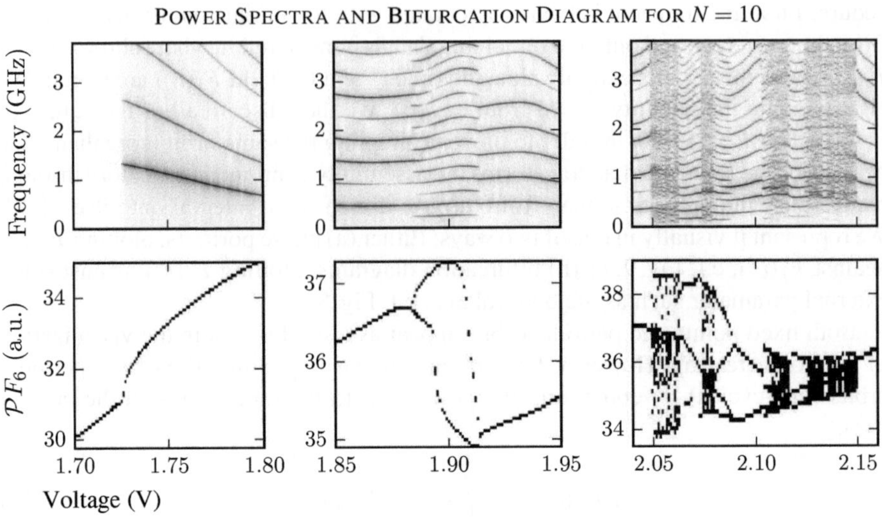

Fig. 8 (Top row) The power spectrum of $J(t)$ plotted against the bias voltage, taken from [42]. (Bottom row) The bifurcation diagram, plotting the Poincaré map against the bias voltage. The Hopf bifurcation from the steady state is shown in the first column. A period doubling "bubble" is shown in the second column. A period-doubling cascade is shown in the third column

are evolved using the parameter values listed in Table 1 for a GaAs/$Al_{0.7}Ga_{0.3}As$ SSL, with the quantized energy levels corresponding to $V_{barr} = 600$ meV. The GaAs/$Al_{0.7}Ga_{0.3}As$ SSL is treated here in order to illustrate bifurcations as clearly as possible, but the same phenomena and instabilities occur in $Al_{0.45}Ga_{0.55}As$ SSLs [21]. Dynamical instabilities are found in two distinct *plateaus*, over which which the local electric fields of the SSL cease to increase monotonically as a function of V_{bias}. The *first plateau* occurs at very low voltages, with tunneling transport between the ground states of adjacent wells that are nearly aligned with one another in energy. The *second plateau* occurs in the region of V_{bias} where the the electric fields bend the potential of the SSL to align the ground state of well i with first excited state of well $i + 1$. We do not observe a third plateau because the third excited state becomes unbound at bias voltages that align it with the first excited state of an adjacent well.

The leading edge, i.e. the lowest value of V_{bias} contained in a plateau, is identified by a supercritical Hopf bifurcation from fixed point to periodic orbit, as shown the leftmost column of Fig. 8. At low temperatures, the Hopf bifurcation may be preceded by bistability, but at higher temperatures this phenomenon is suppressed. Within a plateau, we may observe period-doubling, period-doubling cascades, and chaotic attractors whose locations depend upon on the values of the rest of the parameters, in particular N, the number of wells making up the superlattice. As a general rule, shorter superlattices exhibit a greater variety of dynamical behavior in the second plateau. In SSLs ($N \leq 10$), the Hopf bifurcation in the first plateau disappears and the first plateau is not present. As N increases, the dynamical instabilities appear to move from the second plateau into the first plateau: Near $N = 20$, the Hopf bifurcation appears in the first plateau. In longer SSLs ($N \geq 30$), the second plateau contains only a supercritical Hopf bifurcation to a periodic orbit without any further bifurcations, while the first plateau has gained a period-doubling bubble. In this section, we give a detailed description of the dynamical instabilities of $N = 10$ SSLs, then we point out the effect of increasing N. We close with a discussion of the effects of stochastic terms and disorder on the dynamical instabilities.

13.1 N=10

As mentioned above, the first plateau does not exist for $N = 10$, and all oscillatory behavior takes place in the second plateau. Combining the bifurcation diagram, power spectra and phase portraits shown in Figs. 8 and 9, we characterize the dynamical instabilities of the SRT model for $N = 10$:

Supercritical Hopf Bifurcation: In the leftmost column of Fig. 8, we observe a transition from a stationary state to a periodic orbit. Subsequently, we observe a circle in the phase portrait, similar to the top row of Fig. 9. The Poincaré map consists of a single point, or one-cycle, when visualized. The power spectrum contains peaks falling at integer multiples of a fundamental oscillation frequency as demonstrated in the top row of Fig. 8. In this regime, the SSL acts as a GHz oscillator with a discrete power spectrum involving the frequencies $f_n = n/T,\ n = 1, 2, 3, \ldots$, where T is the period of the lowest-frequency oscillation. The superharmonics $n > 1$ arise due to nonlinearities of the SRT model. We also observe that the fundamental frequency and resulting superharmonics can be continuously tuned by variation of V_{bias} as demonstrated in Fig. 8.

Period Doubling Bifurcation: A period-doubling bifurcation is identified by a doubling of the orbits in phase space and consequent doubling of the number of points in the Poincaré map. We illustrate this phenomena in the transition between rows one and two of Fig. 8 and in the second row of Fig. 9. The power spectrum gains a *subharmonic* peak at half of the former fundamental frequency, and consequently we observe twice as many superharmonics in the power spectrum. A period-doubling bifurcation may be followed by a period-halving bifurcation forming a period-doubling bubble as shown in the second column of Fig. 8.

Period Doubling Cascade: A period doubling cascade is identified when many period-doubling bifurcations occur in rapid succession over some interval of the bias voltage. In principle, an infinite number of doublings may occur over a finite voltage interval. This process terminates when the phase space orbits lose their periodicity altogether and the Poincaré map takes on the characteristics of a chaotic attractor. An example of a period-doubling cascade is illustrated in the last three rows of Fig. 9. The rightmost column of Fig. 8 also shows several period-doubling cascades

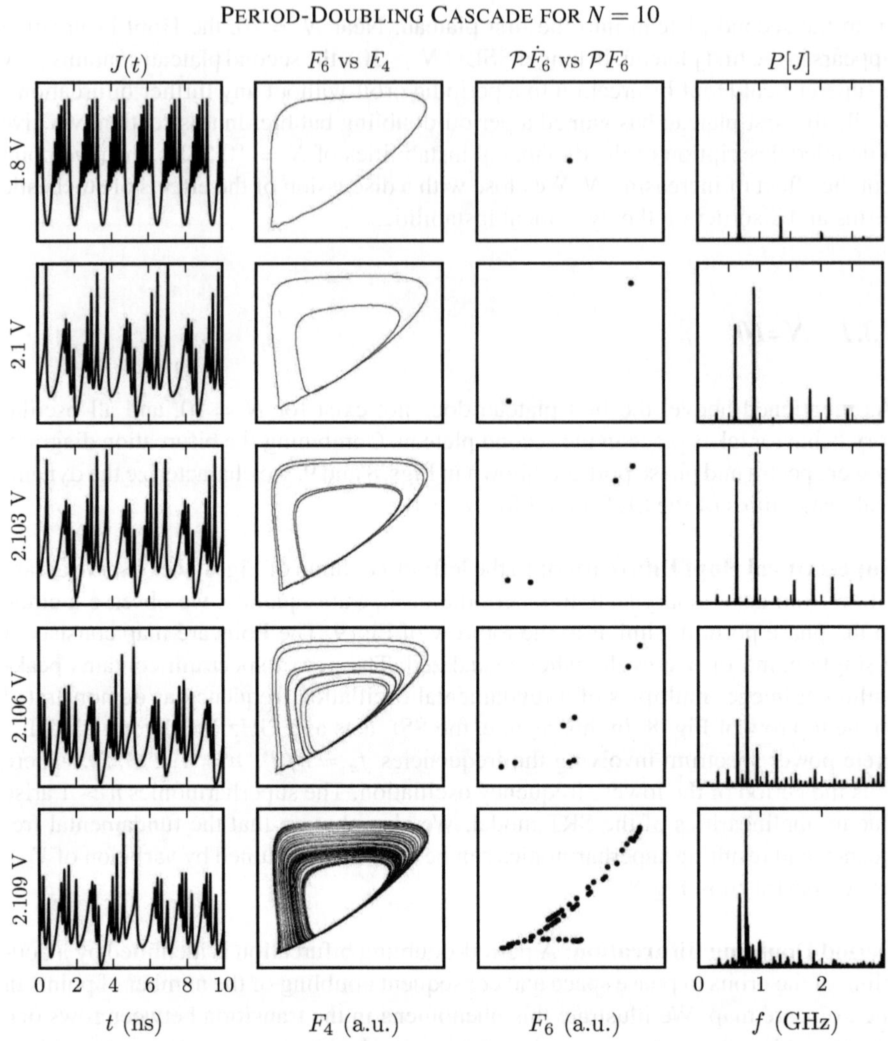

Fig. 9 Representative phase portraits, taken from [42]. The first column shows the average current J plotted against time t. The second column shows the phase portrait $F_6(t)$ plotted against $F_4(t)$. The third column shows the Poincaré map $\mathscr{P}\dot{F}_6(t^*)$ plotted against $\mathscr{P}F_6(t^*)$. The last column shows the power spectrum of $J(t)$. A periodic oscillation is shown in the first row. A period-doubling bifurcation is observed in the second row. The period-doubling cascade to a chaotic attractor is shown in the bottom four rows

connected by regions of orbits with very high periods. The broadening and merging of peaks in the power spectrum is characteristic of a chaotic attractor. We also compute the Feigenbaum constant of the cascade near 2.109 V, shown in Figs. 8 and 9. We introduce the formula

$$\delta_n = \frac{V_{n-1} - V_{n-2}}{V_n - V_{n-1}}, \tag{28}$$

where V_n are the voltages corresponding to the nth doubling in the cascade. For a period-doubling cascade, $\delta_n \to 4.6692\ldots$ as $n \to \infty$. By taking sufficiently small steps (about 10^{-6} V) in the parameter V_{bias}, we have measured the first Feigenbaum constant with less than 1% error. We conclude that the route to chaos in the SRT model is a period-doubling cascade.

13.2 $N > 10$

We next describe the effects of increasing number of periods making up the SSL, keeping all other parameters fixed. In the case of $N = 20$, both the first and second plateaus are present. The supercritical Hopf bifurcation corresponding to the beginning of the first plateau is shown in the first column of Fig. 10. No other bifurcations are observed in the first plateau. The onset of the second plateau is shown in the second column of Fig. 10. In the second plateau, we again find period doubling cascades to chaotic attractors; this behavior is illustrated in the third column of Fig. 10. Comparing the third columns of Figs. 8 and 10, we observe that the period-doubling cascade and the chaotic attractor occur over narrower voltage intervals in the $N = 20$ case compared with the $N = 10$ case. For higher values of N, the voltage intervals containing the period doubling bifurcations become increasingly narrow, and eventually disappear entirely from the second plateau near $N = 30$.

As N is increased, we observe the appearance of more dynamical instabilities in the first plateau. A period-doubling bubble emerges in the first plateau near $N = 25$ and subsequently widens over a larger interval of V_{bias} with increasing N. This bubble is responsible for the period-two orbit illustrated in the top panel of Fig. 12 for the case of $N = 50$. There are no further period-doubling bifurcations present in the first plateau for this value of N. For $N = 100$, simulations by Amann *et al.* showed chaotic dynamics occur in the first plateau [2]. This result fits with the trend of dynamical instabilities moving from the second plateau to the first plateau as N increases.

13.3 Noise

We next consider the effects of the stochastic terms in Eqs. (12) and (13), which model the effects of a noisy voltage source and intrinsically random tunneling processes. We observe that the dynamics become increasingly sensitive to noise with increasing

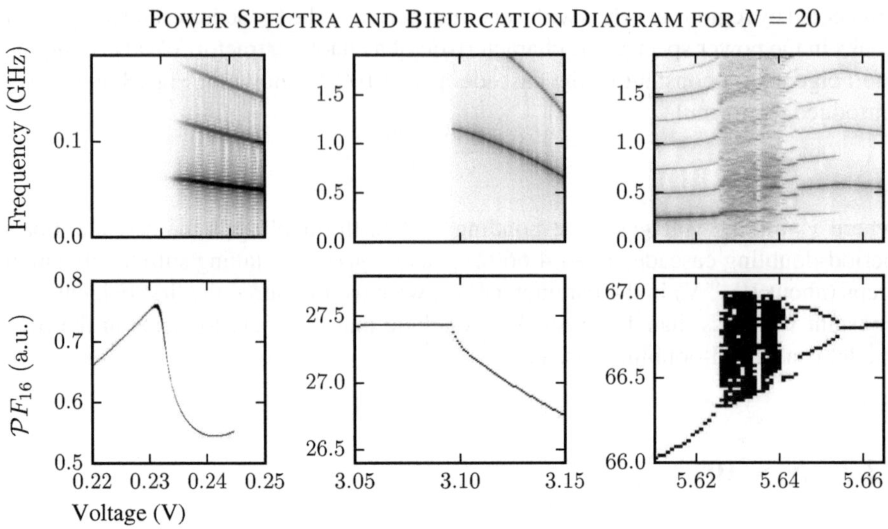

Fig. 10 (Top row) The power spectrum of $J(t)$ plotted against the bias voltage. (Bottom row) The bifurcation diagram, plotting the Poincaré map against the bias voltage. The Hopf bifurcation from the steady state in the first plateau is shown in the first column. The Hopf bifurcation from the steady state in the second plateau is shown in the second column. A narrow region in the second plateau containing a chaotic attractor is shown in the third column

N. For the case of $N = 10$, regions of interest in the bifurcation diagram are plotted in Fig. 11. We have chosen these voltage intervals to be the same as in Fig. 8 for clear comparison. We observe that the addition of noise stimulates the Hopf bifurcation to occur at lower voltages, which widens the second plateau. Noise also has the effect of broadening the peaks in the power spectrum as shown in the first row of Fig. 11.

Upon perturbation by noise, period-doubling behavior may be enhanced and higher period orbits may occur over a particular window of V_{bias} than do in the noise-free case, see Fig. 12. demonstrates the effect of very small perturbations by noise on the phase portrait over the window containing the period-doubling bubble, which occurs in the first plateau for the $N = 50$ case. For higher-period orbits, the broadening effect of perturbations on the power spectrum can cause the narrowly spaced peaks in the spectrum to merge, transforming high-period orbits to chaotic attractors as demonstrated in Fig. 11. This effect may broaden the windows in which chaotic attractors occur, connecting chaotic attractors that are distinct in the noise-free limit as demonstrated by comparison of the chaotic regions in Figs. 8 and 11.

Periodic orbits in the first plateau for $N = 50$ are more sensitive to noise than those which occur in the second plateau for $N = 10$. This difference is evident upon comparison of Fig. 11 with Fig. 12, in particular the power spectra. We see similar results but have used much less noise to produce Fig. 12 than Fig. 11. The bottom panel of Fig. 12 shows that the inclusion of noise terms may cause the Poincaré map and power spectrum to resemble those of a chaotic system. However, we simulate

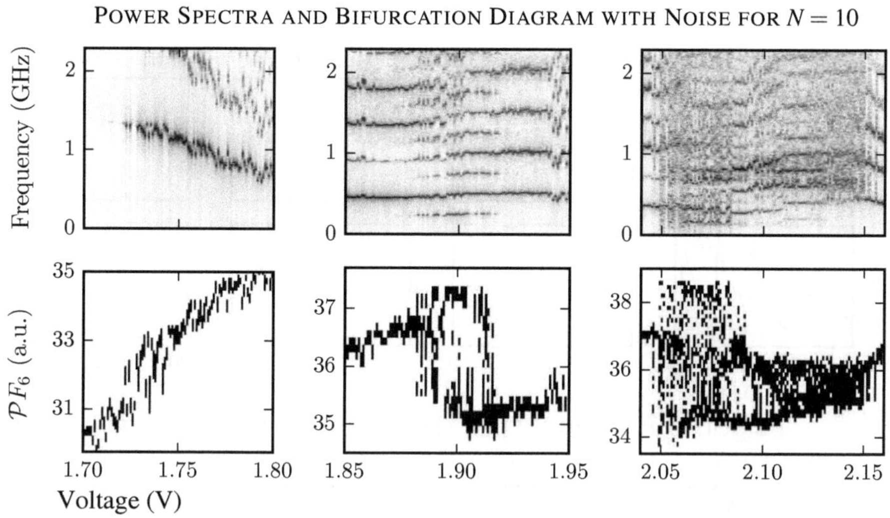

Fig. 11 (Top row) The power spectrum of $J(t)$ plotted against the bias voltage. (Bottom row) The bifurcation diagram, plotting the Poincaré map against the bias voltage. The Hopf bifurcation from the steady state is shown in the first column. A period doubling bubble is shown in the second column. A period-doubling cascade is shown in the third column. $\sigma_\eta = 1.4 \times 10^{-5}$ V

the same situation in the absence of noise in the top panel of Fig. 12. While the SRT model is very sensitive to noise in this regime, the underlying dynamical structure is an orbit of period two, and hence our numerical methods are able discern between noise-sensitivity and true dynamical chaos.

13.4 Disorder

Sensitivity to disorder is independent of N. The model is more sensitive to disorder with correlations between β_i and ζ_i. The introduction of noise decreases the effect of disorder.

The values of V_{bias} at which a particular instability (Hopf bifurcations, period-doublings, period-doubling cascades, and chaotic attractors) occurred in our simulations was shifted upon the introduction of disorder, broadening the voltage windows where it is *possible* for a certain bifurcation to occur. This means that the introduction of disorder makes it possible for chaotic behavior to occur at lower *and* higher values of voltage bias, even for the same value of disorder. Figure 13 shows a chaotic attractor which is centered about the same value of V_{bias} but is much wider in the disordered SSL.

For sufficient values of σ, additional instabilities—including chaotic attractors, may occur which are not observed in the clean (free of disorder) case. Alternatively,

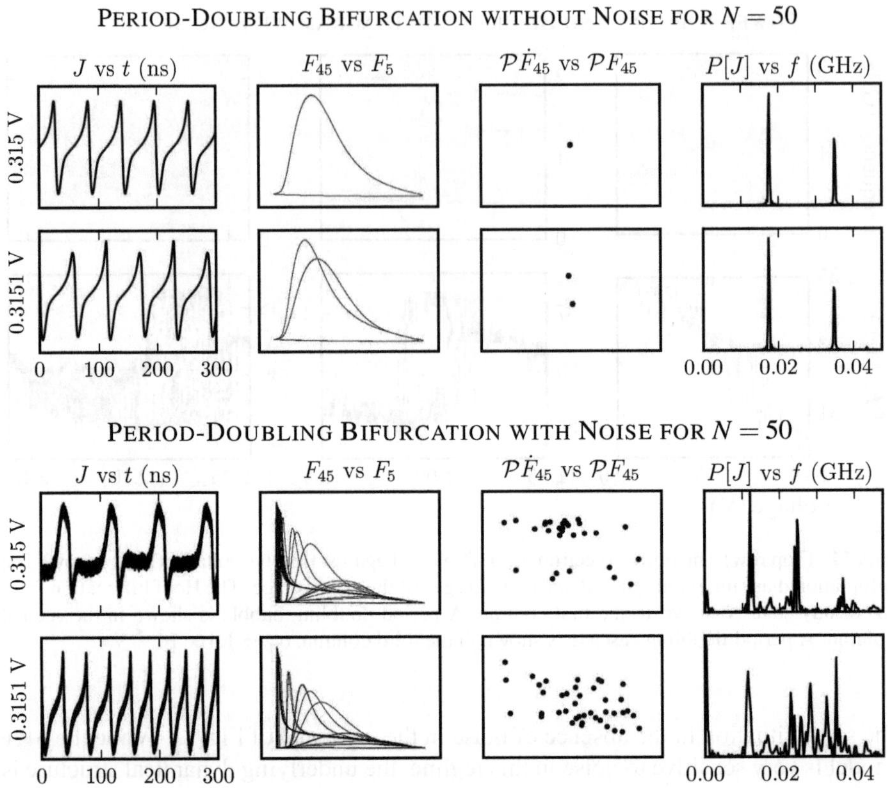

Fig. 12 Period doubling with and without noise. $\sigma_\eta = 2.8 \times 10^{-6}$ V

period doubling bifurcations that are present in the clean case may suppressed by the introduction of disorder, as demonstrated in Fig. 13.

Distinct chaotic attractors which occur in the second plateau for $N = 10$ are robust against disorder to varying degrees, and some have been observed to persist for large values of σ. Over multiple simulations, we have observed that the chaotic region near 2.25 V is the most robust to disorder as demonstrated in Fig. 13.

The enhancement or suppression of chaotic behavior and the voltage regions over which this may occur is dictated by the random numbers β_i and ζ_i so we believe it will be possible to design a superlattice with specific variations in lengths of wells and barriers to enhance the robustness of chaotic attractors.

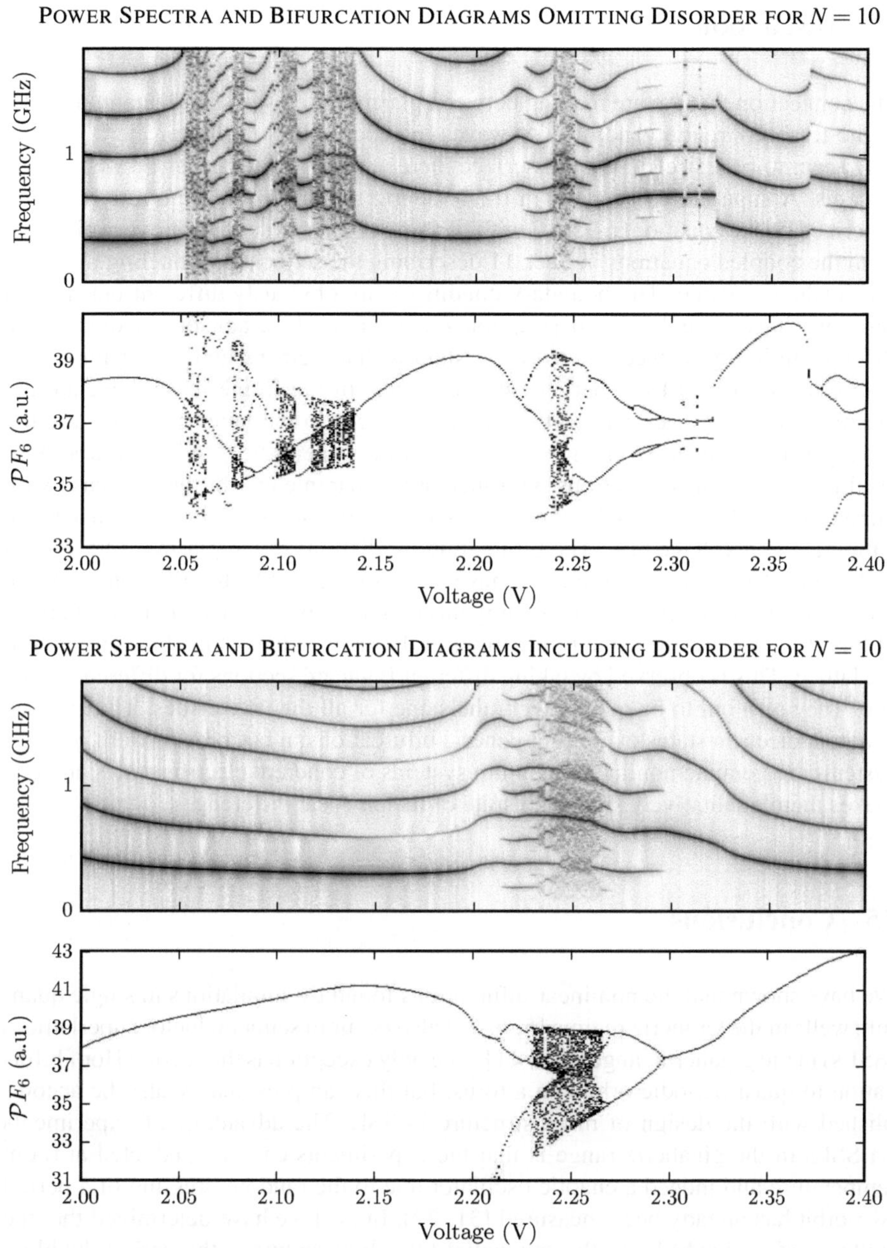

Fig. 13 (Top panel) Bifurcation diagram with $N = 10$ without disorder. (Bottom panel) Bifurcation diagram with disorder of strength $\sigma = 1\%$

14 Discussion

The connection that we are making between plasmon states in single quantum wells in the terahertz regime and density waves in superlattices in the gigahertz regime may seem to be a stretch. On the surface there are many differences between these systems. A superlattice of wells in the terahertz regime would be tightly coupled described by the equation in Sect. 3 with periodic boundary conditions, very different from the coupled equations in Sect. 11 describing the sequential tunneling model in the gigahertz regime. The boundary conditions are obviously different but in both cases we have a qualitative description for a range of parameters involved. The striking similarity between these two system is that their qualitative behavior is in both cases governed by coherent electron states. In the former case these are the plasmons slushing back and forth in quantum well. In the latter case they are the density waves executing Gunn oscillations in the superlattice. In both cases these oscillations exhibit bifurcations with increase in parameters. In the first case with increasing amplitude of the laser drive, in the second case the bifurcations take place with increased voltage bias. The bifurcating oscillations of the plasmons have been understood for a long time, but we have shown in [21, 42] that the density wave form a coherent electron state extending through the superlattice and the oscillations of these states show the same bifurcations at the same values of the bias throughout the lattice. This is observed by taking different Poincaré sections for different lattice site. They turn out to be qualitatively the same for all the lattice sites. Thus we see coherent electron states exhibiting generic bifurcations in both cases. Both of these systems are genuine nonlinear quantum systems of coherent electron states and this makes them qualitatively similar in spite of the physical differences.

15 Conclusions

We have shown that the nonlinear bifurcations found by simulations in single quantum wells in the terahertz regime [5, 6, 25] also occur in semiconductor superlattices (SSLs) in the gigahertz range [2, 3, 11]. The only exception is the second Hopf bifurcation to quasi-periodic orbits on a torus, but this can presumably also be accomplished with the design of more structure in SSL. The advantage of experiments on SSLs in the gigahertz range is that the experiments can be conducted at room temperature and indeed a chaotic oscillator due to the random dressing of a period two-orbit has already been measured [31, 32]. In [42] we have determined that the route to chaos for SSLs in the sequential tunneling regime is the period doubling cascade. Shorter (10-period) superlattices are observed to exhibit faster oscillations compared with longer (50-period) ones. Two plateaus are observed as functions of the voltage bias, and intrinsically chaotic dynamics on the second plateau are possible only for shorter SSLs, while the dynamics in the first plateau contain intrinsic chaos only for longer ($N > 50$) SSLs, see [21]. The robustness of these results to stochastic perturbations in the local tunneling currents and the bias voltage was tested in [21]. It was observed that shorter SSLs are much less sensitive to noise compared with longer SSLs. Therefore two modes of random number generation are possible: Faster, intrinsic chaos in the second plateau for shorter superlattices, and slower, noise-enhanced chaos in the first plateau for longer superlattices, see [21] for more details.

The effects of random variations in doping density and the width of the wells and the barriers was also examined in [21]. It was found that the period-doubling cascade is robust to these perturbations, but the detailed shape of the bifurcation diagram can change significantly. Then these perturbations cannot unfold the period-doubling bifurcation into a second Hopf bifurcation as we initially conjectured. They are simply not strong enough to break the reflection symmetry of the constituent wells. To observe the second Hopf bifurcation it is essential that this symmetry is broken analogous to the work of Batista et al. [5, 6], then two or more states would also exist below the Fermi level. We conjecture this greater number of active states will be successful at exposing the second Hopf bifurcation in a properly designed SSL. It remains an open question whether all of these bifurcations can be extended to SSLs in the terahertz regime. This would signify a nontrivial technological progress since terahertz devices are difficult to make and operate at room temperature. It these bifurcations are found in teraherz range the possibility of making all the devices discussed above opens up and such devices can be operated at significantly faster time-scales.

Acknowledgements This material is based upon work supported by, or in part by, the U.S. Army Research Laboratory and the U. S. Army Research Office under contract/grant number 444045-22682.

References

1. M. Alvaro, M. Carretero, L. Bonilla, Noise-enhanced spontaneous chaos in semiconductor superlattices at room temperature. EPL (Europhys. Lett.) **107**, 37002 (2014)
2. A. Amann, J. Schlesner, A. Wacker, E. Schöll, Chaotic front dynamics in semiconductor superlattices. Phys. Rev. B **65**, 193313 (2002)
3. T. Ando, A.B. Fowler, F. Stern, Electronic properties of two-dimensional systems. Rev. Mod. Phys. **54**, 437 (1982)
4. S. Asmussen, P.W. Glynn. *Stochastic Simulation: Algorithms and Analysis*, vol. 57 (Springer Science & Business Media, 2007)
5. A.A. Batista, P.I. Tamborenea, B. Birnir, M. Sherwin, D.S. Citrin, Nonlinear dynamics in far-infrared driven quantum-well intersubband transitions. Phys. Rev. B **66**, 195325 (2002)
6. A.A. Batista, B. Birnir, P.I. Tamborenea, D.S. Citrin, Period-doubling and Hopf bifurcations in far-infrared driven quantum well intersubband transitions. Phys. Rev. B **68**, 035307 (2003)
7. B. Birnir, Chaotic perturbations of KdV equations: I. rational solutions. Phys. D Nonlinear Phenom. **18**, 464 (1986)
8. B. Birnir, R. Grauer, An explicit description of the global attractor of the damped and driven sine-Gordon equation. Commun. Math. Phys. **162**, 539 (1994)

9. B. Birnir, *Basic Attractors and Control* (To be published by Springer, New York, 2017)
10. K. Binder, D.W. Heermann, *Monte Carlo Simulation in Statistical Physics: An Introduction* (2002)
11. L.L. Bonilla, J. Galán, J. Cuesta, F.C. Martínez, J.M. Molera, Dynamics of electric field domains and oscillations of the photocurrent in a simple superlattice model. Phys. Rev. B **50**, 8644 (1994)
12. L.L. Bonilla, G. Platero, D. Sánchez, Microscopic derivation of transport coefficients and boundary conditions in discrete drift-diffusion models of weakly coupled superlattices. Phys. Rev. B **62**, 2786 (2000)
13. L.L. Bonilla, Theory of nonlinear charge transport, wave propagation and self-oscillations in semiconductor superlattices. J. Phys. Condens. Matter **14**, R341 (2002)
14. L.L. Bonilla, H.T. Grahn, Non-linear dynamics of semiconductor superlattices. Rep. Prog. Phys. **68**, 577 (2005)
15. L.L. Bonilla, S.W. Teitsworth, *Nonlinear Wave Methods for Charge Transport* (Wiley VCH, Weinheim, 2009)
16. L.L. Bonilla, R. Escobedo, G. Dell'Acqua, Voltage switching and domain relocation in semiconductor superlattices. Phys. Rev. B **73**, 115341 (2006)
17. L.L. Bonilla, M. Alvaro, M. Carretero, Chaos-based true random number generators. J. Math. Ind. **7**, 1 (2016)
18. P. Collet, J. Eckmann, *Iterated Maps on the Interval as Dynamical Systems*, Modern Birkhäuser Classics (Birkhäuser Boston, 2009)
19. K. Craig et al., Undressing a collective intersubband excitation in quantum well. Phys. Rev. Lett. **76**, 2382 (1996)
20. G. Dell'Acqua, L.L. Bonilla, R. Escobedo, Hopf Bifurcation in a Superlattice Model, in *Proceedings of the International Conference on Computational and Mathematical Methods in Science and Engineering* (2006)
21. J. Essen, M. Ruiz-Garcia, I. Jenkins, M. Carretero, L.L. Bonilla, B. Birnir, Parameter dependence of high-frequency nonlinear oscillations and intrinsic chaos in short GaAs/(Al,Ga)As superlattices (2017) (Submitted)
22. J. Essen, Simulations of open ballistic quantum systems and semiclassical semiconductor superlattices. Ph.D. thesis, University of California, Santa Barbara, 2017
23. R.G. Gallager, *Principles of Digital Communication* (Cambridge University Press, Cambridge, UK, 2008)
24. B. Galdrikian, M. Sherwin, B. Birnir, Self-consistent Floquet states for periodically driven quantum wells. Phys. Rev. B **49**, 13744 (1994)
25. B. Galdrikian, B. Birnir, Period Doubling and Strange Attractors in Quantum Wells. Phys. Rev. Lett. **76**, 3308 (1996)
26. R. Graham, Squeezing and frequency changes in harmonic oscillations. J. Mod. Optics **34**, 873 (1987)
27. R. Grauer, B. Birnir, The center manifold and bifurcations of damped and driven sine-Gordon breathers. Phys. D Nonlinear Phenom. **56**, 165 (1992)
28. J. Guckenheimer, P. Holmes, *Nonlinear oscillations, dynamical systems, and bifurcations of vector fields* (Springer, New York, 1983)
29. J.N. Heyman et al., Resonant harmonic generation and dynamics screening in a double quantum well. Phys. Rev. Lett. **72**, 2183 (1994)
30. J.N. Heyman et al., Temperature and intensity dependence of intersubband relaxation rates from photovolgate and absorption. Phys. Rev. Lett. **74**, 2682 (1995)
31. Y. Huang, W. Li, W. Ma, H. Qin, Y. Zhang, Experimental observation of spontaneous chaotic current oscillations in GaAs/Al0.45Ga0.55As superlattices at room temperature. Chin. Sci. Bull. **57**, 2070 (2012)
32. Y. Huang, W. Li, W. Ma, H. Qin, H.T. Grahn, Y. Zhang, Spontaneous quasi-periodic current self-oscillations in a weakly coupled GaAs/(Al, Ga)As superlattice at room temperature. Appl. Phys. Lett. **102**, 242107 (2013)
33. I. Kanter, Y. Aviad, I. Reidler, E. Cohen, M. Rosenbluth, An optical ultrafast random bit generator. Nat. Photonics **4**, 58 (2010)

34. J. Kastrup, R. Hey, K.H. Ploog, H.T. Grahn, L.L. Bonilla, M. Kindelan, M. Moscoso, A. Wacker, J. Galán, Electrically tunable GHz oscillations in doped GaAs-AlAs superlattices. Phys. Rev B. **55**, 2476 (1997)
35. E.W. Laedke, K.H. Spatschek, On localized solutions in nonlinear faraday resonance. J. Fluid Mech. **223**, 589–601 (1991)
36. K. Leo, P.H. Bolivar, F. Brüggemann, R. Schwedler, K. Klaus, Köhler, Observation of Bloch oscillations in a semiconductor superlattice. Solid State Commun. **10**, 943–946 (1992)
37. W. Li, I. Reidler, Y. Aviad, Y.Y. Huang, H. Song, Y.H. Zhang, M. Rosenbluh, I. Kanter, Fast physical random-number generation based on room-temperature chaotic oscillations in weakly coupled superlattices. Phys. Rev. Lett. **111**, 044102 (2013)
38. C.M. Morris, Interplay of quantum confinement, electron-electron interactions, and terahertz radiation in indium gallium arsenide quantum posts and gallium arsenide quantum wells. Ph.D. thesis, University of California, Santa Barbara, 2011
39. T.E. Murphy, R. Roy, The world's fastest dice. Nat. Photonics **2**, 714–715 (2008)
40. M.A. Nielsen, I.L. Chuang, *Quantum Computation and Quantum Information* (Cambridge University Press, Cambridge, UK, 2000)
41. I. Reidler, Y. Aviad, M. Rosenbluh, I. Kanter, Ultrahigh-speed random number generation based on a chaotic semiconductor laser. Phys. Rev. Lett. **103**, 024102 (2009)
42. M. Ruiz-Garcia, J. Essen, M. Carretero, L.L. Bonilla, B. Birnir, Enhancing chaotic behavior at room temperature in GaAs/(Al, Ga)As superlattices. Phys. Rev. B **95**, 085204 (2017)
43. A.K. Saxena, The conduction band structure and deep levels in Ga1-xAlxAs alloys from a high-pressure experiment. J. Phys. C Solid State Phys. **23**, 4323 (1980)
44. M. Sciamanna, K.A. Shore, Physics and applications of laser diode chaos. Nat. Photonics **9**, 151–162 (2015)
45. D.R. Stinson, *Cryptography: Theory and Practice*, 3rd edn. (CRC Press, Boca Raton, 2006)
46. A. Uchida, K. Amano, M. Inoue, K. Hirano, S. Naito, H. Someya, I. Oowada, T. Kurashige, M. Shiki, S. Yoshimori, K. Yoshimura, P. Davis, Fast physical random bit generation with chaotic semiconductor lasers. Nat. Photonics **2**, 728–732 (2008)
47. J.Q. Wu, D.S. Jiang, B.Q. Sun, Room-temperature microwave oscillation in alas/gaas superlattices. Phys. E Low-dimensional Syst. Nanostruct. **4**(2), 137–141 (1999)
48. Z. Yin, Y. Zhang, M. Ruiz-García, M. Carretero, L.L. Bonilla, K. Biermann, H.T. Grahn, Noise-enhanced chaos in a weakly coupled GaAs/(Al,Ga)As superlattice (unpublished)
49. M. Zalużny, Influence of the depolarization effect on the nonlinear intersubband absorption spectra of quantum wells. Phys. Rev. B **47**, 3995 (1993)
50. M. Zalużny, Saturation of intersubband absorption and optical rectification in asymmetric quantum wells. J. Appl. Phys. **74**, 4716 (1993)

Chaotic Current Self-oscillations in Doped, Weakly Coupled Semiconductor Superlattices for True Random Number Generation

Yaohui Zhang, Yaara Aviad and Holger T. Grahn

Abstract A weakly coupled semiconductor superlattice (SL) represents an almost ideal one-dimensional nonlinear dynamical system, the nonlinearity of which is due to sequential resonant tunneling between adjacent quantum wells. A great richness of nonlinear transport behavior has been observed in weakly coupled SLs, including periodic as well as quasi-period current self-oscillations and even driven as well as undriven chaos. Recently, spontaneous chaotic and quasi-periodic self-oscillations have been observed in doped GaAs/(Al,Ga)As SLs with 45% Al at room temperature. Based on this type of SLs, an *all-electronic* true random number generator has been demonstrated at room temperature with achievable bit rates of up to 80 Gbit/s, about two orders of magnitude larger than typical bit rates for currently available all-electronic true random number generators. The synchronization of chaos using these SLs has been demonstrated as a useful building block for various tasks in secure communications. The realization of chaotic SLs without external feedback and the synchronization among differently structured SLs open up the possibility for advanced secure multi-user communication methods based on large networks of coupled SLs.

Keywords Semiconductor superlattice · Spontaneous chaos · True random number generator · Synchronization

---

Y. Zhang
Key Laboratory of Nanodevices and Applications, Suzhou Institute of Nano-Tech and Nano-Bionics, Chinese Academy of Sciences, Suzhou 215125, China
e-mail: yhzhang2006@sinano.ac.cn

Y. Aviad
Department of Applied Mathematics, Jerusalem College of Technology, 93721 Jerusalem, Israel
e-mail: aviaday@gmail.com

Y. Aviad
Department of Physics, Bar-Ilan University, 52900 Ramat-Gan, Israel

H. T. Grahn (✉)
Paul-Drude-Institut für Festkörperelektronik, Leibniz-Institut im Forschungsverbund Berlin e. V., Hausvogteiplatz 5–7, 10117 Berlin, Germany
e-mail: htgrahn@pdi-berlin.de

L. L. Bonilla et al. (eds.), *Coupled Mathematical Models for Physical and Biological Nanoscale Systems and Their Applications*, Springer Proceedings in Mathematics & Statistics 232, https://doi.org/10.1007/978-3-319-76599-0_2

1 Introduction

Undriven chaotic current self-oscillations in semiconductor superlattices (SLs) were discovered in the mid 1990s in doped, weakly coupled GaAs/AlAs SLs at low temperatures [1]. Because of the indirect nature of the band gap in AlAs together with the rather low energy of the X valley, Γ-X transfer from the GaAs wells through the AlAs barriers can easily occur by thermal excitation at elevated temperatures. However, in 2012, Huang et al. [2] demonstrated that, by using the highest direct barrier of $Al_{0.45}Ga_{0.55}As$ in GaAs/(Al,Ga)As SLs, spontaneous chaotic current self-oscillations are observable at room temperature. The frequency spectrum of the chaotic current self-oscillations ranges from 0 to 4 GHz. The corresponding devices are of sub-millimeter size so that they can be easily integrated in existing devices.

This type of SL was subsequently used to demonstrate *all-electronic* true random number generators with bit rates up to 80 Gbit/s [3]. In comparison, the highest bit rates of currently available all-electronic true random number generators are typically below 1 Gbit/s [4] so that this type of true random number generator would result in an increase of the current bit rate by about two orders of magnitude. These devices do not require external feedback or conversion to an electronic signal prior to digitization. The method is robust and insensitive to external perturbations. At the same time, its fully electronic implementation suggests scalability and minimal postprocessing in comparison to existing optical implementations of true random number generators.

More recently, chaos synchronization has been demonstrated as a useful building block for various tasks in secure communications, including a source of all-electronic ultrafast true random number generators based on room temperature spontaneous chaotic oscillations in this type of SL [5]. The emergence of several types of chaos synchronization has been experimentally demonstrated, e.g. leader-laggard, face-to-face, and zero-lag synchronization in network patterns of coupled SLs consisting of unidirectional and mutual coupling as well as self-feedback coupling. Chaotic SL oscillators without external feedback and the synchronization among differently structured SLs open up the possibility for advanced secure multi-user communication methods based on large networks of coupled SLs.

2 Electric-Field Domain Formation and Chaos in Semiconductor Superlattices

In this section, we will briefly review the formation of static electric-field domains in doped, weakly coupled SLs and present spontaneous current self-oscillations due to the formation of dynamic domains. Spontaneous current self-oscillations can be driven with an incommensurate sinusoidal voltage for a fixed bias into a chaotic regime, i.e. transitions between synchronization and chaos are observed via pattern-forming bifurcations. However, it is also possible for doped, weakly coupled SLs to exhibit undriven chaos, i.e. spontaneous chaotic current self-oscillations. For a detailed review of the work performed at temperatures well below room temperature, see Ref. [6].

2.1 Electric-Field Domains

Applying an electric field to a weakly coupled SL results in nonlinear transport along the SL axis due to resonant tunneling between different subbands of adjacent quantum wells. The resulting drift velocity-electric field characteristic becomes nonlinear with resonances at field strengths corresponding to the energetic separation of the involved subbands divided by the SL period d and the electron charge e. Between these resonances, the current density will exhibit regions of negative differential conductivity (NDC) and positive differential conductivity (PDC) as a function of the applied electric field. If carriers are added by for example doping the SL, the regions of negative differential conductivity become unstable, when the SL is biased in this region. For large carrier densities, static electric-field domains will form, where a region of a lower electric field F_{LFD} is separated by a charge accumulation layer (CAL), i.e. the domain boundary, from a region with a larger electric field F_{HFD} as shown in Fig. 1. LFD and HFD refer to the low-field and high-field domain, respectively. The values of F_{LFD} and F_{HFD} are in the PDC regions of the first and second resonance, respectively. The carrier density of the accumulation layer is determined by the difference of the electric field strengths of the two domains $F_{\mathrm{HFD}} - F_{\mathrm{LFD}}$ multiplied by the product of the relative permittivity ε of the material and the permittivity of the vacuum ε_0 divided by e.

2.2 Spontaneous Current Self-oscillations and Chaos

When the carrier density is reduced, the field distribution becomes unstable, resulting in spontaneous current self-oscillations, the frequency of which depends on the

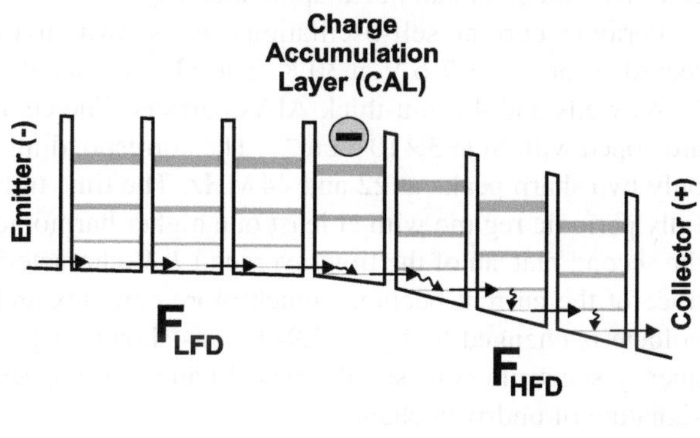

Fig. 1 Schematic diagram of the distribution of the low- (F_{LFD}) and high-field domain (F_{HFD}) separated by a charge accumulation layer (CAL) in a doped, weakly coupled SL

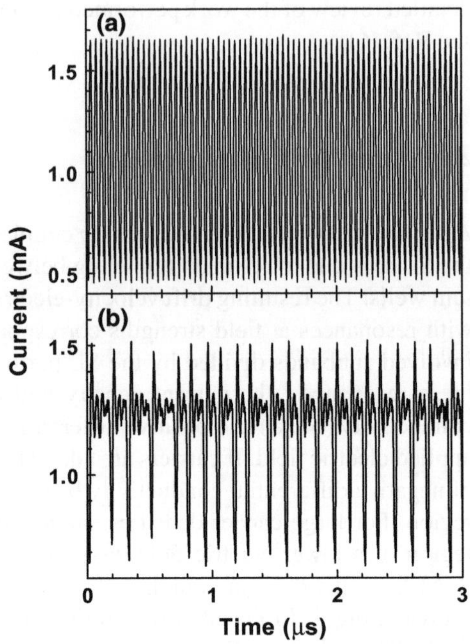

Fig. 2 Current oscillations in **a** the periodic regime at $V_{dc} = 7.6$ V and **b** the chaotic regime at $V_{dc} = 7.941$ V for the GaAs/AlAs SL recorded at 30 K [7]

resonant coupling between adjacent wells. There are three types of current self-oscillations:

- spontaneous periodic,
- spontaneous quasi-periodic, and
- driven as well as spontaneous chaotic.

For spontaneous periodic current self-oscillations, either a charge monopole or a dipole, which represents the domain boundary, performs a recycling motion through the whole or part of the SL. For spontaneous quasi-periodic current self-oscillations, there is a competition between the dipole and monopole oscillation modes. Finally, chaotic current self-oscillations can be induced by driving the SL with an appropriate external voltage or can occur spontaneously.

Periodic current self-oscillations are shown in Fig. 2a for a GaAs/AlAs SL recorded for $V_{dc} = 7.6$ V at 30 K. The SL consists of 40 periods with 9.0-nm-thick GaAs wells and 4.0-nm-thick AlAs barriers. The central 5 nm of each GaAs well are doped with Si at 3×10^{17} cm^{-3}. The corresponding frequency spectrum contains only two sharp peaks at 22 and 44 MHz. The time trace in Fig. 2a corresponds to a fully periodic regime with at least one higher harmonic. These oscillations occur in the second plateau of the time-averaged I-V characteristic in this sample. The time trace of the current becomes much more complex and irregular, when the applied voltage is changed to $V_{dc} = 7.941$ V as shown in Fig. 2b. The corresponding frequency spectrum consists of a broad-band noise spectrum. This time trace has the signature of undriven chaos.

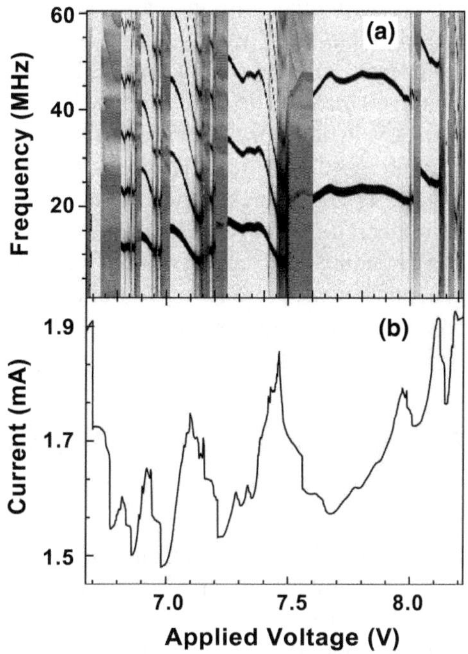

Fig. 3 **a** Frequency spectra of the spontaneous current oscillations and **b** time-averaged current versus applied voltage for the GaAs/AlAs SL recorded at 30 K. In the power spectra, dark areas correspond to large amplitudes [7]

The existence of chaotic windows in the frequency spectra as a function of applied voltage becomes more evident when the power spectra for the GaAs/AlAs SL recorded at 30 K are plotted versus the applied voltage on a grey scale as in Fig. 3a. Dark areas correspond to a large amplitude of the oscillations. The corresponding time-averaged I-V characteristic is shown in Fig. 3b on the same voltage scale at the same temperature [1, 7]. With increasing bias, several chaotic windows exist, which are separated by periodic windows containing only the fundamental frequency and several higher harmonics. The transition from the chaotic to the periodic window occurs over a very narrow voltage range, e.g. at $V_{dc} = 7.6$ V. However, the transition from a periodic to a chaotic window usually consists of a random-enhancing process covering a broader voltage range. Comparing the voltage ranges of the chaotic windows with the time-averaged I-V characteristic in Fig. 3b, there is a clear correlation between the existence of a chaotic window and a large NDC in the time-averaged current. In contrast, the periodic windows appear for PDC.

The sharp transitions from the chaotic to the periodic window correspond to the minimum value of the average current, i.e. the onset of PDC. The regions with a PDC are characterized by attractive coupling between the different quantum wells. The SL behaves as a self-synchronized unit with spontaneous periodic current oscillations. However, in the regions with a large NDC, the coupling becomes strongly repulsive. With increasing repulsion, the synchronized oscillations become more and more destabilized, until the SL enters a chaotic state. With increasing temperature, the chaotic windows become smaller and smaller, until they completely disappear at 60 K

[7]. As the temperature increases, the thermal energies increase so that thermionic emission into the X valley of the AlAs barriers becomes more easily possible, since the energy separation between the Γ valley in GaAs and the X valley in AlAs is about 160 meV, while the difference between the Γ valleys of both materials is about 980 meV. In addition, other scattering mechanisms such as optical phonon scattering become more and more likely, thereby reducing the effective non-linearity of the system. The structures in the time-averaged I-V characteristic disappear completely for temperatures above 60 K in GaAs/AlAs SLs so that a structureless current plateau is a signature of a large periodic window. The undriven chaos itself is probably generated by the presence of the finite time scale of the sequential resonant tunnelling process between adjacent wells.

2.3 *Spontaneous Current Self-oscillations at Room Temperature*

In order to achieve spontaneous current self-oscillations in a weakly coupled SL at room temperature, the choice of the barrier material is crucial. To achieve the highest direct barrier in a GaAs/Al_xGa_{1-x}As SL, the Al content x should be chosen with a value of about 0.45. In this case, the Γ and X valleys of the $Al_{0.45}Ga_{0.55}As$ barrier are separated by about 390 meV from the Γ valley of GaAs so that the thermal excitation of electrons in the GaAs quantum wells over the (Al, Ga)As barrier can be effectively reduced. This suppression may provide the existence of the nonlinearity in SLs up to higher temperature so that spontaneous current self-oscillations can occur in GaAs/$Al_{0.45}Ga_{0.55}As$ SLs even at room temperature. Indeed, spontaneous chaotic current self-oscillations have been observed experimentally in a GaAs/$Al_{0.45}Ga_{0.55}As$ SL at 300 K in 2012 [2]. The SL consists of 50 periods with 7.0-nm-thick GaAs wells and 4.0-nm-thick $Al_{0.45}Ga_{0.55}As$ barriers. The central 3.0 nm of each GaAs well are doped with Si at 3×10^{17} cm^{-3}.

When this GaAs/$Al_{0.45}Ga_{0.55}As$ SL is biased in the voltage range between 6.21 and 6.80 V, spontaneous chaotic current self-oscillations are observed. The corresponding frequency spectrum is very wide and very noisy with a frequency distribution ranging up to 4 GHz, providing a clear signature of spontaneous chaos [2].

Figure 4a shows the I-V characteristic of a GaAs/$Al_{0.45}Ga_{0.55}As$ SL with similar parameters as in Ref. [2], but a different doping density of 2×10^{17} cm^{-3} in the GaAs well, recorded at 300 K. When the dc bias voltage ranges between 4 and 4.3 V as well as 6.2 and 6.8 V marked by the gray-shaded regions in Fig. 4a, the SL exhibits spontaneous chaotic current self-oscillations. The oscillation amplitude exceeds 0.4 V into a 50 Ω load resistor, corresponding to 8 mA, and the bandwidth amounts to several hundreds of MHz as shown in Fig. 4b. The frequency spectrum of these spontaneous chaotic current self-oscillations is so wide that it can be used as wideband noise source. The inset of Fig. 4b displays a 100-ns-long time trace, which was digitized at 40 GHz.

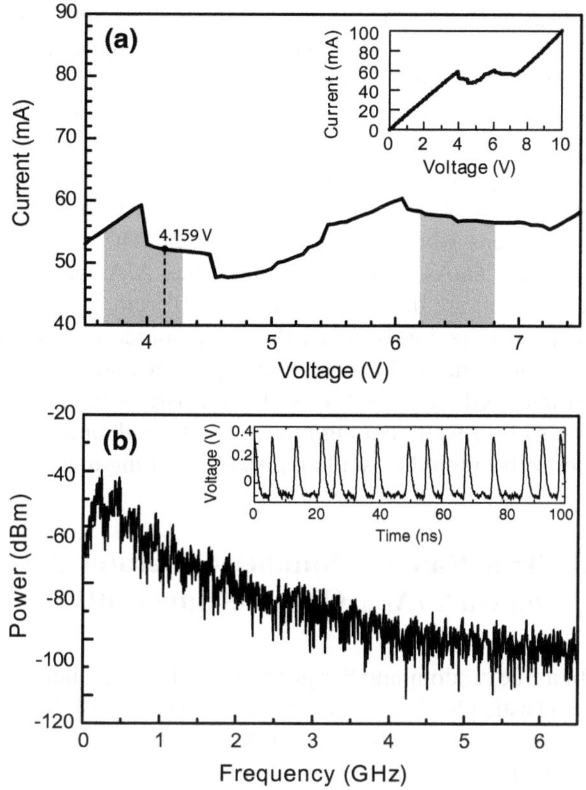

Fig. 4 **a** Plateau region of the I-V characteristic of the $GaAs/Al_{0.45}Ga_{0.55}As$ SL consisting of two voltage segments (gray regions) characterized by chaotic current oscillations. Most of the measurements discussed in Sect. 3 were carried out at 4.159 V (dashed vertical line). The inset shows the I-V characteristic over the entire voltage range. **b** Spectrum of the chaotic oscillations [2]. Inset: a 100-ns-long time trace of the SL current oscillations [3]

Bonilla et al. [8] have discussed a mathematical model to describe spontaneous chaos in idealized SLs with identical wells and barriers. Their numerical simulations show that spontaneous chaos may possibly appear directly from a two-frequency quasi-periodic attractor. In addition, they demonstrated that the unavoidable shot and thermal noise, which exist in the SL, both enhance existing deterministic chaos and induce chaos in nearby voltage intervals. They also discussed that the differences between numerical and experimental results may be due to imperfections in the doping density, the Ga content in the barriers, and the size thereof. A better model has to be developed in order to discuss the imperfections and their effect on the chaotic oscillations.

Spontaneous quasi-periodic current self-oscillations have also been observed at room temperature in this $GaAs/Al_{0.45}Ga_{0.55}As$ SL [9]. The underlying physical mechanism may be related to the interaction of two oscillation modes, namely, the collective dipole motion mode through the whole SL and the well-to-well hopping mode within a few SL periods. Furthermore, noise-enhanced chaos in a $GaAs/Al_{0.45}Ga_{0.55}As$ SL has been observed at room temperature in experiments [10] as well as in the results of the simulation of nonlinear transport based on a discrete tunneling model [10, 11]. The investigated sample has nominally the same parameters as in Ref. [3], but it was grown in a different place and a different time. When external noise is added, both the measured and simulated current-versus-time traces contain irregularly spaced spikes for particular applied voltages, which separate a regime of periodic current oscillations from a region of no current oscillations at all. Increased noise can consequently enhance chaotic current oscillations in SLs.

Finally, the presence of a coherence resonance has been experimentally identified in a doped, weakly coupled GaAs/AlAs SL at 77 K, which exhibits current self-oscillations when driven by noise [12]. The SL consists of 28 periods with 9.0-nm-thick GaAs wells and 4.0-nm-thick AlAs barriers. The central 5.0 nm of each GaAs well are doped with Si at 3×10^{17} cm^{-3}. The physical mechanism is related to the interaction between the noise source and two oscillation modes of the SL oscillator: the well-to-well hopping mode and the dipole-motion mode. For a range of intermediate noise levels, the intrinsic oscillation mode of the SL oscillator, i.e. the well-to-well hopping mode, is activated, and the dipole motion is forced into a collective periodic oscillating state, resulting in a coherence resonance.

3 True Random Number Generator Based on GaAs/$A_{0.45}Ga_{0.55}As$ Superlattices

In a true random number generator (TRNG), random numbers are produced from a stochastic physical process rather than a deterministic computer program. Currently, the bit rate of *all-electronic* TRNGs is typically below 1 Gbit/s [4]. These devices are often based on microscopic phenomena that generate low-level, statistically random noise signals such as thermal noise, the photoelectric effect, or quantum phenomena. In contrast, nonlinear electrical transport in SLs generates spontaneous chaotic current self-oscillations with GHz frequencies at room temperature, which may be used for TRNG devices. It is very important to increase the current bit rate of *all-electronic* TRNGs by several orders of magnitude for applications such as encryption, secure communication, stochastic computational codes, and electronic commerce.

Two methods have been used to generate a random bit sequence from the spontaneous chaotic current self-oscillations in GaAs/$A_{0.45}Ga_{0.55}As$ SLs schematically shown in Fig. 5. The first method is similar to the already demonstrated *optical* TRNG scheme, where the generation of the random bit sequence consists of the following two steps [13]. In the first step, a dynamical buffer of the last $n + 1$ successively digitized electrical current values of the SL is used to calculate the nth discrete derivative as exemplified for $n = 3$ in Fig. 5b. In the second step, the m least significant bits (LSBs) of the resulting nth derivative are appended to the random bit sequence as shown in Fig. 5a. The second method consists of a linear combination of the analog chaotic current oscillations of several uncorrelated SLs as shown in Fig. 5c.

The time trace of the spontaneous chaotic current self-oscillations shown in Fig. 6a was digitized at 40 and 1.25 GHz as indicated by the blue line and red dots, respectively. Figure 6b shows the 1st and the 4th derivatives of the signal presented in Fig. 6a sampled at 1.25 GHz. Whereas for the first derivative [red line in Fig. 6b]

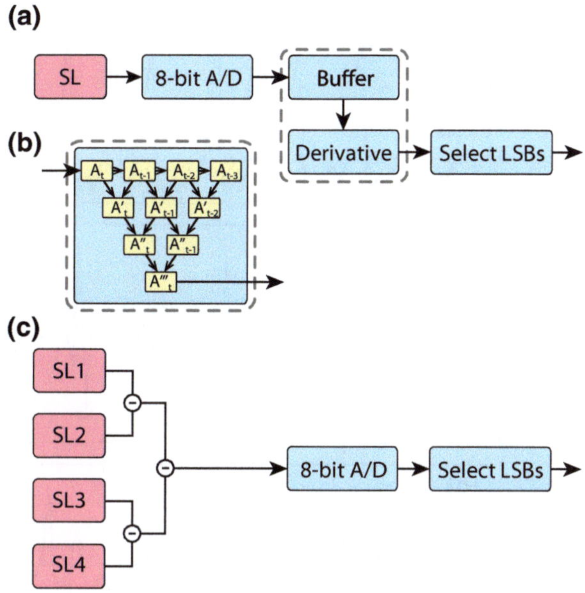

Fig. 5 **a** Schematic diagram of the TRNG method based on discrete time derivatives of the digitized current signal and retention of only a number of least significant bits (LSBs). **b** An example of the 3rd discrete derivative implementation of the method described in (**a**), where A_t stands for the digitized signal. **c** Schematic diagram of the parallel combination TRNG method where a minus sign stands for the subtraction, i.e., signal of SL1 minus the signal of SL2 [3]

there are many time points with near zero value, the fourth derivative values [blue line in Fig. 6b] fluctuate on all time scales. Retaining only the LSBs is still required to guarantee the generation of random sequences with verified randomness using the National Institute of Standards and Technology (NIST) statistical test suite [14]. Higher sampling rates than the 1.25 GHz used here, with the retention of 5 LSBs, require a much higher-order derivative; for example, a sampling rate of 5 GHz requires the 10th discrete derivative so as to pass the NIST statistical test. The implementation of higher derivatives requires a longer buffer and does not result in much improved bit rate generation. Using the first method, a bit rate of 6.25 Gbit/s was achieved using a sampling rate of 1.25 GHz, the 4th derivative, and a retention of 5 LSBs out of 8 bits.

The second method for implementing an SL-based TRNG overcomes these difficulties. Since the spike timings of independent SL devices are uncorrelated, one can linearly combine several such signals and "fill" the interspike intervals as shown in Fig. 6c for a linear combination of four recorded SL time traces digitized at 40 GHz. Because of the unavailability of several such SL devices in our experiment, this TRNG method was tested by combining segments with a large separation in time of the recorded current oscillations of a single device. The chaotic nature of the device ensures absence of any correlation between these segments. This method, as illustrated in Fig. 5c, requires only a single analog-to-digital converter regardless of the number of combined SL devices. To minimize the possible emergence of bias in the combined analog signals, each pair of signals is combined by subtraction. For example in Fig. 5c, the combined signal consists of SL1 + SL4 − SL2 − SL3. A 40 Gbit/s TRNG with verified randomness was obtained using a linear combination

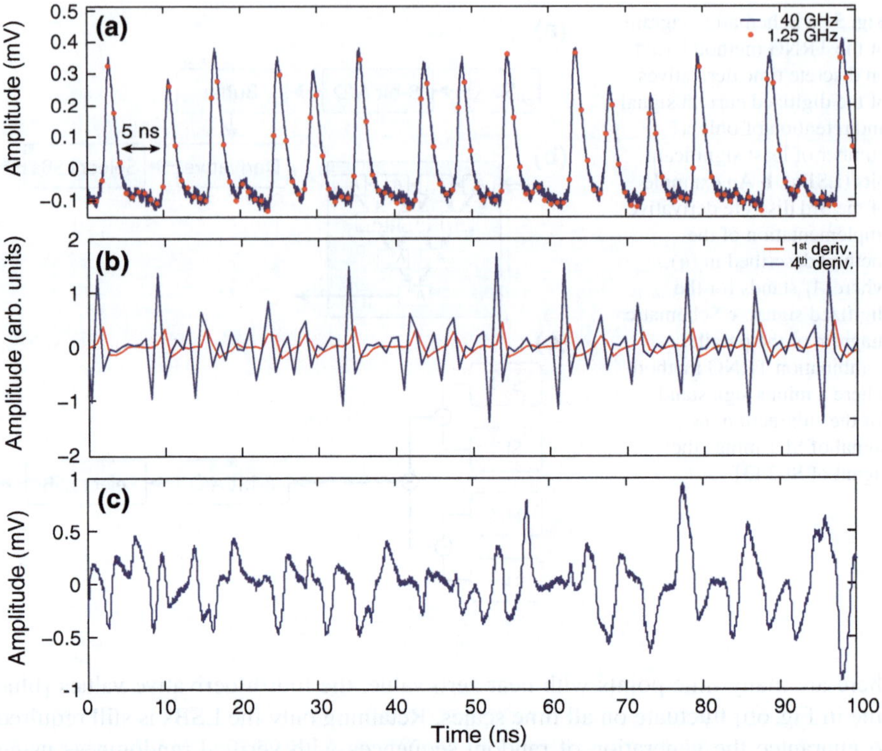

Fig. 6 **a** A 100 ns time trace of SL oscillations digitized at 40 (blue line) and 1.25 GHz (red dots). **b** 1st and 4th order discrete derivatives of the SL oscillations sampled at 1.25 GHz presented in panel (**a**). **c** Linear combination of four recorded SL time traces digitized at 40 GHz [3]

of four signals, a sampling rate of 10 GHz, and 4 LSBs. At a sampling rate of 20 GHz, the statistical NIST tests for randomness of this combination failed. However, a combination of six signals passed the statistical tests, resulting in a TRNG with a bit rate of 80 Gbit/s with verified randomness. This value is about two orders of magnitude larger than the bit rates for currently available *all-electronic* TRNGs. Because of the submillimeter dimensions and all-electronic nature of the SL devices, they can be easily integrated into existing devices.

Table 1 summarizes several key points of the tested parameter space according to the order of the derivative and the number of combined independent SL signals for the retention of 5 or 4 LSBs. For a single SL sampled at 1.25 GHz and 5 LSBs retained, at least the 4th order derivative is required to generate a verified random sequence resulting in a bit rate of 6.25 Gbit/s for the TRNG. When the sampling rate is increased to 5 GHz, higher derivatives become necessary even for fewer than 5 LSBs. Alternatively, a combination of both methods may be used at 5 GHz sampling rate with two SLs, a 3rd order derivative, and retention of 4 LSBs for a bit rate of 20 Gbit/s. The previously mentioned parallel combination of 4 and 6 SL devices with

Table 1 For a given number of combined independent signals, the TRNG bit rate is fixed by the sampling rate and the number of retained LSBs. The minimum order of derivative necessary for verified randomness is minimized to zero for four or more combined signals.

Number of combined SL devices	1	2	4	6
Derivative	4	3	…	…
Maximum sampling rate (GHz)	1.25	5	10	20
Retained LSBs	5	4	4	4
TRNG bit rate (Gbit/s)	6.25	20	40	80

a retention of 4 LSBs allows for a further increase of the sampling rate, resulting in bit rates of 40 and 80 Gbit/s, respectively, for the TRNG without the use of derivatives. The results demonstrate the interplay between the sampling rate, order of derivative, and number of combined SL signals. For a given number of retained LSBs, the rate of the TRNG can be enhanced either by increasing the number of combined SL signals or by increasing the order of the derivative.

4 Chaos Synchronization in Networks of $GaAs/A_{0.45}Ga_{0.55}As$ Superlattices

The emergent synchronization of chaotic systems is a fundamental phenomenon in nonlinear dynamics with interesting applications for private- and public-channel secure-communication protocols, since messages can be concealed and transmitted by the synchronized chaotic signal. A variety of types of synchronization have been already observed between coupled chaotic oscillators including leader-laggard synchronization using unidirectional coupling and face-to-face synchronization using mutual coupling. In addition, zero-lag synchronization (ZLS) has been observed based on mutual coupling with an additional delayed self-feedback to each one of the chaotic lasers or by using multiple mutual couplings. In ZLS, chaotic oscillations are simultaneously synchronized in spite of an arbitrarily large physical distance between the chaotic oscillators. ZLS between two or more chaotic semiconductor lasers is of great current interest because of promising applications to public-channel cryptography without relying on the number theory, as in other existing methods.

The investigated SL devices have been obtained from the same wafer as the ones in Ref. [10]. We will first discuss the leader-laggard configuration. A chaotic SL A drives unidirectionally a nonchaotic SL B having a similar structure, applying no or a moderate bias voltage. The distance between the two SLs was set to 16 m as indicated in Fig. 7a. The unidirectionality is attained by inserting a broadband amplifier into the connecting link that enhances the signal traveling from SL A to SL B, but not vice versa. Although the two SLs might differ somewhat in their I-V characteristics as well as in their chaotic voltage regions, their waveforms become highly synchronized with a delay of about 65 ns as shown in Fig. 7b. This delay originates

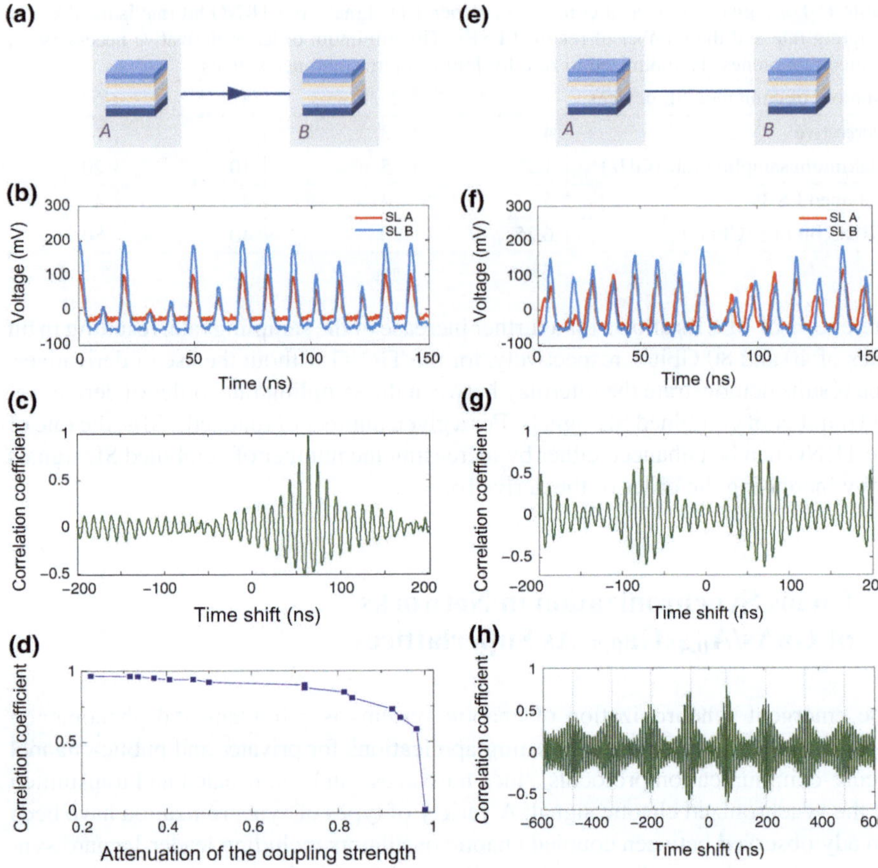

Fig. 7 **a** Schematic diagram of a leader-laggard setup with unidirectional coupling, which is realized by a broadband amplifier enhancing signals from SL A and transmitting them to SL B. **b** A snapshot of a 150-ns-long time trace of spontaneous chaotic current self-oscillations digitized at 5 GHz for SL A (blue) and SL B (red) shifted by 65 ns. **c** The cross-correlation between the current waveforms of the two SLs in (**a**), indicating a nearly-unity correlation peak at about 65 ns, where any additional delay due to the amplifier is negligible. **d** Correlation coefficient of the leader-laggard synchronization as a function of attenuation of the coupling strength. **e** Schematic representation of a face-to-face setup similar to (**a**), but without an amplifier. **f** Similar to (**b**) with the same shift to SL B. **g** The cross-correlation between the current waveforms of the two SLs indicating fairly good symmetry around zero delay and peaks at ± 65 ns. **h** Magnification of the cross-correlation in (**g**) indicating additional fading revival peaks at $\pm(2n+1)\times 65$ ns (vertical dashed lines) with n being an integer [5]

from the 16- m-long unidirectional coupling corresponding to the velocity of the signal-propagating electric device. Quantitatively, the almost perfect leader-laggard synchronization, both in spike timings and in relative amplitudes of the spikes, is marked by a cross-correlation peak approaching one, which rapidly vanishes on a shifted time scale of several inter-spike-intervals as shown in Fig. 7c. Interestingly, the quality of the leader-laggard synchronization is found to be nearly insensitive to the coupling strength, which was measured by the height of the peaks in their correlation as a function of the coupling strength. The results indicate a high degree of synchronization even for relatively weak coupling as indicated in Fig. 7d. Moreover, unlike synchronization among coupled lasers which requires temperature control, the synchronization of coupled SLs is insensitive to the variation at room temperature.

Changing the unidirectional coupling to bidirectional by removing the amplifier results in a face-to-face configuration as shown in Fig. 7e. A snapshot of a 150-ns-long time trace of spontaneous chaotic current self-oscillations digitized at 5 GHz is shown in Fig. 7f for the face-to-face configuration. The symmetry between the two similarly structured SLs is now partially restored, where both SLs are now simultaneously leaders and laggards as well. To completely restore the symmetry, the voltage of the second SL B is enhanced in such a way that each isolated SL without the mutual connection is chaotic. This complete symmetry is now reflected in the symmetric cross-correlation function around the origin shown in Fig. 7g. Specifically, there are two similar dominant peaks around ± 65 ns related to the delay of the electric signal propagation between the two SLs. Since the output current of a SL typically increases with the applied voltage across the device (cf. Fig. 4a), one expects a quasi-periodic structure for the output current waveform due to the mutual coupling, which is also reflected by the cross-correlation function (cf. Fig. 7g). Indeed, the cross-correlation function exhibits fading revivals at about $(2n + 1) \times 65$ ns, where n is an integer as can be seen in Fig. 7g, h. Such revivals are absent in the leader-laggard setup (cf. Fig. 7c). Note that ZLS is another possible solution maintaining the symmetry between the two SLs. However, our experimental results indicate that ZLS becomes unstable in a face-to-face configuration.

The ability to achieve ZLS between two SLs, but without a relay mechanism, requires additional self-couplings as indicated in Fig. 8a. Here, the two SL devices, which both consist of 60-μm-square mesas, are mutually connected with a 16-m-long cable between the top contact of SL A and the bottom contact of SL B. In addition, an 8-m-long cable was connected to the bottom contact of SL A, while its other terminal was open-circuited, and similarly an 8-m-long cable was connected to the bottom contact of SL B. Each of these two cables reflects from its open end a fraction of the chaotic signal back into its SL, hence functioning as a 16-m-long self-feedback connection. Consequently, the face-to-face solution shown in Fig. 7e is now an inconsistent solution, and the stable solution becomes ZLS with a dominant cross-correlation peak of about 0.84 at zero time shift as shown in Fig. 8b. The quasi-periodicity of the waveform is represented by fading revivals at about $n \times 65$ ns, where n is an integer, as opposed to about 2×65 ns time lags between revival peaks in the case of the face-to-face configuration (cf. Fig. 7h).

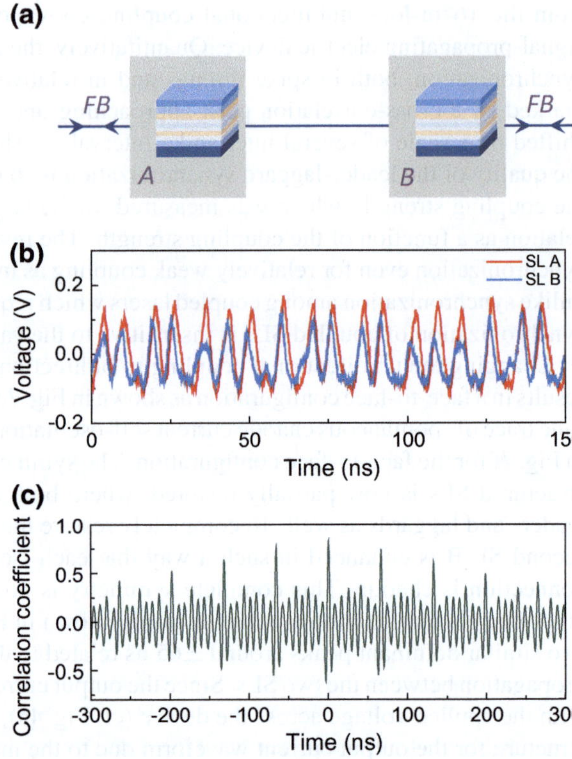

Fig. 8 **a** Schematic diagram of two mutually coupled SLs with a 16 m cable with an additional self-feedback (FB) coupling of 8 m for each SL. **b** A snapshot of a 150-ns-long time trace of spontaneous chaotic current self-oscillations of the SLs A (red) and B (blue). **c** Cross-correlation between the two SLs with a dominant ZLS peak of about 0.84 and fading revival peaks at about $n \times 65$ ns, where n is an integer [5]

Leader-laggard, face-to-face, and zero-lag types of synchronization were experimentally demonstrated among mutually coupled chaotic SLs at room temperature. In addition, the synchronization of different mesa shapes and sizes has been demonstrated in a star configuration using five SL devices [5]. Each type of synchronization clearly reflects the symmetry of the topology of its network pattern. The emergence of a chaotic SL without external feedback as well as the effective interaction between coupled, but differently structured SLs are distinguishing features of SLs as opposed to typical chaotic lasers. These features are expected to lead to the ability of synchronization of much larger scale patterns and hint on reliable as well as advanced secure communication protocols based on chaos synchronization.

5 Summary and Conclusions

Spontaneous chaotic current self-oscillations with a bandwidth of several GHz have been experimentally demonstrated in doped, weakly coupled $GaAs/Al_{0.45}Ga_{0.55}As$ SLs at room temperature. The use of an (Al,Ga)As barrier with 45% Al content instead of an AlAs barrier strongly reduces leakage currents induced by thermally excited Γ-X intervalley transfer. At the same time, the nonlinearity of these SLs is sufficiently strong to stimulate spontaneous chaos up to room temperature.

A high-speed, *all-electronic* true random number generator based on spontaneous chaotic current self-oscillations of GaAs/$Al_{0.45}Ga_{0.55}As$ SL devices has been demonstrated experimentally operating at room temperature. The randomness of the generated sequences has been verified using a statistical test suite. The all-electronic method uses the parallel combination of multiple, independent SL signals, a single SL with high-order derivatives, or a combination of both methods, demonstrating a large degree of scalability and customization options, depending on TRNG rate requirements and complexity restrictions of the setup. The achieved bit rates are 80 Gbit/s, which are about two orders of magnitude larger than for currently commercially available all-electronic TRNGs. Further developments of the proposed TRNG methods may lead to a miniaturized, on-chip, high-speed TRNG with verified randomness.

The synchronization of spontaneous chaotic current self-oscillations of weakly coupled GaAs/$Al_{0.45}Ga_{0.55}As$ SL devices has been demonstrated as a useful building block for various tasks in secure communications. The emergence of several types of chaos synchronization, e.g. leader-laggard, face-to-face, and zero-lag synchronization in network patterns of coupled SLs consisting of unidirectional and mutual coupling as well as self-feedback coupling has been experimentally demonstrated. Each type of synchronization clearly reflects the symmetry of the topology of its network pattern. The emergence of a chaotic SL oscillator without external feedback and synchronization among differently structured SLs open up the possibility for advanced secure multi-user communication methods based on large networks of coupled SLs.

Acknowledgements The authors would like to thank Y. Y. Huang, W. Li, H. Qin, H. Song, as well as Z. Z. Yin from the Key Laboratory of Nanodevices and Applications, Suzhou Institute of Nano-Tech and Nano-Bionics, Chinese Academy of Sciences, Suzhou, China, I. Kanter, I. Reidler, as well as M. Rosenbluh from the Department of Physics, Bar-Ilan University, Ramat-Gan, Israel, K. Biermann from the Paul-Drude-Institut für Festkörperelektronik, Leibniz-Institut im Forschungsverbund Berlin e. V., Berlin, Germany, and Luis L. Bonilla, M. Carretero, as well as M. Ruiz-García from the Gregorio Millán Institute for Fluid Dynamics, Nanoscience and Industrial Mathematics, Universidad Carlos III de Madrid, 28911 Leganés, Spain, for their continuous and fruitful contributions as well as L. Schrottke from the Paul-Drude-Institut für Festkörperelektronik, Leibniz-Institut im Forschungsverbund Berlin e. V., Berlin, Germany, and Stephen Teitsworth from the Department of Physics, Duke University, Durham, North Carolina, USA, for a critical reading of the manuscript.

References

1. Y.H. Zhang, J. Kastrup, R. Klann, K.H. Ploog, H.T. Grahn, Phys. Rev. Lett. **77**, 3001–3004 (1996)
2. Y.Y. Huang, W. Li, W.Q. Ma, H. Qin, Y.H. Zhang, Chin. Sci. Bull. **57**, 2070–2072 (2012)

3. W. Li, I. Reidler, Y. Aviad, Y.Y. Huang, H. Song, Y.H. Zhang, M. Rosenbluh, I. Kanter, Phys. Rev. Lett. **111**, 044102, 5 (2013)
4. https://en.wikipedia.org/wiki/Comparison_of_hardware_random_number_generators
5. W. Li, Y. Aviad, I. Reidler, H. Song, Y.Y. Huang, K. Biermann, M. Rosenbluh, Y.H. Zhang, H.T. Grahn, I. Kanter, Europhys. Lett. **112**, 30007, 5 (2015)
6. L.L. Bonilla, H.T. Grahn, Rep. Prog. Phys. **68**, 577–683 (2005)
7. Y.H. Zhang, R. Klann, H.T. Grahn, K.H. Ploog, Superlattices Microstruct. **21**, 565–568 (1997)
8. L.L. Bonilla, M. Alvaro, M. Carretero, J. Math. Ind. **7**(1), 1–17 (2017)
9. Y.Y. Huang, W. Li, W.Q. Ma, H. Qin, H.T. Grahn, Y.H. Zhang, Appl. Phys. Lett. **102**, 242107, 3 (2013)
10. Z.Z. Yin, H. Song, Y.H. Zhang, M. Ruiz-García, M. Carretero, L.L. Bonilla, K. Biermann, H.T. Grahn, Phys. Rev. E **95**, 012218, 7 (2017)
11. M. Alvaro, M. Carretero, L.L. Bonilla, Europhys. Lett. **107**, 37002, 6 (2014)
12. Y.Y. Huang, H. Qin, W. Li, S.L. Lu, J.R. Dong, H.T. Grahn, Y.H. Zhang, Europhys. Lett. **105**, 47005, 5 (2014)
13. I. Kanter, Y. Aviad, I. Reidler, E. Cohen, M. Rosenbluh, Nat. Photon. **4**, 58–61 (2009)
14. NIST Statistical Test Suite at http://csrc.nist.gov/groups/ST/toolkit/rng/stats_tests.html

Transport Out of Locally Broken Detailed Balance

Rafael Sánchez

Abstract Electrons move along potential or thermal gradients. In the presence of a global gradient, applied e.g. to the two terminals of a conductor, this induces electric charge and heat currents. They can also flow between two equilibrated terminals (at the same voltage and temperature) if detailed balance is broken in some part of the system. A minimal model involving two metallic islands in series is introduced whose internal potential and temperatures can be externally modulated. The conditions for a finite electric flow are discussed.

Keywords Mesoscopic thermoelectrics · Interactions · Fluctuations

1 Introduction

An electronic conductor responds to a nonequilibrium situation in the form of charge and heat currents. It can be due to the presence of electric or thermal gradients applied to the two terminals of the system, $V_L - V_R$ and $T_L - T_R$. Transport is however not restricted to this situation. The role of fluctuations in the intermediate region (the system) cannot be overlooked. If the two terminals are in thermal equilibrium (for being at the same temperature and potential) a current can nevertheless flow due to rectified noise.

The relevance of noisy states was pointed out by Landauer for modifying the relative occupation of bistable potentials [1]. This is the case for instance if temperature is locally increased (with an ideal blowtorch) on one side of the barrier only [1]. These ideas were later applied to transport configurations in a series of papers by Büttiker [2], van Kampen [3] and Landauer [4]. They predict classical particles to flow along symmetric potentials which are subject to state-dependent noise. As an example, particles overcome a potential barrier from the side where the diffusion

R. Sánchez (✉)
Departamento de Física Teórica de la Materia Condensada,
Universidad Autónoma de Madrid, 28049 Madrid, Spain
e-mail: rafael.sanchez@uam.es

L. L. Bonilla et al. (eds.), *Coupled Mathematical Models for Physical and Biological Nanoscale Systems and Their Applications*, Springer Proceedings in Mathematics & Statistics 232, https://doi.org/10.1007/978-3-319-76599-0_3

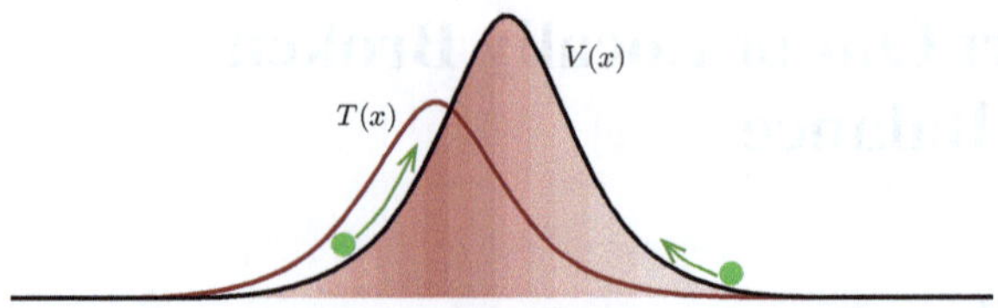

Fig. 1 Transport in the presence of state-dependent noise. Symmetric potential and temperature distributions give nevertheless a finite particle current if their respective maxima are displaced [2–4]

coefficient is higher (due e.g. to a locally increased temperature, cf. Fig. 1), leading to a broken detailed balance situation. The resulting currents depend on the separation of the potential and temperature fields: symmetric state-dependent potential and temperature distributions give no transport [2]. Of course Onsager relations are still fulfilled [5] if one takes into account the heat injected (by the blowtorch) in order to keep the temperature stationary.

Similar ideas have motivated proposals of thermal ratchets [6, 7] and thermoelectric metamaterials [9], and the measurement of radiation induced currents [8, 10] or transverse rectification in semiconductor 2DEGs [11, 12]. They are closely related to the recent field of mesoscopic three-terminal thermoelectrics [13, 14] which discusses the conversion of a heat current injected from a thermal reservoir into a charge current in an equilibrium conductor [15–28]. Also noise generated either in a coupled conductor [29–31] or externally [32, 33] is used for this purpose. Most of these cases use discrete levels in quantum dot systems where a temperature cannot be properly defined [34]. Another possibility is a mesoscopic thermocouple configuration [35–39] where the current is due to the separation of electron-hole excitations at the two contacts of a hot cavity. Common to all of them is the need of a conductor with broken left-right inversion and electron-hole symmetries (usually due to energy-dependent transport coefficients).

Here a minimal model of a mesoscopic conductor is proposed where all the necessary ingredients can be engineered and controlled experimentally. It is based on the discretization of a metallic conductor by means of at least three tunnel junctions, cf. Fig. 2. This way an array of two small metallic islands are formed whose level spacing is much smaller than the thermal energy, $\Delta E \ll k_{\mathrm{B}}T$. The electron-electron relaxation rate is fast such that they equilibrate to a Fermi distribution function. Hence, each island has a well defined temperature, T_i. A physical *mesoscopic blowtorch* can be defined that modulates the temperature of the system locally. This can be done either by introducing time dependent drivings [40–42], or by using either on-chip refrigerators [43–47] or the noise generated in a coupled conductor [29, 49]. In the first case, the frequency of the driving must be much larger than the energy relaxation rate such that electrons have an increased effective temperature in the island. In the second case, it is important that the refrigerator exchanges heat but no charge with the conductor [15]. Additionally, the internal potential of each island can be externally modulated by means of gate voltages. This way, both temperature and potential profiles can be spacially resolved and tuned in a simple configuration.

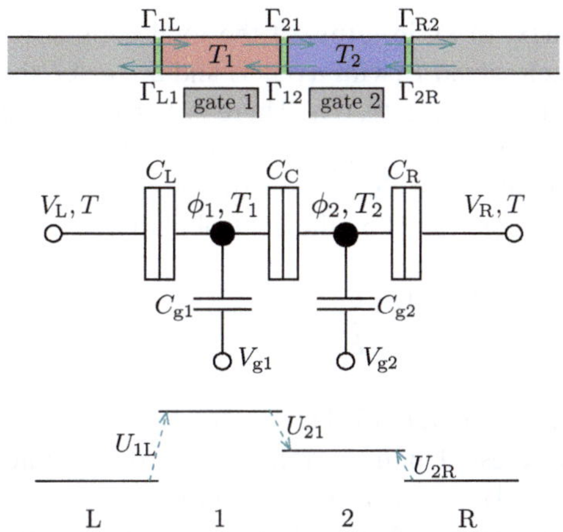

Fig. 2 Sketch of our system. Two metallic islands are tunnel-coupled in series in a two-terminal configuration. Electrons are assumed to thermalise at a different temperature T_i in each piece of the conductor. The island internal potentials can be tuned by gate voltages $V_{g,i}$. The two terminals are considered at the same voltage, $V_L = V_R = 0$, and temperature, T. The piecewise Coulomb gap U_{ji} experienced by tunneling electrons can be modulated with the gate voltages V_{gk}

2 Two Coulomb Islands in Series

Our model is based on the well known physics of single-electron tunneling at small-capacitance tunnel junctions [50]. Metallic islands can be separated by insulating barriers such that the energy cost for adding an extra electron (the charging energy, E_i) is larger than the thermal energy $E_i \gg k_B T$. For islands of a micrometer size, the charging energy is of the order of 0.1 meV. This introduces electron-electron correlations that suppress the low voltage transport, what is called Coulomb blockade. The extra energy can be supplied by a gate voltage V_{gi} such that electrons can flow one by one through the island giving rise to conductance [51–53] and thermopower [54–58] oscillations. Electronic cooling based on these properties has been recently proposed [59] and realized experimentally [60] in single-electron transistors.

Consider an array of three tunnel junctions forming two islands [61–64], as sketched in Fig. 2. Each barrier is described by a capacitance C_i and a resistance R_i in parallel. The former determine the electrostatic potential of each island, ϕ_i, which is obtained self-consistently from the equations for the charge accumulated in each island:

$$Q_1 = C_L(\phi_1 - V_L) + C_C(\phi_1 - \phi_2) + C_{g1}V_{g1} \tag{1}$$

$$Q_2 = C_R(\phi_2 - V_R) + C_C(\phi_2 - \phi_1) + C_{g2}V_{g2}. \tag{2}$$

The electrostatic potential in the coupled system is $U(Q_1, Q_2) = \sum_i \int dQ_i\phi_i$. It defines the charging energies $E_i = e^2/2\tilde{C}_i$, with $\tilde{C}_i = \left(C_{\Sigma_1}C_{\Sigma 2} - C_C^2\right)/C_{\Sigma\bar{i}}$ (where

$\bar{i}$ denotes the island next to i), in terms of the total geometric capacitance $C_{\Sigma i} = C_{\alpha_i} + C_\mathrm{C} + C_{gi}$ of each island i (with $\alpha_1 = \mathrm{L}$ and $\alpha_2 = \mathrm{R}$), the centre capacitance C_C, and the one coupling to the gate, C_{gi}. There is a crossed charging energy related to the occupation of the next island, parametrized as $J = 2e^2 C_\mathrm{C}/\left(C_{\Sigma_1} C_{\Sigma 2} - C_\mathrm{C}^2\right)$. It corresponds to the repulsion of electrons in different islands. The internal potential of the islands is furthermore tuned by the control parameters

$$n_{gi} = \frac{1}{e}\frac{4E_1E_2}{J^2 - 4E_1E_2}\left(Q_i^{\mathrm{ext}} - \frac{J}{4E_i}Q_{\bar{i}}^{\mathrm{ext}}\right), \tag{3}$$

with the externally induced charges [52] $Q_1^{\mathrm{ext}} = C_\mathrm{L} V_\mathrm{L} + C_{g1} V_{g1}$ and $Q_2^{\mathrm{ext}} = C_\mathrm{R} V_\mathrm{R} + C_{g2} V_{g2}$. Here we are interested in the current generated in an unbiased configuration, such that $V_\mathrm{L} = V_\mathrm{R} = 0$. Hence, n_{gi} depends only on the gate voltages. Note that each gate affects the two islands. A time-dependent modulation of the gates has been used for single-electron pumping with metrological precision [62, 63] and provides a feed-back control mechanism for Maxwell's demon proposals [65].

All these can be cast into the electrostatic term of the Hamiltonian of the system:

$$H = \sum_i E_i(n_i - n_{gi})^2 + J(n_1 - n_{g1})(n_2 - n_{g2}), \tag{4}$$

where $n_i = Q_i/e$ is the number of electrons in each island. Let us restrict here to the simplest configuration with up to one extra electron in the system. This is the case at low temperatures, $k_\mathrm{B}T \ll E_i, J$, which for typical experimental situations is around 100 mK. The charge configuration of the system is then described by the three states (n_1, n_2): (0, 0), (1, 0) and (0, 1).

Tunneling events are characterized by the energy cost U_{ji} (the Coulomb gap) for electrons tunneling from region i to j, which read:

$$U_{1\mathrm{L}} = 2E_1\left(\frac{1}{2} - n_{g1}\right) - Jn_{g2} \tag{5}$$

$$U_{2\mathrm{R}} = 2E_2\left(\frac{1}{2} - n_{g2}\right) - Jn_{g1} \tag{6}$$

$$U_{21} = U_{2\mathrm{R}} - U_{1\mathrm{L}} \tag{7}$$

in the absence of a bias voltage. Obviously, $U_{ij} = -U_{ji}$. We emphasize that the energetics of the mesoscopic junction is fully tunable with the gate voltages. The rates for the corresponding tunneling transitions are given by the well-known expression:

$$\Gamma_{ji} = \frac{1}{e^2 R_{ji}}\int dE f(E, T_i)[1 - f(E - U_{ji}, T_j)], \tag{8}$$

where $R_{ji} = R_\mathrm{L}$, R_C or R_R, is the tunneling resistance of the involved junction. They are quite generally energy-independent (typically around 10–100 kΩ). Here

$f(E,T)=1/\left(1+e^{E/k_BT}\right)$ is the Fermi-Dirac distribution for a region at temperature T.

With these, one can write the rate equations:

$$\begin{aligned}\dot{P}_{(0,0)} &= \Gamma_{\rm L1}P_{(1,0)}+\Gamma_{\rm R2}P_{(0,1)}-(\Gamma_{\rm 1L}+\Gamma_{\rm 2R})\,P_{(0,0)}\\ \dot{P}_{(1,0)} &= \Gamma_{\rm 1L}P_{(0,0)}+\Gamma_{12}P_{(0,1)}-(\Gamma_{\rm L1}+\Gamma_{21})\,P_{(1,0)}\\ \dot{P}_{(0,1)} &= \Gamma_{21}P_{(1,0)}+\Gamma_{\rm 2R}P_{(0,0)}-(\Gamma_{12}+\Gamma_{\rm R2})\,P_{(0,1)},\end{aligned} \tag{9}$$

for the dynamics of the occupation probability of each state, $P_m(t)$. The stationary occupation of the different states are obtained by solving $\dot{P}=0$, giving:

$$\begin{aligned}P_{(0,0)} &= \Gamma_{\rm T}^{-2}\left(\Gamma_{21}\Gamma_{\rm R2}+\Gamma_{\rm L1}\Gamma_{12}+\Gamma_{\rm L1}\Gamma_{\rm R2}\right)\\ P_{(1,0)} &= \Gamma_{\rm T}^{-2}\left(\Gamma_{\rm 1L}\Gamma_{\rm R2}+\Gamma_{12}\Gamma_{\rm 1L}+\Gamma_{12}\Gamma_{\rm 2R}\right)\\ P_{(0,1)} &= \Gamma_{\rm T}^{-2}\left(\Gamma_{\rm L1}\Gamma_{\rm 2R}+\Gamma_{21}\Gamma_{\rm 1L}+\Gamma_{21}\Gamma_{\rm 2R}\right),\end{aligned} \tag{10}$$

with $\Gamma_{\rm T}^2$ given by the normalization condition, $\sum_m P_m=1$. The stationary state obeys detailed balance if tunneling transitions satisfy: $\Gamma_{\rm 1L}P_{(0,0)}=\Gamma_{\rm L1}P_{(1,0)}$, for the left $\Gamma_{12}P_{(1,0)}=\Gamma_{21}P_{(0,1)}$, for the center, and $\Gamma_{\rm 1R}P_{(0,0)}=\Gamma_{\rm R1}P_{(0,1)}$, for the right junction. Hence, the current in the right terminal:

$$I_{\rm R}=e\left(\Gamma_{R2}P_{(0,1)}-\Gamma_{\rm 2R}P_{(0,0)}\right), \tag{11}$$

gives a measure of the deviation from detailed balance for processes at the right barrier. Injected currents from either terminal are defined as positive. Using Eqs. (10) results in the expression:

$$I_{\rm R}=e\Gamma_{\rm T}^{-2}\left(\Gamma_{\rm 1L}\Gamma_{21}\Gamma_{\rm R2}-\Gamma_{\rm L1}\Gamma_{12}\Gamma_{\rm 2R}\right). \tag{12}$$

From charge conservation, we obtain the current through the left junction: $I_{\rm L}=-I_{\rm R}$. Note that the first term in the right hand side of Eq. (12) is proportional to the rate for an electron to be transported from the left to the right terminal (via a sequence L→1→2→R). The second term is proportional to the rate of the reversed process. It is then clear that the case of having transitions satisfying local detailed balance at every junction results in no net current: In the unbiased and isothermal configuration, this translates to having tunneling rates related by $\Gamma_{ij}=\Gamma_{ji}e^{U_{ij}/k_BT}$. This is not the case if one of the leads or islands is at a different temperature, as discussed in the next section.

2.1 Broken Detailed Balance. Linear Response

Let us first emphasize the importance of the Coulomb gap introduced by the electronic confinement in the island. If $U_{ji}=0$, tunneling is electron-hole symmetric and a

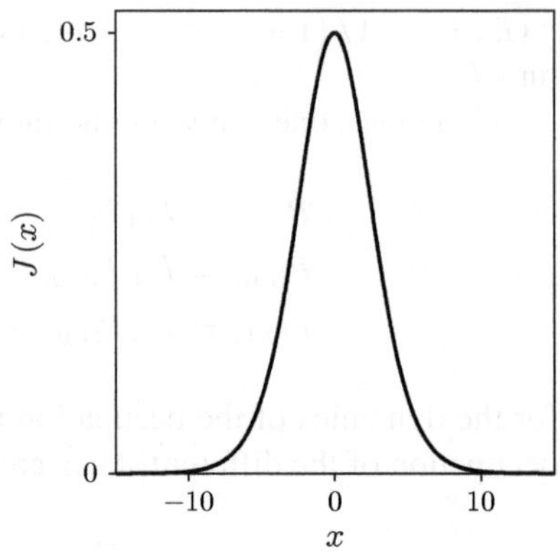

Fig. 3 Function $J(x)$ defined in Eq. (17)

temperature drop across the junction is not sufficient to break detailed balance. It can be easily shown from symmetries of the Fermi function that the integral

$$\int_{-\infty}^{\infty} dEf(E,T)\left[1-f(E,T')\right] \tag{13}$$

is invariant under the exchange of temperatures $T \leftrightarrow T'$. Hence $\Gamma_{ji} = \Gamma_{ij}$, for $U_{ji} = 0$, independently of the temperatures T_i and T_j.

In the following, the case with finite U_{ji} is assumed, unless explicitly mentioned. Some of the rates Eq. (8) can be analytically calculated by using the relation:

$$f(E,T)[1-f(E',T')] = n_{\mathrm{B}}\left(\frac{E}{k_{\mathrm{B}}T}-\frac{E'}{k_{\mathrm{B}}T'}\right)[f(E',T')-f(E,T)], \tag{14}$$

where $n_{\mathrm{B}}(x) = (e^x - 1)^{-1}$ is the Bose-Einstein distribution function. For transitions between regions at the same temperature one gets:

$$\Gamma_{ji}^{(0)} = -\frac{U_{ji}}{e^2 R_{ji}} n_{\mathrm{B}}\left(\frac{U_{ji}}{k_{\mathrm{B}}T}\right) e^{U_{ji}/k_{\mathrm{B}}T} \tag{15}$$

and, on the other hand, using $n_{\mathrm{B}}(-x) = -e^x n_{\mathrm{B}}(x)$:

$$\Gamma_{ij}^{(0)} = -\frac{U_{ji}}{e^2 R_{ji}} n_{\mathrm{B}}\left(\frac{U_{ji}}{k_{\mathrm{B}}T}\right), \tag{16}$$

for the reversed process. It is straightforward to check that local detailed balance is satisfied at such a junction. If furthermore the gates are tuned such that there is no energy cost, $U_{ji} = 0$, tunneling is governed by thermal fluctuations: $\Gamma_{ji}^{(0)} = \Gamma_{ij}^{(0)} = k_{\mathrm{B}}T/(e^2 R_{ji})$.

This is not the case when the junction separates two pieces of the metal which are at different temperatures. This effect shows up in the linear regime. Considering a small temperature difference $\delta T = T_j - T_i$, one can linearize the involved tunneling rates, $\Gamma_{ji} \approx \Gamma_{ji}^{(0)} + \Gamma_{ji}^{(1)}\delta T/T$, with $\Gamma_{ji}^{(1)} = k_B T(e^2 R_{ji})^{-1} J(U_{ji}/k_B T)$ and:

$$J(x) = e^x \left[n_B(x)\right]^2 \left[\frac{\pi^2}{6} + x^2 + \mathrm{Li}_2\left(-e^x\right) + \mathrm{Li}_2\left(-e^{-x}\right)\right], \quad (17)$$

with the dilogarithm function $\mathrm{Li}_2(z) = \sum_{k=1}^{\infty} z^k/k^2$. The function $J(x)$ is a peak centered at the origin, as shown in Fig. 3. That is, the response will be larger close to the Fermi energy. From the above expression, we can verify that $J(x) = J(-x)$ and therefore $\Gamma_{ij}^{(1)} = \Gamma_{ji}^{(1)}$. Hence the rates do not satisfy detailed balance:

$$\frac{\Gamma_{ji}}{\Gamma_{ij}} \approx e^{U_{ji}/k_B T} + \frac{\delta T}{T} F(U_{ji}/k_B T), \quad (18)$$

where $F(x) = x^{-1} J(x)[n_B(x)]^{-2}$. Note that the effect appears in the simultaneous occurence of a temperature drop and a Coulomb gap.

Using the above relations in the expression for the charge current, Eq. (12), one finds that breaking detailed balance in a single barrier is enough to have a finite current, even if the two terminals are at the same voltage and temperature.

To be more specific, let us consider the case where only the first island is at a different temperature, $T_1 = T + \delta T$, with $T_2 = T$. Then a current is generated:

$$I_R \propto \frac{\delta T}{T}\left[e^{-U_{21}/k_B T} F(U_{21}/k_B T) - e^{U_{1L}/k_B T} F(-U_{1L}/k_B T)\right], \quad (19)$$

in terms of the energy cost for tunneling to the hot island from the left lead and from the other island, U_{1L} and $-U_{21}$, respectively. The avoided prefactor depends on the equilibrium rates, $\Gamma_{ji}^{(0)}$. It is clear from the above expression that no current will flow if detailed balance is broken symmetrically at the two barriers of the hot island, i.e. if the energy cost is equal: $U_{1L} = -U_{21}$. This way, tuning the gate voltages allows one to control the current.

3 Transport from a Hotspot

As discussed in the previous section, broken detailed balance occurs in tunneling through a junction separating regions at different temperatures. This is however not a sufficient condition to generate transport in a conductor in global equilibrium. For example, no current will occur in a system consisting on a single hot island, as broken

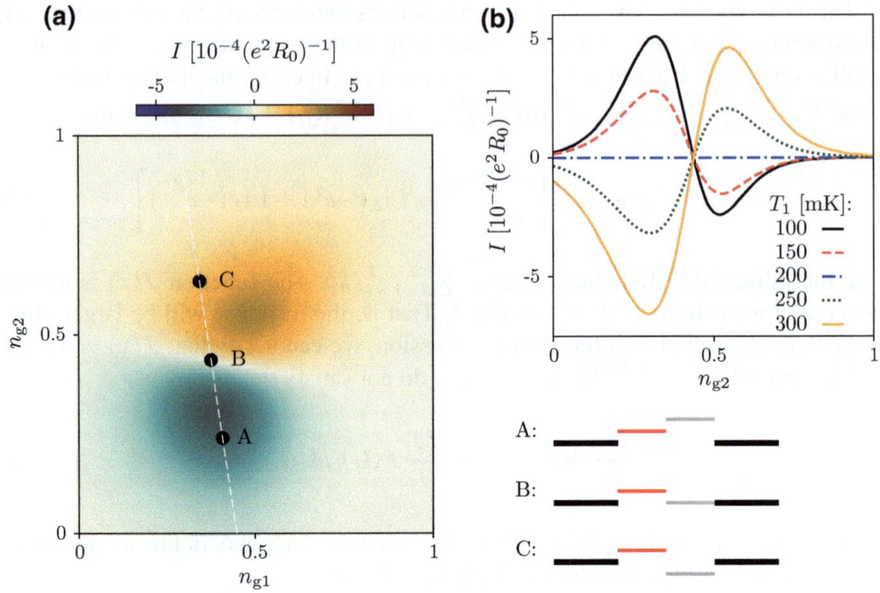

Fig. 4 Transport from a hot spot. **a** Current as a function of the island control parameters, n_{gi} for an unbiased configuration $V_L = V_R = 0$ and $T = T_2 = 200$ mK. The temperature of island 1 is increased at $T_1 = 150$ mK, with $R_L = R_C = R_R = R_0$, $E_1 = E_2 = 0.15$ meV, and $J = 0.05$ meV. The configurations in points A, B and C are sketched on the right bottom side. **b** Cut along the white dashed line in (**a**) as a function of n_{g2}. Island 1 is tuned such that $U_{1L} = 0.015$ meV is constant. Reversing the temperature gradient or the sign of U_{2R} changes the sign of the current

detailed balance is symmetric in the two junctions.[1] In order to have a current, an asymmetry needs to be introduced, e.g. by making the energy cost for tunneling through the two barriers different. In our case, this is the role of the second island. Its charging energy lifts the asymmetry making U_{1L} and U_{12} different. Furthermore, this asymmetry can be tuned with gate voltages V_{g1} and V_{g2}, as discussed above. If the second island is at the same temperature than the two terminals, $T_2 = T$, detailed balance is satisfied at the third junction, with $\Gamma_{2R} = \Gamma_{R2} e^{U_{2R}/k_B T}$.

The current generated in such a configuration with $T_1 \neq T$ is plotted in Fig. 4a as a function of the control parameters n_{g1} and n_{g2}. Fixing the gate voltage in island 1, the current changes sign when the gate voltage of island 2 is tuned. As depicted in the inset, the sign of the current depends on the sign of $U_{1L} + U_{21}$: electrons tunneling

[1]Remember that tunneling resistances are energy independent in metallic islands. This is not necessarily the case in other related systems with energy-dependent tunneling junctions, e.g. in semiconductor quantum dots.

out from the hot island into the region i giving the largest U_{1i} have a larger rate. The response is restricted to a region of gate voltages such that $U_{li} \lesssim k_B T$. Far from this region, the difference of the two rates is exponentially small and therefore current is suppressed.

Reversing the sign of the local temperature gradient, $T_1 - T$, changes the sign of the current, cf. Fig. 4b. Being an obvious statement, it has practical consequences: rather than by heating one of the islands, the effect can be experimentally detected by cooling it. This can be done locally and non-invasively by a coupled refrigerator system within nowadays technology [45, 47, 48].

A particularly interesting configuration is when n_{g2} is tuned such that $U_{2R} = 0$. This is configuration B in Fig. 4. Then, the rates $\Gamma_{R2} = \Gamma_{2R} = k_B T/(e^2 R_R)$ only contribute to the prefactor of the current and the second island plays no role. In this case, $U_{1L} = U_{12}$, i.e. the temperature gradient and the energy cost are the same for tunneling through the two tunneling junctions of island 1. Hence electrons in the hot island have no preferred direction to tunnel out. Thus detailed balance is symmetrically broken in the two junctions, with the overall rate through the island being the same in both directions: $\Gamma_{1L}\Gamma_{21} = \Gamma_{12}\Gamma_{L1}$, and the current is zero, see Eq. (12).

4 Multijunction Arrays

One can envision to extend the above results to longer arrays, cf. Fig. 5. This relaxes the requirement to increase the temperature in a micrometer-size single island. The size of the islands can be controlled in the sample growth, which introduces a natural way to spacially modulate the Coulomb gap along the conductor. This way the need to gate the system can also be avoided. The current is then induced by the interplay of local non-equilibrium (only due to a piecewise-constant temperature distribution) and electron-electron interactions, emphasizing the mesoscopic nature of the device.

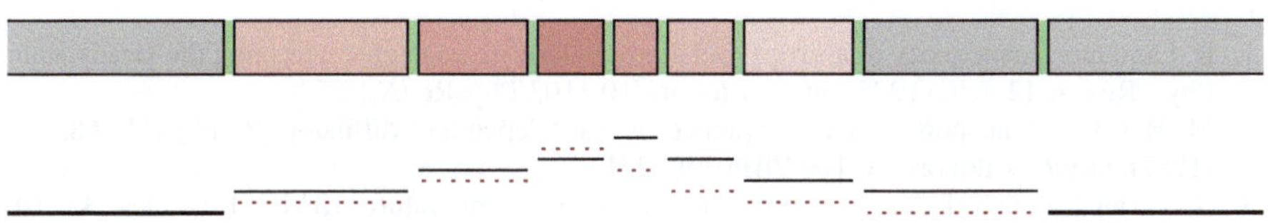

Fig. 5 An array of islands of different size in the presence of a local temperature increase. The charging energy of each island depends on its size, hence introducing an spacial modulation of the Coulomb gap (black solid lines). The piecewise-constant temperature distribution (red dotted lines) allows for a wider extension due to heat leaking to neighbour islands

5 Summary

A simple model of locally broken detailed balance giving rise to transport in an electronic conductor is presented. A system of two metallic islands, one of which experiences a different temperature, can be interpreted as a mesoscopic analogue of transport in state-dependent diffusion at a single scatterer. Dividing the conductor in an array of metallic islands allows for the local control of voltages and temperatures. The cooperative occurrence of a local temperature difference and a Coulomb gap introduces a preferred direction for tunneling electrons. This asymmetry, which determines the sign of the current, can furthermore be tuned by means of gate voltages applied to each island. Broken detailed balance is of relevance e.g. for the investigation of the Jarzynski equality and fluctuation theorems [29, 41, 66–69], and can be detected in higher-order cummulants of the full counting statistics of Coulomb-blockade systems [70].

The electron-hole asymmetry required for having a thermoelectric response is introduced by inhomogeneous energy costs in the tunneling processes. This is the case of islands with different charging energies or which are modulated by gate voltages. Thus energy-dependence of the barriers is not necessary. The involved technology is within nowadays reach [41, 45, 47, 62–64] and could readily be tested in an experiment. This proposal contributes to the field of interaction-induced thermoelectric properties [59, 60] and the control of thermal flows [71, 72] in low-dimensional metallic conductors.

Acknowledgements I acknowledge Markus Büttiker and Jukka P. Pekola for discussions and comments on an earlier version of the results presented here. This work was supported by the Spanish Ministerio de Economía y Competitividad via grants No. MAT2014-58241-P, No. FIS2015-74472-JIN (AEI/FEDER/UE) and RYC-2016-20778. I also thank the COST Action MP1209 "Thermodynamics in the quantum regime".

References

1. R. Landauer, Inadequacy of entropy and entropy derivatives in characterizing the steady state. Phys. Rev. A **12**, 636 (1975), https://doi.org/10.1103/PhysRevA.12.636
2. M. Büttiker, Transport as a consequence of state-dependent diffusion. Z. Phys. B **68**, 161 (1987), http://dx.doi.org/10.1007/BF01304221
3. N.G. van Kampen, Relative stability in nonuniform temperature. IBM J. Res. Dev. **32**, 107 (1988), http://dx.doi.org/10.1147/rd.321.0107
4. R. Landauer, Motion out of noisy states. J. Stat. Phys. **53**, 233 (1988), http://dx.doi.org/10.1007/BF01011555
5. N.G. van Kampen, Onsager relations for transport in inhomogeneous media. J. Stat. Phys. **63**, 1019 (1991), http://dx.doi.org/10.1007/BF01029996
6. Y.M. Blanter, M. Büttiker, Rectification of fluctuations in an underdamped ratchet. Phys. Rev. Lett. **81**, 4040 (1998), http://dx.doi.org/10.1103/PhysRevLett.81.4040
7. R. Benjaminm, R. Kawai, Inertial effects in Büttiker-Landauer motor and refrigerator at the overdamped limit. Phys. Rev. E **77**, 051132 (2008), http://dx.doi.org/10.1103/PhysRevE.77.051132

8. P. Olbrich, E.L. Ivchenko, R. Ravash, T. Feil, S.D. Danilov, J. Allerdings, D. Weiss, D. Schuh, W. Wegscheider, S.D. Ganichev, Ratchet effects induced by terahertz radiation in heterostructures with a lateral periodic potential. Phys. Rev. Lett. **103**, 090603 (2009), http://dx.doi.org/10.1103/PhysRevLett.103.090603
9. T.E. Humphrey, H. Linke, Reversible thermoelectric nanomaterials. Phys. Rev. Lett. **94**, 096601 (2005), https://doi.org/10.1103/PhysRevLett.94.096601
10. P. Olbrich, J. Karch, E.L. Ivchenko, J. Kamann, B. März, M. Fehrenbacher, D. Weiss, S.D. Ganichev, Classical ratchet effects in heterostructures with a lateral periodic potential. Phys. Rev. B **83**, 165320 (2011), http://dx.doi.org/10.1103/PhysRevB.83.165320
11. A. Ganczarczyk, S. Rojek, A. Quindeau, M. Geller, A. Hucht, C. Notthoff, J. König, A. Lorke, D. Reuter, A.D. Wieck, Transverse rectification in density-modulated two-dimensional electron gases. Phys. Rev. B **86**, 085309 (2012), https://doi.org/10.1103/PhysRevB.86.085309
12. S. Rojek, J. König, Mesoscopic diffusion thermopower in two-dimensional electron gases. Phys. Rev. B **90**, 115403 (2014), https://doi.org/10.1103/PhysRevB.90.115403
13. B. Sothmann, R. Sánchez, A.N. Jordan, Thermoelectric energy harvesting with quantum dots. Nanotechnology **26**, 032001 (2015), http://iopscience.iop.org/0957-4484/26/3/032001/
14. G. Benenti, G. Casati, K. Saito, R.S. Whitney, Fundamental aspects of steady-state conversion of heat to work at the nanoscale. Phys. Rep. **694**, 1 (2017), https://doi.org/10.1016/j.physrep.2017.05.008
15. R. Sánchez, M. Büttiker, Optimal energy quanta to current conversion. Phys. Rev. B **83**, 085428 (2011), http://link.aps.org/doi/10.1103/PhysRevB.83.085428
16. H. Thierschmann, R. Sánchez, B. Sothmann, F. Arnold, C. Heyn, W. Hansen, H. Buhmann, L.W. Molenkamp, Three-terminal energy harvester with coupled quantum dots. Nat. Nanotechnol. **10**, 854 (2015), http://dx.doi.org/10.1038/nnano.2015.176
17. H. Thierschmann, R. Sánchez, B. Sothmann, H. Buhmann, L.W. Molenkamp, Thermoelectrics with Coulomb coupled quantum dots. C. R. Physique, **17**, 1109 (2016), http://dx.doi.org/10.1016/j.crhy.2016.08.001
18. O. Entin-Wohlman, Y. Imry, A. Aharony, Three-terminal thermoelectric transport through a molecular junction. Phys. Rev. B **82**, 115314 (2010), http://dx.doi.org/10.1103/PhysRevB.82.115314
19. T. Ruokola, T. Ojanen, Theory of single-electron heat engines coupled to electromagnetic environments. Phys. Rev. B **86**, 035454 (2012), http://dx.doi.org/10.1103/PhysRevB.86.035454
20. B. Sothmann, R. Sánchez, A.N. Jordan, M. Büttiker, Rectification of thermal fluctuations in a chaotic cavity heat engine. Phys. Rev. B **85**, 205301 (2012), http://link.aps.org/doi/10.1103/PhysRevB.85.205301
21. B. Roche, P. Roulleau, T. Jullien, Y. Jompol, I. Farrer, D.A. Ritchie, D.C. Glattli, Harvesting dissipated energy with a mesoscopic ratchet. Nature Comm. **6**, 6738 (2015), http://www.nature.com/ncomms/2015/150401/ncomms7738/abs/ncomms7738.html
22. J.-H. Jiang, O. Entin-Wohlman, Y. Imry, Thermoelectric three-terminal hopping transport through one-dimensional nanosystems. Phys. Rev. B **85**, 075412 (2012), https://doi.org/10.1103/PhysRevB.85.075412
23. J.-H. Jiang, O. Entin-Wohlman, Y. Imry, Three-terminal semiconductor junction thermoelectric devices: improving performance. New J. Phys. **15**, 075021 (2013), http://dx.doi.org/10.1088/1367-2630/15/7/075021
24. J.-H. Jiang, O. Entin-Wohlman, Y. Imry, Hopping thermoelectric transport in finite systems: boundary effects. Phys. Rev. B **87**, 205420 (2013), https://doi.org/10.1103/PhysRevB.87.205420
25. F. Mazza, R. Bosisio, S. Valentini, G. Benenti, V. Giovannetti, R. Fazio, F. Taddei, Thermoelectric efficiency of three-terminal quantum thermal machines. New J. Phys. **16**, 085001 (2014), http://dx.doi.org/10.1088/1367-2630/16/8/085001
26. O. Entin-Wohlman, Y. Imry, A. Aharony, Enhanced performance of joint cooling and energy production. Phys. Rev. B **91**, 054302 (2015), http://dx.doi.org/10.1103/PhysRevB.91.054302
27. F. Mazza, S. Valentini, R. Bosisio, G. Benenti, V. Giovannetti, R. Fazio, F. Taddei, Separation of heat and charge currents for boosted thermoelectric conversion. Phys. Rev. B **91**, 245435 (2015), https://doi.org/10.1103/PhysRevB.91.245435

28. R. Bosisio, G. Fleury, J.-L. Pichard, C. Gorini, Nanowire-based thermoelectric ratchet in the hopping regime. Phys. Rev. B **93**, 165404 (2016), https://doi.org/10.1103/PhysRevB.93.165404
29. R. Sánchez, R. López, D. Sánchez, M. Büttiker, Mesoscopic Coulomb drag, broken detailed balance and fluctuation relations. Phys. Rev. Lett. **104**, 076801 (2010), http://link.aps.org/doi/10.1103/PhysRevLett.104.076801
30. D. Bischoff, M. Eich, O. Zilberberg, C. Rössler, T. Ihn, K. Ensslin, Measurement back-action in stacked graphene quantum dots. Nano Lett. **15**, 6003 (2015), http://dx.doi.org/10.1021/acs.nanolett.5b02167
31. A.J. Keller, J.S. Lim, D. Sánchez, R. López, S. Amasha, J.A. Katine, H. Shtrikman, D. Goldhaber-Gordon, Cotunneling drag effect in Coulomb-coupled quantum dots. Phys. Rev. Lett. **117**, 066602 (2016), https://doi.org/10.1103/PhysRevLett.117.066602
32. F. Hartmann, P. Pfeffer, S. Höfling, M. Kamp, L. Worschech, Voltage fluctuation to current converter with coulomb-coupled quantum dots. Phys. Rev. Lett. **114**, 146805 (2015), http://dx.doi.org/10.1103/PhysRevLett.114.146805
33. P. Pfeffer, F. Hartmann, S. Höfling, M. Kamp, L. Worschech, Logical stochastic resonance with a coulomb-coupled quantum-dot rectifier. Phys. Rev. Appl. **4**, 014011 (2015), http://dx.doi.org/10.1103/PhysRevApplied.4.014011
34. R.S. Whitney, R. Sánchez, F. Haupt, J. Splettstoesser, Thermoelectricity without absorbing energy from the heat sources. Physica E **75**, 257 (2016), http://dx.doi.org/10.1016/j.physe.2015.09.025
35. T.E. Humphrey, R. Newbury, R.P. Taylor, H. Linke, Reversible quantum Brownian heat engines for electrons. Phys. Rev. Lett. **89**, 116801 (2002), https://doi.org/10.1103/PhysRevLett.89.116801
36. A.N. Jordan, B. Sothmann, R. Sánchez, M. Büttiker, Powerful and efficient energy harvester with resonant-tunneling quantum dots. Phys. Rev. B **87**, 075312 (2013), http://link.aps.org/doi/10.1103/PhysRevB.87.075312
37. B. Sothmann, R. Sánchez, A.N. Jordan, M. Büttiker, Powerful energy harvester based on resonant-tunneling quantum wells. New J. Phys. **15**, 095021 (2013), http://iopscience.iop.org/1367-2630/15/9/095021
38. R. Sánchez, B. Sothmann, A.N. Jordan, Chiral thermoelectrics with quantum Hall edge states. Phys. Rev. Lett. **114**, 146801 (2015), http://dx.doi.org/10.1103/PhysRevLett.114.146801
39. R.S. Whitney, Quantum coherent three-terminal thermoelectrics: maximum efficiency at given power output. Entropy **18**, 208 (2016), http://dx.doi.org/10.3390/e18060208
40. D.V. Averin, J.P. Pekola, Statistics of the dissipated energy in driven single-electron transitions. EPL **96**, 67004 (2011), http://dx.doi.org/10.1209/0295-5075/96/67004
41. J.P. Pekola, A. Kutvonen, T. Ala-Nissila, Dissipated work and fluctuation relations for non-equilibrium single-electron transitions. J. Stat. Mech. P02033 (2013), http://dx.doi.org/10.1088/1742-5468/2013/02/P02033
42. O.-P. Saira, Y. Yoon, T. Tantu, M. Möttönen, D.V. Averin, J.P. Pekola, Test of the Jarzynski and Crooks fluctuation relations in an electronic system. Phys. Rev. Lett. **109**, 180601 (2012), https://doi.org/10.1103/PhysRevLett.109.180601
43. M. Nahum, T.M. Eiles, J.M. Martinis, Electronic microrefrigerator based on a normal insulator superconductor tunnel junction. Appl. Phys. Lett. **65**, 3123 (1994), http://dx.doi.org/10.1063/1.112456
44. M.M. Leivo, J.P. Pekola, D.V. Averin, Efficient Peltier refrigeration by a pair of normal metal/insulator/superconductor junctions. Appl. Phys. Lett. **68** (1996), http://dx.doi.org/10.1063/1.115651
45. A.V. Timofeev, M. Helle, M. Meschke, M. Möttönen, J.P. Pekola, electronic refrigeration at the quantum limit. Phys. Rev. Lett. **102**, 200801 (2009), http://dx.doi.org/10.1103/PhysRevLett.102.200801
46. J.V. Koski, A. Kutvonen, I.M. Khaymovich, T. Ala-Nissila, J.P. Pekola, On-chip Maxwell's demon as an information-powered refrigerator. Phys. Rev. Lett. **115**, 260602 (2015), http://dx.doi.org/10.1103/PhysRevLett.115.260602

47. M. Partanen, K.Y. Tan, J. Govenius, R.R. Lake, M.K. Mäkelä, T. Tanttu, M. Möttönen, Quantum-limited heat conduction over macroscopic distances. Nature Phys. **12**, 460 (2016), http://dx.doi.org/10.1038/nphys3642
48. R. Sánchez: Correlation-induced refrigeration with superconducting single-electron transistors. Appl. Phys. Lett. **111**(22), 223103 (2017), https://doi.org/10.1063/1.5008481
49. B. Kubala, G. Johansson, J. König, Transport in metallic multi-island Coulomb blockade systems: a systematic perturbative expansion in the junction transparency. Phys. Rev. B 165316 (2006), https://doi.org/10.1103/PhysRevB.73.165316
50. D.V. Averin, K.K. Likharev, Coulomb blockade of single-electron tunneling, and coherent oscillations in small tunnel junctions. J. Low. Temp. Phys. **62**, 345 (1986), http://dx.doi.org/10.1007/BF00683469
51. L.I. Glazman, R.I. Shekhter, Coulomb oscillations of the conductance in a laterally confined heterostructure. J. Phys. Condens. Matter **1**, 5811 (1989), http://dx.doi.org/10.1088/0953-8984/1/33/027
52. C.W.J. Beenakker, Theory of Coulomb-blockade oscillations in the conductance of a quantum dot. Phys. Rev. B **44**, 1646 (1991), https://doi.org/10.1103/PhysRevB.44.1646
53. H. Grabert, M.H. Devoret (eds.), *Single Charge Tunneling: Coulomb Blockade Phenomena In Nanostructures* (Plenum Press, New York, 1992)
54. C.W.J. Beenakker, A.A.M. Staring, Theory of the thermopower of a quantum dot. Phys. Rev. B **46**, 9667 (1992), https://doi.org/10.1103/PhysRevB.46.9667
55. A.A.M. Staring, L.W. Molenkamp, B.W. Alphenaar, H. van Houten, O.J.A. Buyk, M.A.A. Mabesoone, C.W.J. Beenakker, C.T. Foxon, Coulomb-Blockade oscillations in the thermopower of a quantum dot. Europhys. Lett. **22**, 57 (1993), http://stacks.iop.org/0295-5075/22/i=1/a=011
56. A.S. Dzurak, C.G. Smith, M. Pepper, D.A. Ritchie, J.E.F. Frost, G.A.C. Jones, D.G. Hasko, Observation of Coulomb blockade oscillations in the thermopower of a quantum dot. Solid State Commun. **87**, 1145 (1993), http://dx.doi.org/10.1016/0038-1098(93)90819-9
57. S. Möller, H. Buhmann, S.F. Godijn, L.W. Molenkamp, Charging energy of a chaotic quantum dot. Phys. Rev. Lett. **81**, 5197 (1998), https://doi.org/10.1103/PhysRevLett.81.5197
58. M. Turek, K.A. Matveev, Cotunneling thermopower of single electron transistors. Phys. Rev. B **65**, 115332 (2002), https://doi.org/10.1103/PhysRevB.65.115332
59. J.P. Pekola, J.V. Koski, D.V. Averin, Refrigerator based on the Coulomb barrier for single-electron tunneling. Phys. Rev. B **89**, 081309(R) (2014), https://doi.org/10.1103/PhysRevB.89.081309
60. A.V. Feshchenko, J.V. Koski, J.P. Pekola, Experimental realization of a Coulomb blockade refrigerator. Phys. Rev. B **90**, 201407(R) (2014), https://doi.org/10.1103/PhysRevB.90.201407
61. H. Pothier, P. Lafarge, P.F. Orfila, C. Urbina, D. Esteve, M.H. Devoret, Single electron pump fabricated with ultrasmall normal tunnel junctions. Physica B **169**, 573 (1991), http://dx.doi.org/10.1016/0921-4526(91)90332-9
62. H. Pothier, P. Lafarge, C. Urbina, D. Esteve, M.H. Devoret, Single-electron pump based on charging effects. Europhys. Lett. **17**, 249 (1992), http://dx.doi.org/10.1209/0295-5075/17/3/011
63. S.V. Lotkhov, S.A. Bogoslovsky, A.B. Zorin, J. Niemeyer, Operation of a three-junction single-electron pump with on-chip resistors. Appl. Phys. Lett. **78**, 946 (2001), http://dx.doi.org/10.1063/1.1347017
64. B. Limbach, P. vom Stein, C. Wallisser, R. Schäfer, Coulomb blockade in two-island systems with highly conductive junctions. Phys. Rev. B **72**, 045319 (2005), https://doi.org/10.1103/PhysRevB.72.045319
65. D.V. Averin, M. Möttönen, J.P. Pekola, Maxwell's demon based on a single-electron pump. Phys. Rev. B **84**, 245448 (2011), https://doi.org/10.1103/PhysRevB.84.245448
66. G. Bulnes Cuetara, M. Esposito, P. Gaspard, Fluctuation theorems for capacitively coupled electronic currents. Phys. Rev. B **84**, 165114 (2011), https://doi.org/10.1103/PhysRevB.84.165114

67. R. López, J.S. Lim, D. Sánchez, Fluctuation relations for spintronics. Phys. Rev. Lett. **108**, 246603 (2012), https://doi.org/10.1103/PhysRevLett.108.246603
68. R. Sánchez, M. Büttiker, Detection of single-electron heat transfer statistics. Europhys. Lett. **100**, 47008 (2012); ibid.**104**, 49901 (2013), http://stacks.iop.org/0295-5075/100/i=4/a=47008, http://dx.doi.org/10.1209/0295-5075/104/49901
69. R. Hussein, S. Kohler, Capacitively coupled nano conductors: Ratchet currents and exchange fluctuation relations. Ann. Phys. (Berlin) **527**, 610 (2015), https://doi.org/10.1002/andp.201500141
70. P. Stegmann, J. König, Violation of detailed balance for charge-transfer statistics in Coulomb-blockade systems. Phys. Status Solidi B. **254**, 1600507 (2017), https://doi.org/10.1002/pssb.201600507
71. T. Ruokola, T. Ojanen, Single-electron heat diode: asymmetric heat transport between electronic reservoirs through Coulomb islands. Phys. Rev. B **83**, 241404(R) (2011), http://dx.doi.org/10.1103/PhysRevB.83.241404
72. R. Sánchez, H. Thierschmann, L.W. Molenkamp, All-thermal transistor based on stochastic switching. Phys. Rev. B **95**, 241401(R) (2017), https://doi.org/10.1103/PhysRevB.95.241401

Non-perturbative Approaches in Nanoscience and Corrections to Finite-Size Scaling

J. Kaupužs and R. V. N. Melnik

Abstract Non-perturbative approaches in nanoscience are discussed. Traditional applications of these approaches cover description of charge transport and optical phenomena in nano-scale systems. We focus on finite-size effects in spin systems near the critical point, based on Monte Carlo (MC) method and some analytical arguments. We have performed MC simulations of the 3D Ising model for small, as well as large linear lattice sizes up to $L = 2560$, providing a numerical evidence for a recent challenging prediction, according to which the asymptotic decay of corrections to finite-size scaling is remarkably slower than it was expected before. Our approach along with several other non-perturbative approaches, like, e.g., the non-perturbative nonequilibrium Greens functions (NEGF) method, reveals a potential application of non-perturbative methods to nanoscience and nanotechnology through condensed matter physics, including semiconductor physics and physics of disordered systems like spin glasses.

Keywords Ising model · Non-perturbative methods · Finite-size effects
Corrections to scaling · Critical exponents · Monte Carlo simulation

J. Kaupužs (✉)
Faculty of Materials Science and Applied Chemistry, Institute of Technical Physics, Riga Technical University, Paula Valdena Str. 3/7, Riga LV–1048, Latvia
e-mail: kaupuzs@latnet.lv

J. Kaupužs
Institute of Mathematical Sciences and Information Technologies, University of Liepaja, 14 Liela Street, Liepaja LV–3401, Latvia

J. Kaupužs · R. V. N. Melnik
The MS2 Discovery Interdisciplinary Research Institute, Wilfrid Laurier University, Waterloo, Ontario N2L 3C5, Canada
e-mail: rmelnik@wlu.ca

R. V. N. Melnik
BCAM - Basque Center for Applied Mathematics, E48009 Bilbao, Spain

L. L. Bonilla et al. (eds.), *Coupled Mathematical Models for Physical and Biological Nanoscale Systems and Their Applications*, Springer Proceedings in Mathematics & Statistics 232, https://doi.org/10.1007/978-3-319-76599-0_4

1 Introduction

Non-perturbative approaches like Monte Carlo (MC) simulation [1, 2] and non-perturbative nonequilibrium Greens functions (NEGF) method [3, 4] become increasingly important in nanoscience. Particular applications cover the description of charge transport and optical phenomena in nano-scale systems such as microelectronic devices, graphene layers, etc.

Perturbative approaches not always provide a satisfactory description of physical phenomena in such systems. For example, the recent results of [5] show that the many-body localization (MBL) in translation-invariant systems with two or more very different energy scales is less robust than perturbative arguments suggest. It possibly points to the importance of non-perturbative effects, which induce delocalization in the thermodynamic limit [5]. The importance of non-perturbative effects in laser-illuminated graphene nanoribbons has been demonstrated in [6].

Here we consider another application—behavior of small and large spin systems near the phase transition point, based on non-perturbative analytical evaluation of **k**-space integrals, as well as MC method. Lattice spin models are considered, where certain value of the spin variable is related to each lattice site. In the Ising model, the spin variable σ can take only one of two possible values ± 1. In the scalar φ^4 model, the spin variable φ can take any value within $-\infty < \varphi < \infty$. Such models exhibit second-order phase transition in the thermodynamic limit, where the linear lattice size L tends to infinity. The behavior of these models on finite lattices is described by the finite-size scaling. Moreover, the scaling behavior near the critical point can be remarkably varied depending on whether small or large lattices are considered. This effect is described by corrections to the leading scaling behavior.

Our analytical and MC results serve as a basis for a challenging prediction that the correction-to-scaling exponent ω has a remarkably smaller value $\omega \leq \omega_{\max} \approx 0.38$ than the usually accepted ones of about 0.83 [7] in the 3D φ^4 and Ising models. It implies a much slower decay of corrections to scaling than it was usually expected. We have performed MC simulations of the 3D Ising model for small, as well as large linear lattice sizes up to $L = 2560$, providing a numerical evidence for this challenging prediction.

The vicinity of critical point is not the natural domain of validity of any perturbation theory, therefore one should prefer non–perturbative approaches. Among them are:

- Exact and rigorous analytical solution methods (transfer matrix methods, combinatorial methods, Bethe-ansatz)
- Conformal field theory (CFT) analysis
- Non–perturbative renormalization group (RG) analysis
- Numerical transfer–matrix calculations
- Molecular dynamics simulations
- Monte Carlo (MC) simulations.

Recent study of hierarchical Edwards-Anderson model of spin glasses [8] has shown that non-perturbative effects can be really important in critical phenomena.

Namely, it has been found that the perturbative approach (ε-expansion) correctly predicts the existence of the critical point only in certain mean-field region of parameters. At the same time, non-perturbative calculations show that the critical point exists also in the non-mean-field region, where no phase transition is predicted by the ε-expansion. From the point of view of the renormalization group (RG) theory, the critical behavior in this region is described by certain non-perturbative fixed point [8].

The results of [8] refer to the phase transitions in spin glasses, whereas our analytical and MC arguments, discussed in this paper, allow us to question the validity and/or accuracy of the perturbative treatments even in the apparently very well studied case of the 3D Ising model.

Our approach along with several other non-perturbative approaches reveals a potential application of non-perturbative methods to nanoscience and nanotechnology through condensed matter physics as an example.

2 Finite-Size Effects

Finite-size effects are very important in nano-scaled systems. For example, the charge density in metal nanoparticles after absorption of oxygen shows a very interesting patterns, which essentially depend on the size of these nanoparticles [9]. As another example, the importance of finite-size effects in the many-body localization has been shown in [5].

In our study, it is demonstrated the finite-size effects are pronounced via the finite-size scaling. We consider behavior of spin systems near the phase transition point depending on the lattice size, using the standard arguments of the finite-size scaling theory. For example, magnetic susceptibility χ at the critical temperature scales as $\chi \propto L^{2-\eta}\left(1+\mathscr{O}\left(L^{-\omega}\right)\right)$ at $L \to \infty$, where η is the critical exponent, describing the leading asymptotic behavior, whereas the exponent ω describes the leading correction to scaling.

Finite-size scaling and corrections to scaling are important to understand the difference in the behavior of very small nano-scale systems and large systems in the thermodynamic limit.

3 Non-perturbative Analytical Arguments

First, we consider the continuous φ^4 model as one of the simplest basic models in a hierarchy of spin models. The Hamiltonian $\mathscr{H}$ of this model is given by

$$\frac{\mathscr{H}}{k_B T} = \int \left(r_0 \varphi^2(\mathbf{x}) + c(\nabla \varphi(\mathbf{x}))^2 + u \varphi^4(\mathbf{x})\right) d\mathbf{x}, \tag{1}$$

where k_B is the Boltzmann constant, T is the temperature, $\varphi(\mathbf{x})$ is the local order parameter—an n-component vector, which depends on the coordinate $\mathbf{x}$, and r_0, c, u are Hamiltonian parameters. Configurations of $\varphi(\mathbf{x})$ obey a constraint, represented by certain upper cut-off Λ for its Fourier components. Namely, the Hamiltonian in the Fourier representation reads

$$\frac{\mathscr{H}}{k_B T} = \sum_{i,\mathbf{k}} \left(r_0 + c\,\mathbf{k}^2\right) \mid \varphi_{i,\mathbf{k}} \mid^2 + uV^{-1} \sum_{i,j,\mathbf{k}_1,\mathbf{k}_2,\mathbf{k}_3} \varphi_{i,\mathbf{k}_1}\varphi_{i,\mathbf{k}_2}\varphi_{j,\mathbf{k}_3}\varphi_{j,-\mathbf{k}_1-\mathbf{k}_2-\mathbf{k}_3} , \quad (2)$$

where $\varphi_{j,\mathbf{k}} = V^{-1/2} \int \varphi_j(\mathbf{x}) e^{-i\mathbf{k}\mathbf{x}} d\mathbf{x}$ and $\varphi_j(\mathbf{x}) = V^{-1/2} \sum_{k<\Lambda} \varphi_{j,\mathbf{k}} e^{i\mathbf{k}\mathbf{x}}$. Here V is the volume.

The Fourier-transformed two-point correlation function $G(\mathbf{k}) = \langle \mid \varphi_{j,\mathbf{k}} \mid \rangle^2$ is important in our following consideration. The case where r_0 is a linear function of T has been studied in [10], showing that, in this case, the leading singularity of specific heat C_V is given by an integral of $G(\mathbf{k})$ over wave vectors, i.e.,

$$C_V^{sing} = B\xi^{1/\nu} \left(\int_{k<\Lambda} [G(\mathbf{k}) - G^*(\mathbf{k})] d\mathbf{k} \right)^{sing} , \quad (3)$$

where B is a constant, $G^*(\mathbf{k})$ is the critical correlation function, and the superscript "sing" implies the leading singular part, represented in terms of the correlation length ξ. It diverges as $\xi \propto \mid T - T_c \mid^{-\nu}$, approaching the critical temperature T_c.

Non-perturbative analytical calculation of this integral has been performed in [10], based on the well known scaling hypothesis (see, e.g., [11]),

$$G(\mathbf{k}) = \sum_{i \ge 0} \xi^{(\gamma - \theta_i)/\nu} g_i(k\xi) , \quad (4)$$

where γ and ν are the critical exponents of susceptibility and correlation length, the term with $i = 0$ (at $\theta_0 = 0$) is the leading term, the terms with $i > 0$ are corrections to scaling with positive correction-to-scaling exponents $\theta_i = \omega_i \nu$, and $g_i(k\xi)$ are scaling functions. The singularity of specific heat in the general form of $C_V^{sing} \propto (\ln \xi)^\lambda \xi^{\alpha/\nu}$ has also been accounted for.

In addition, it has been assumed that the contribution of small-k region $k < \Lambda'$ is relevant in the limit $\lim_{\Lambda' \to 0} \lim_{\xi \to \infty}$. It has been verified by MC simulation tests in the lattice φ^4 model [10]. From a physics point of view, this assumption represents the usual statement about the importance of long-wavelength fluctuations in critical phenomena.

Based on these scaling assumptions, a theorem has been formulated and proven in [10], according to which the two-point correlation function contains a correction exponent $\theta_\ell = \gamma + 1 - \alpha - d\nu = \gamma - 1$, provided that $\gamma > 1$ holds. According to the current knowledge, the latter condition is strictly satisfied in two and three dimensions. The exponent θ_ℓ corresponds to the correction exponent $\omega_\ell = (\gamma - 1)/\nu$ in the finite-size scaling. Since ω_ℓ is not necessarily the leading correction-to-scaling exponent, $\omega \leq (\gamma - 1)/\nu$ is expected for the leading correction exponent. In particular, we have $\omega_\ell = 3/4$ and $\omega \leq 3/4$ for the scalar 2D φ^4 model.

The scalar φ^4 model belongs to the same universality class as the Ising model. Although corrections with $\omega = 3/4$ tend to cancel in the 2D Ising model [10], there is no reason for such a cancellation in general. Hence, corrections with $\omega \leq (\gamma - 1)/\nu$ are expected both in φ^4 and Ising models in three dimensions. Using the widely accepted estimates $\gamma \approx 1.24$ and $\nu \approx 0.63$ [12] for the 3D Ising model, we obtain the upper bound $\omega_{\max} = (\gamma - 1)/\nu \approx 0.38$ for ω. The prediction of the grouping of Feynman diagrams (GFD) theory [13] is $\gamma = 5/4$, $\nu = 2/3$ and, therefore, $\omega_{\max} = 0.375$. Thus, we can state that in any case $\omega_{\max}$ is about 0.38. The value of ω is expected to be $1/8$ according to the GFD theory considered in [13, 14].

The above arguments represent a very challenging prediction, since the currently widely known estimations of ω give essentially larger than 0.38 values, e.g., 0.782(5) [15] and 0.832(6) [7].

4 MC Tests for the 3D Ising Model

We consider the 3D Ising model on a simple cubic lattice with periodic boundary conditions. The Hamiltonian H is given by

$$H/T = -\beta \left(\sum_{\langle ij \rangle} \sigma_i \sigma_j + h \sum_i \sigma_i \right), \tag{5}$$

where T is the temperature in energy units, β is the coupling constant, h is the dimensionless (normalized) external field, and $\langle ij \rangle$ denotes the pairs of neighboring spins $\sigma_i = \pm 1$.

We have performed a detailed MC analysis based on the available simulation data for linear lattice sizes L, ranging from a small value $L = 8$ to a relatively very large value $L = 2560$, with the aim to test the challenging prediction $\omega \leq \omega_{\max} \approx 0.38$, discussed in Sect. 3. These data have been summarized in our papers [16, 17], where the simulations at $h = 0$ have been performed with the Wolff single cluster algorithm [18], using its parallel implementation [19].

The quantities of interest are the magnetization per spin $m = N^{-1} \sum_i \sigma_i$ and the magnetic susceptibility at zero external field, i.e., $\chi = \lim_{h \to 0} \frac{\partial \langle m \rangle}{\partial (\beta h)} = N \langle m^2 \rangle$, where $N = L^3$ is the total number of lattice sites. We consider χ and magnetization moments at certain pseudocritical coupling, i.e., at $\beta = \widetilde{\beta}_c(L)$. This coupling corresponds to a constant value of $\langle m^4 \rangle / \langle m^2 \rangle^2 = U$.

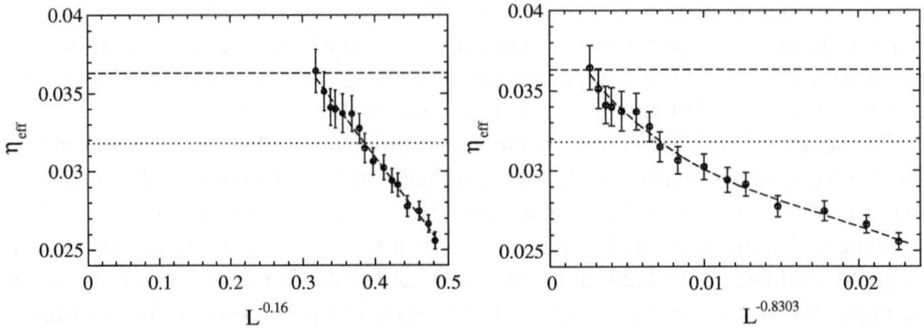

Fig. 1 The effective exponent $\eta_{\mathrm{eff}}(L)$ depending on $L^{-0.16}$ (left) and $L^{-0.8303}$ (right). The dashed straight lines represent the linear fit (left) and a guide to eye (right). The known estimates $\eta = 0.03631(3)$ [20] and $\eta = 0.318(3)$ [15] are shown by dashed and dotted horizontal lines, respectively

The pseudocritical coupling converges to the true critical coupling β_c as $\widetilde{\beta}_c(L) = \beta_c + \mathcal{O}\left(L^{-1/\nu}\right)$ at $L \to \infty$ for $1 < U < 3$ [19]. The constant $U = 1.6$ has been chosen, which is close to the critical value of $\langle m^4 \rangle / \langle m^2 \rangle^2$ at $\beta = \beta_c$ and $L \to \infty$.

According to the finite-size scaling theory, the susceptibility scales as

$$\chi \propto L^{2-\eta}\left(1 + \mathcal{O}\left(L^{-\omega}\right)\right) \tag{6}$$

at $\beta = \beta_c$ or $\beta = \widetilde{\beta}_c(L)$. It allows to estimate the critical exponent η by fitting the MC data. However, we do not know precisely the correction term, and a fit without this term gives us an effective value of the exponent rather than the true critical exponent.

We define the effective exponent $\eta_{\mathrm{eff}}(L)$ as the average slope of $-\ln\left(\chi/L'^2\right)$ versus $\ln L'$ plot, evaluated by fitting the data within $L' \in [L/2, 2L]$ at $\beta = \widetilde{\beta}_c(L')$. According to (6), it scales as

$$\eta_{\mathrm{eff}}(L) = \eta + \mathcal{O}\left(L^{-\omega}\right) \tag{7}$$

at $L \to \infty$. Hence, the $\eta_{\mathrm{eff}}(L)$ versus $L^{-\omega}$ plot is asymptotically linear at $L \to \infty$. The best linearity of this plot within $L \in [96, 1280]$ (extracted from the susceptibility data within $L \in [48, 2560]$) is observed at $\omega = 0.16(36)$. This plot looks, indeed, much more linear at $\omega = 0.16$ than at $\omega = 0.8303$, as it can be seen from Fig. 1. The latter value comes from the estimate of $\omega = 0.8303(18)$, obtained by the conformal bootstrap method in [20], which agrees with the MC value of 0.832(6) obtained in [7], but is claimed to be more accurate. Other MC values, usually reported in literature, are between 0.82 and 0.87 (see [7, 21]). The perturbative RG estimates are somewhat smaller, e.g., $\omega = 0.799 \pm 0.011$ [12] and $\omega = 0.782(5)$ [15].

The critical exponent $\eta = 0.03631(3)$, estimated in [20], and $\eta = 0.318(3)$, estimated in [15], are also indicated in Fig. 1 for comparison. The MC value of [7], $\eta = 0.03627(10)$, is very similar to that of [20].

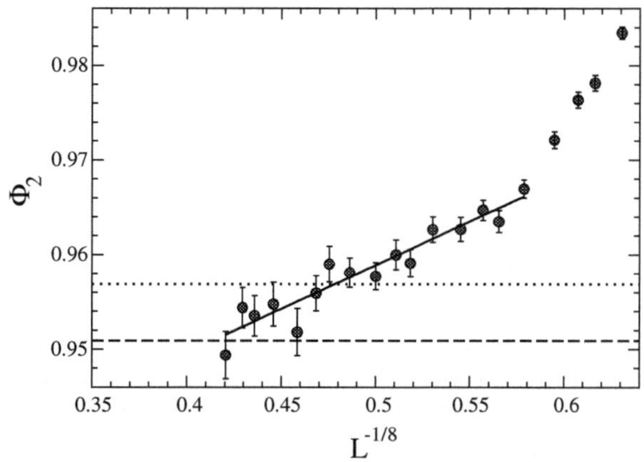

Fig. 2 The ratio $\Phi_2(L) = 2^{-4}\chi(2L)/\chi(L/2)$, evaluated from MC data at $\beta = \widetilde{\beta}_c(L)$, depending on $L^{-1/8}$. The asymptotic values of Φ_2 for $\eta = 0.03631(3)$ [20] and $\eta = 0.318(3)$ [15] are shown by dashed and dotted horizontal lines, respectively

Next, in the following, we consider a more direct method, which gives similar, but slightly more accurate results for ω. We consider the ratio $\Phi_b(L) = b^{-4}\chi(bL)/\chi(L/b)$ at $\beta = \widetilde{\beta}_c(L)$, where b is a constant. According to (6), $\Phi_b(L)$ behaves as

$$\Phi_b(L) = A + BL^{-\omega} \tag{8}$$

at $L \to \infty$, where $A = b^{-2\eta}$ and $B = ab^{-2\eta}\left(b^{-\omega} - b^{\omega}\right)$. The choice $b = 2$ is found to be optimal for the analysis of our data. The $\Phi_2(L)$ versus $L^{-1/8}$ plot can be well approximated by a straight line for large enough lattice sizes $L \in [80, 1280]$, as shown in Fig. 2. It confirms that ω could be as small as $1/8$, in agreement with [13, 14]. Considering ω as a fit parameter in (8), the fit over this region of sizes gives $\omega = 0.21(29)$. Taking into account that the asymptotic value of ω is positive, we can judge from this estimation that ω, most probably, lies between 0 and 0.5. The asymptotic value of Φ_2 is $2^{-2\eta}$. The estimates, corresponding to the same η values as in Fig. 1, are indicated in Fig. 2 for comparison.

Summarizing the results of this section, our MC estimation suggests that $0 < \omega < 0.5$ most probably holds for the correction-to-scaling exponent ω, in agreement with the analytical arguments in Sect. 2. Our values of 0.16(36) and 0.21(29) are remarkably smaller than the earlier estimates, e.g., 0.832(6) [7]. Although the latter MC estimate of [7] is claimed to be very accurate, it has been obtained from relatively small lattice sizes, i.e., $L \leq 360$, as compared to $L \leq 2560$ in our current study. Various estimations from lattice sizes $L \leq 384$ have been discussed in [17], clearly demonstrating that these sizes are still too small for a reliable estimation of ω.

Concerning the exponent η, the plots in Fig. 1 show that η could be similar or larger than the already mentioned estimates of 0.03631(3) [20] and 0.03627(10) [7]. The effective exponent increases well above the perturbative RG value of 0.318(3) obtained in [15] (dotted line).

5 Remarks and Discussion

It should be emphasized that our MC analysis is based on the simulation results for substantially large lattice sizes as compared to those typically considered in literature, e.g., $L \leq 360$ in [7] and even smaller sizes in other studies. It reveals a possibility to explain the inconsistency of our actual findings with earlier numerical results by relating the previous numerical results to a "perturbative" region, whereas our findings—to a "nonperturbative" region, as explained further on.

The "perturbative" region is a region of not–too–small reduced temperatures in the thermodynamic limit and not–too–large lattice sizes in the finite–size scaling regime. It is, very likely, true that the perturbative RG methods describe just only this region, due to the fundamental problems recovered in [22] (see Sect. 5.4.6 there). The "perturbative" region is easily accessible by various numerical methods. It explains the fact why these methods easily produce the results, which are approximately consistent with the estimates of the perturbative RG method and, therefore, also with each other. To the contrary, our findings refer to the "non-perturbative" asymptotic region, which corresponds to substantially larger lattice sizes in the finite–size scaling regime and extremely small reduced temperatures in the thermodynamic limit. This region cannot be easily accessed and is not yet properly investigated by numerical methods.

From the perspective of the non-perturbative RG approach, the "non-perturbative" asymptotic region is described by certain non-perturbative fixed point. We assume that the non-perturbative fixed point is not described by the usual perturbative treatments, just like in the case of spin glasses [8], discussed in Sect. 1. Hence, the critical exponents can be inconsistent with the perturbative RG estimates.

The results of the ε–expansion coincide with a particular treatment of the conformal field theory (CFT) in [20, 23]. This, however, only implies that the perturbative fixed point is, indeed, conformally symmetric and consistent with CFT at specific constraints, assumed in [20, 23].

6 Conclusions

In summary, we conclude the following:

- Non-perturbative approaches become increasingly important in study of critical phenomena, including their applications in nanoscience and nanotechnology.
- Non-perturbative analytical arguments have been provided for the existence of certain correction to scaling with exponent $\omega_\ell = (\gamma - 1)/\nu$ in the φ^4 model, and the leading finite-size correction-to-scaling exponent being $\omega \leq (\gamma - 1)/\nu$.
- The above statement has been supported by MC simulation results for small and very large lattices in 2D φ^4 and 3D Ising models.
- Our results emphasize the importance of corrections to scaling in nano-scale systems, pointing to a slow crossover in finite-size scaling behavior when the

system size is varied from small (nano-scale) to large (thermodynamic limit). This crossover is related to small ω value—remarkably smaller than it was expected before.

Acknowledgements This work was made possible by the facilities of the Shared Hierarchical Academic Research Computing Network (SHARCNET: www.sharcnet.ca). The authors acknowledge the use of resources provided by the Latvian Grid Infrastructure and High Performance Computing centre of Riga Technical University. R. M. acknowledges the support from the NSERC and CRC program.

References

1. M.E.J. Newman, G.T. Barkema, *Monte Carlo Methods in Statistical Physics* (Clarendon Press, Oxford, 1999)
2. C. Riddet, A. Asenov, *Proceedings Simulation of Semiconductor Processes and Devices* (2008), pp. 261–264
3. A. Martiner, M. Bescond, J.R. Barker, A. Svizhenko, M.P. Anantram, C. Millar, A. Asenov, IEEE Trans. Electron Devices **54**, 2213–2222 (2007)
4. D.A. Areshkin, B.K. Nikolic, Phys. Rev. B **81**, 155450 (2010)
5. Z. Papić, E.M. Stoudenmire, D.A. Abanin, Ann. Phys. **362**, 714–725 (2015)
6. H.L. Calve, P.M. Perez-Piskunov, H.M. Pastawski, S. Roche, L.E.F.F. Torres, J. Phys. Condens. Matter **25**, 144202 (2013)
7. M. Hasenbusch, Phys. Rev. B **82**, 174433 (2010)
8. M. Castellana, G. Parisi, Sci. Rep. **5**, 12367 (2015)
9. L. Lin, A.H. Larsen, N.A. Romero, V.A. Morozov, C. Glinsvad, F. Abild-Pedersen, J. Greeley, K.W. Jacobsen, J.K. Nørskov, J. Phys. Chem. Lett. **4**, 222–226 (2013)
10. J. Kaupužs, R.V.N. Melnik, J. Rimšāns, Int. J. Mod. Phys. C **27**, 1650108 (2016)
11. S.K. Ma, *Modern Theory of Critical Phenomena* (W. A. Benjamin Inc, New York, 1976)
12. J. Zinn-Justin, *Quantum Field Theory and Critical Phenomena* (Clarendon Press, Oxford, 1996)
13. J. Kaupužs, Ann. Phys. (Berlin) **10**, 299–331 (2001)
14. J. Kaupužs, Canadian. J. Phys. **9**, 373 (2012)
15. A.A. Pogorelov, I.M. Suslov, J. Exp. Theor. Phys. **106**, 1118 (2008)
16. J. Kaupužs, J. Rimšāns, R.V.N. Melnik, Ukr. J. Phys. **56**, 845 (2011)
17. J. Kaupužs, R.V.N. Melnik, J. Rimšāns, Int. J. Mod. Phys. C **28**, 1750044 (2017)
18. U. Wolff, Phys. Rev. Lett. **62**, 361 (1989)
19. J. Kaupužs, J. Rimšāns, R.V.N. Melnik, Phys. Rev. E **81**, 026701 (2010)
20. S. El-Showk, M.F. Paulos, D. Poland, S. Rychkov, D. Simmons-Duffin, A. Vichi, J. Stat. Phys. **157**, 869 (2014)
21. M. Hasenbusch, Int. J. Mod. Phys. C **12**, 911 (2001)
22. J. Kaupužs, Int. J. Mod. Phys. A **27**, 1250114 (2012)
23. F. Kos, D. Poland, D. Simmons-Duffin, JHEP **1406**, 091 (2014)

Continuum Model for Coupled Acousto-Optical Phonons in Piezoelectric Materials

Morten Willatzen and Zhong Lin Wang

Abstract A 3D continuum model is used to find optical and acoustic phonon fields in zincblende GaAs. We start out using continuum elastic differential equations and the Maxwell-Poisson equation to describe dynamic lattice strain and internal strain effects accounting for the full crystal symmetry of zincblende GaAs. The analytical model is derived in detail in a first-principles analysis. Our results reveal that for a slab of crystal GaAs grown along the [001] direction the mechanical displacements along the x and z directions u_x, u_z couple while the mechanical displacement u_y couple solely to the electric field E by virtue of piezoelectricity. As a consequence optical and acoustic phonon fields are inherently coupled due to piezoelectricity and acoustic and optical phonon modes must be found by solving simultaneously the full elastic solid and electric governing equations and the relevant (elastic and electric) boundary conditions. We then derive phonon dispersion curves for a GaAs slab and compare cases with and without anisotropy and piezoelectricity and show that neglecting the latter in the description of both acoustic and optical modes of GaAs, as is done in many classical descriptions, is a too crude approximation. We finally discuss two novel results: (i) confined coupled acousto-optical $u_y - u_\phi$ modes cannot exist in piezoelectric media except at certain discrete q_x wavenumber values, and (ii) piezoelectricity prohibits the existence of optical phonon fields at the LO phonon frequency. The model presented is general and can be applied to other materials and other crystal structures.

M. Willatzen (✉)
Beijing Institute of Nanoenergy and Nanosystems, Chinese Academy of Sciences, Beijing, China
e-mail: mortenwillatzen@binn.cas.cn

Z. Lin Wang
School of Materials Science & Engineering, Georgia Institute of Technology, Atlanta, GA 30332, USA

Z. Lin Wang
Beijing Institute of Nanoenergy and Nanosystems, Chinese Academy of Sciences, Beijing, China

L. L. Bonilla et al. (eds.), *Coupled Mathematical Models for Physical and Biological Nanoscale Systems and Their Applications*, Springer Proceedings in Mathematics & Statistics 232, https://doi.org/10.1007/978-3-319-76599-0_5

Keywords Piezoelectricity · Continuum elastic and electric model · Optical and acoustic phonons · Coupling · Anisotropy · Symmetry

1 Introduction

Whether discussing thermal, transport, or optical properties of solids it is of major importance to include dissipation mechanisms and account accurately for the lattice vibrational coupling to electronic states. Usually dissipation mechanisms have degrading effects on device performance where the active components consist of bulk and lower-dimensional solids and it is therefore essential to be able to control their strength and tune phonon properties and device characteristics through geometry, composition of materials, and/or external excitation conditions. Some recent examples of phonon tuning applications include the realization of population inversion in InGaAs quantum dot systems [1–3] and phonon-assisted gain in semiconductor quantum dot masers [4–6]. Coupling to phonon reservoirs significantly influences dephasing mechanisms [7–10] in quantum dot-cavity systems which is a central question and obstacle to address for applications in quantum information technology (QIT) [11]. Dephasing is a limiting factor for QIT since it destroys the entanglement between light and matter [7].

Optical and acoustic phonon interactions are also central to the fundamental understanding of dissipation mechanisms in the novel class of 2D materials, such as graphene, silicene, phospherene, molybdenum disulfide, and boron nitride, and to optimize devices where 2D materials function as active components [12–14].

A unified continuum approach to compute acoustic and optical phonon modes is described applicable to all crystals solids. The main purpose of this work is to demonstrate that acoustic and optical phonons mix in cases where inversion symmetry of the host crystal is broken such that piezoelectric effects play a role. Hence phonon modes must be described using a combination of elastic and electric equations and appropriate mechanical and electric boundary conditions. The combined theory of acoustic and optical phonon modes presented is shown to simplify, in the absence of piezoelectricity, to the classical phonon picture of separate acoustic and optical phonon models. However, even in the case of weak piezoelectricity of a solid lacking inversion symmetry, we demonstrate qualitative and significant differences due to piezoelectricity in the phonon density of states that eventually lead to important changes in phonon interaction strengths.

2 Elastic Equations

For zincblende GaAs, using Voigt notation, the constitutive equations between stress and strain are [15]

$$T_1 = c_{11}S_1 + c_{12}S_2 + c_{12}S_3, \tag{1}$$
$$T_2 = c_{12}S_1 + c_{11}S_2 + c_{12}S_3, \tag{2}$$
$$T_3 = c_{12}S_1 + c_{12}S_2 + c_{11}S_3, \tag{3}$$
$$T_4 = c_{44}S_4 - e_{14}E_1, \tag{4}$$
$$T_5 = c_{44}S_5 - e_{14}E_2, \tag{5}$$
$$T_6 = c_{44}S_6 - e_{14}E_3, \tag{6}$$
$$D_1 = \epsilon_{11}E_1 + e_{14}S_4, \tag{7}$$
$$D_2 = \epsilon_{11}E_2 + e_{14}S_5, \tag{8}$$
$$D_3 = \epsilon_{11}E_3 + e_{14}S_6, \tag{9}$$

where

$$S_1 = \frac{\partial u_x}{\partial x}, \tag{10}$$
$$S_2 = \frac{\partial u_y}{\partial y}, \tag{11}$$
$$S_3 = \frac{\partial u_z}{\partial z}, \tag{12}$$
$$S_4 = \frac{\partial u_y}{\partial z} + \frac{\partial u_z}{\partial y}, \tag{13}$$
$$S_5 = \frac{\partial u_x}{\partial z} + \frac{\partial u_z}{\partial x}, \tag{14}$$
$$S_6 = \frac{\partial u_x}{\partial y} + \frac{\partial u_y}{\partial x}. \tag{15}$$

Here, u_x, u_y, u_z are the displacements along the x, y, z directions and T, S, D, E are the stress, strain, electric displacement, and the electric field, respectively. Further we introduced the stiffness tensor c_{IJ}, the permittivity tensor ϵ_{ij}, and the piezoelectric e tensor e_{iJ}.

Let us compute coupled phonon modes in a slab defined by its two boundaries at $z = -\frac{a}{2}$ and $z = \frac{a}{2}$ assuming the in-plane coordinates x, y range from $-\infty$ to $+\infty$. We write for the lattice displacement

$$\mathbf{u}(\mathbf{r}, t) = \sum_n \int d\mathbf{q}_{\|} e^{i\mathbf{q}_{\|}\cdot\mathbf{r}_{\|}} \mathbf{u}_n(\mathbf{q}_{\|}, z) \exp(i\omega t). \tag{16}$$

The elastic equations now read

$$\frac{\partial T_1}{\partial x} + \frac{\partial T_6}{\partial y} + \frac{\partial T_5}{\partial z} = -\rho\omega^2 u_x, \tag{17}$$

$$\frac{\partial T_6}{\partial x} + \frac{\partial T_2}{\partial y} + \frac{\partial T_4}{\partial z} = -\rho\omega^2 u_y, \tag{18}$$

$$\frac{\partial T_5}{\partial x} + \frac{\partial T_4}{\partial y} + \frac{\partial T_3}{\partial z} = -\rho\omega^2 u_z, \tag{19}$$

where ρ is the mass density, and the Maxwell-Poisson equation becomes

$$\frac{\partial D_1}{\partial x} + \frac{\partial D_2}{\partial y} + \frac{\partial D_3}{\partial z} = 0, \tag{20}$$

if free electric charges are neglected.

Further we shall consider low frequencies such that the quasi-static approximation applies and contributions from magnetic fields can be neglected. In this approximation, a scalar potential suffices to determine the electric field:

$$\mathbf{E} = -\nabla\phi. \tag{21}$$

We shall need a similar relation for the scalar potential as Eq. (16) for the displacement

$$\phi(\mathbf{r}, t) = \sum_n \int d\mathbf{q}_\parallel e^{i\mathbf{q}_\parallel \cdot \mathbf{r}_\parallel} \phi_n(\mathbf{q}_\parallel, z) \exp(i\omega t). \tag{22}$$

In the following we will suppress the trivial time factor $\exp(i\omega t)$ for notational simplicity.

Let the phonon in-plane wavevector be directed along the x axis, i.e., $\mathbf{q}_\parallel = (q_x, 0)$. Insertion of Eqs. (1)–(9) in Eqs. (17)–(20) then yields [16]

$$\left(-q_x^2 c_{11} + c_{44}\frac{\partial^2}{\partial z^2} + \rho\omega^2\right) u_x + iq_x\left(c_{12} + c_{44}\right)\frac{\partial u_z}{\partial z} = 0, \tag{23}$$

$$\left(-q_x^2 c_{44} + c_{44}\frac{\partial^2}{\partial z^2} + \rho\omega^2\right) u_y + 2iq_x e_{14}\frac{\partial \phi}{\partial z} = 0, \tag{24}$$

$$iq_x\left(c_{12} + c_{44}\right)\frac{\partial u_x}{\partial z} + \left(-q_x^2 c_{44} + c_{11}\frac{\partial^2}{\partial z^2} + \rho\omega^2\right) u_z = 0, \tag{25}$$

$$2iq_x e_{14}\frac{\partial u_y}{\partial z} + \epsilon_{11}\left(q_x^2 - \frac{\partial^2}{\partial z^2}\right)\phi = 0. \tag{26}$$

Evidently, couplings exist between u_x and u_z only and between u_y and ϕ only.

Next, assuming solutions in the form $\frac{\partial}{\partial z} = i\gamma$, the detrimental equation stemming from the two equations in u_x and u_z reads

$$\left(-q_x^2 c_{11} - c_{44}\gamma^2 + \rho\omega^2\right)\left(-q_x^2 c_{44} - c_{11}\gamma^2 + \rho\omega^2\right) - q_x^2\left(c_{12} + c_{44}\right)^2 \gamma^2 = 0. \tag{27}$$

Similarly, assuming solutions in the form $\frac{\partial}{\partial z} = i\delta$, the determinental equation stemming from the two equations in u_y and ϕ reads

$$\left(-q_x^2 c_{44} - c_{44}\delta^2 + \rho\omega^2\right)\epsilon_{11}\left(q_x^2 + \delta^2\right) - 4q_x^2 e_{14}^2 \delta^2 = 0. \tag{28}$$

Both determinental equations are second-order polynomial equations, in γ^2 and δ^2, respectively. We find

$$\gamma^4 + \frac{q_x^2\left(c_{11}^2 + c_{44}^2\right) - \left(c_{11} + c_{44}\right)\rho\omega^2 - q_x^2\left(c_{12} + c_{44}\right)^2}{c_{11}c_{44}}\gamma^2 + \left(\frac{\omega^2}{c_{11}/\rho} - q_x^2\right)\left(\frac{\omega^2}{c_{44}/\rho} - q_x^2\right) = 0, \tag{29}$$

with solutions $\gamma = \pm\gamma_1, \pm\gamma_2$, and

$$\delta^4 + \left[2q_x^2 - \frac{\omega^2}{c_{44}/\rho} + 4q_x^2\frac{e_{14}^2}{c_{44}\epsilon_{11}}\right]\delta^2 + \left(q_x^2 - \frac{\omega^2}{c_{44}/\rho}\right)q_x^2 = 0, \tag{30}$$

with solutions $\delta = \pm\delta_1, \pm i\delta_2$. Note the appearance of the imaginary unit multiplied on δ_2. This choice reflects that in the absence of piezoelectricity, Eq. (26) gives $\frac{\partial}{\partial z} = i\delta = q_x$ combined with the fact that piezoelectricity is a rather weak coupling in typical zincblende materials.

It also follows from the structure of the differential equations that two solution types (I and II) for u_x and u_z are possible. For the type-I modes, we have

$$u_x = \sum_{j=1}^{2} A_{x,j}\cos\gamma_j z, \tag{31}$$

$$u_z = \sum_{j=1}^{2} A_{z,j}\sin\gamma_j z, \tag{32}$$

and for type-II modes

$$u_x = \sum_{j=1}^{2} A_{x,j} \sin \gamma_j z, \tag{33}$$

$$u_z = \sum_{j=1}^{2} A_{z,j} \cos \gamma_j z. \tag{34}$$

Similarly, two solution forms exist for u_y and ϕ. For type-I modes

$$u_y = A_{y,1} \cos(\delta_1 z) + A_{y,2} \cosh(\delta_2 z), \tag{35}$$

$$\phi = A_{\phi,1} \sin(\delta_1 z) + A_{\phi,2} \sinh(\delta_2 z), \tag{36}$$

and for type-II modes

$$u_y = A_{y,1} \sin(\delta_1 z) + A_{y,2} \sinh(\delta_2 z), \tag{37}$$

$$\phi = A_{\phi,1} \cos(\delta_1 z) + A_{\phi,2} \cosh(\delta_2 z). \tag{38}$$

From Eqs. (23) and (31)–(32), we obtain (type-I modes)

$$A_{z,j} = \frac{c_{44}\gamma_j^2 + q_x^2 c_{11} - \rho\omega^2}{i\gamma_j q_x (c_{12} + c_{44})} A_{x,j} \ (j = 1, 2), \tag{39}$$

and from Eqs. (23) and (33)–(34) (type-II modes)

$$A_{z,j} = -\frac{c_{44}\gamma_j^2 + q_x^2 c_{11} - \rho\omega^2}{i\gamma_j q_x (c_{12} + c_{44})} A_{x,j} \ (j = 1, 2). \tag{40}$$

Similarly, from Eqs. (26) and (35)–(36), we obtain (type-I modes)

$$2iq_x e_{14}\left(-\delta_1 A_{y,1} \sin(\delta_1 z) + \delta_2 A_{y,2} \sinh(\delta_2 z)\right) \\ +\epsilon_{11}\left(q_x^2 + \delta_1^2\right) A_{\phi,1} \sin(\delta_1 z) + \epsilon_{11}\left(q_x^2 - \delta_2^2\right) A_{\phi,2} \sinh(\delta_2 z) = 0 \ (j = 1, 2), \tag{41}$$

and from Eqs. (26) and (37)–(38) (type-II modes)

$$2iq_x e_{14}\left(\delta_1 A_{y,1} \cos(\delta_1 z) + \delta_2 A_{y,2} \cosh(\delta_2 z)\right) \\ +\epsilon_{11}\left(q_x^2 + \delta_1^2\right) A_{\phi,1} \cos(\delta_1 z) + \epsilon_{11}\left(q_x^2 - \delta_2^2\right) A_{\phi,2} \cosh(\delta_2 z) = 0 \ (j = 1, 2). \tag{42}$$

2.1 Complete Set of Equations for the Coupled (u_x, u_z) Acoustic Modes

A complete set of equations in the coefficients $A_{i,j}$ $(i = x, y, z, \phi)$ is obtained by imposing boundary conditions. For the coupled system u_x, u_z they are

$$T_5(z = \pm a/2) = 0, \tag{43}$$

$$T_3(z = \pm a/2) = 0, \tag{44}$$

subject to vacuum environment giving, in the case of Eqs. (31)–(32) (type-I modes),

$$\sum_{j=1}^{2} -\gamma_j A_{x,j} \sin\left(\gamma_j \frac{a}{2}\right) + iq_x A_{z,j} \sin\left(\gamma_j \frac{a}{2}\right) = 0, \tag{45}$$

$$\sum_{j=1}^{2} iq_x c_{12} A_{x,j} \cos\left(\gamma_j \frac{a}{2}\right) + \gamma_j c_{11} A_{z,j} \cos\left(\gamma_j \frac{a}{2}\right) = 0. \tag{46}$$

For type-II modes, using Eqs. (33)–(34), we find

$$\sum_{j=1}^{2} \gamma_j A_{x,j} \cos\left(\gamma_j \frac{a}{2}\right) + iq_x A_{z,j} \cos\left(\gamma_j \frac{a}{2}\right) = 0, \tag{47}$$

$$\sum_{j=1}^{2} iq_x c_{12} A_{x,j} \sin\left(\gamma_j \frac{a}{2}\right) - \gamma_j c_{11} A_{z,j} \sin\left(\gamma_j \frac{a}{2}\right) = 0. \tag{48}$$

Solving the 4 × 4 determinental equation in the four unknowns $A_{x,1}$, $A_{x,2}$, $A_{z,1}$, $A_{z,2}$ obtained from Eqs. (39) and (45)–(46) leads to the dispersion relations (q_x, ω_n) for type-I acoustic phonons as well as the coupled (u_x, u_z) phonon modes.

Similarly, solving the 4 × 4 determinental equation in the four unknowns $A_{x,1}$, $A_{x,2}$, $A_{z,1}$, $A_{z,2}$ obtained from Eqs. (40) and (47)–(48) leads to the dispersion relations (q_x, ω_n) for type-II acoustic phonons as well as the coupled (u_x, u_z) phonon modes.

Note that the subscript n added to ω indicates the discrete set of ω solutions that results from solving the determinental equations.

2.2 Complete Set of Equations for the Coupled (u_y, u_ϕ) Acousto-Optical Phonon Modes

In the case of (u_y, u_ϕ) modes, coupling between electric and elastic fields makes the calculation more involved. We start out by invoking the elastic boundary conditions

$$T_4(z = \pm a/2) = 0, \tag{49}$$

giving (type-I modes)

$$-c_{44}\delta_1 A_{y,1} \sin\left(\delta_1 \frac{a}{2}\right) + c_{44}\delta_2 A_{y,2} \sinh\left(\delta_2 \frac{a}{2}\right) + e_{14} i q_x A_{\phi,1} \sin\left(\delta_1 \frac{a}{2}\right) + e_{14} i q_x A_{\phi,2} \sinh\left(\delta_2 \frac{a}{2}\right) = 0, \tag{50}$$

and (type-II modes)

$$c_{44}\delta_1 A_{y,1} \cos\left(\delta_1 \frac{a}{2}\right) + c_{44}\delta_2 A_{y,2} \cosh\left(\delta_2 \frac{a}{2}\right) + e_{14} i q_x A_{\phi,1} \cos\left(\delta_1 \frac{a}{2}\right) + e_{14} i q_x A_{\phi,2} \cosh\left(\delta_2 \frac{a}{2}\right) = 0, \tag{51}$$

Next, consider the electric boundary conditions. Here, the electric field outside the slab must be taken into account as well. Since we assume vacuum environment, the Maxwell-Poisson equation simply reads, if $|z| > a/2$,

$$-\epsilon_0 \nabla^2 \phi = \epsilon_0 \left(q_x^2 - \frac{\partial^2}{\partial z^2}\right) \phi = 0, \tag{52}$$

with the solution

$$\phi(z) = \phi_+ e^{-|\mathbf{q}_\parallel| z} e^{i\mathbf{q}_\parallel \cdot \mathbf{r}} \quad \text{if } z > a/2, \tag{53}$$

$$\phi(z) = \phi_- e^{|\mathbf{q}_\parallel| z} e^{i\mathbf{q}_\parallel \cdot \mathbf{r}} \quad \text{if } z < -a/2. \tag{54}$$

Continuity of the transverse electric field at $z = \pm a/2$ yields (type-I modes)

$$-\frac{\partial \phi}{\partial x}|_{z=a/2} = -i q_x \left(A_{\phi,1} \sin\left(\delta_1 \frac{a}{2}\right) + A_{\phi,2} \sinh\left(\delta_2 \frac{a}{2}\right)\right) = -i q_x \phi_+ e^{-|\mathbf{q}_\parallel| a/2}, \tag{55}$$

$$-\frac{\partial \phi}{\partial x}|_{z=-a/2} = -i q_x \left(A_{\phi,1} \sin\left(-\delta_1 \frac{a}{2}\right) + A_{\phi,2} \sinh\left(-\delta_2 \frac{a}{2}\right)\right) = -i q_x \phi_- e^{-|\mathbf{q}_\parallel| a/2}. \tag{56}$$

Hence

$$\phi_+ = \frac{A_{\phi,1} \sin\left(\delta_1 \frac{a}{2}\right) + A_{\phi,2} \sinh\left(\delta_2 \frac{a}{2}\right)}{e^{-|\mathbf{q}_\parallel| a/2}}, \tag{57}$$

$$\phi_- = -\phi_+. \tag{58}$$

Finally, we impose continuity of the normal electric displacement at the slab boundaries (type-I modes)

$$\left[-\epsilon_{11}\frac{\partial \phi}{\partial z}+e_{14}\frac{\partial u_y}{\partial x}\right]|_{z=a/2} = -\epsilon_0\frac{\partial \phi}{\partial z}|_{z=a/2}, \tag{59}$$

$$\left[-\epsilon_{11}\frac{\partial \phi}{\partial z}+e_{14}\frac{\partial u_y}{\partial x}\right]|_{z=-a/2} = -\epsilon_0\frac{\partial \phi}{\partial z}|_{z=-a/2}, \tag{60}$$

i.e.,

$$\begin{aligned}
&-\epsilon_{11}\delta_1 A_{\phi,1}\cos\left(\delta_1\frac{a}{2}\right)-\epsilon_{11}\delta_2 A_{\phi,2}\cosh\left(\delta_2\frac{a}{2}\right)\\
&+e_{14}iq_x A_{y,1}\cos\left(\delta_1\frac{a}{2}\right)+e_{14}iq_x A_{y,2}\cosh\left(\delta_2\frac{a}{2}\right)\\
&=\epsilon_0\phi_+|\mathbf{q}_\parallel|e^{-|\mathbf{q}_\parallel|a/2},
\end{aligned} \tag{61}$$

$$\begin{aligned}
&-\epsilon_{11}\delta_1 A_{\phi,1}\cos\left(-\delta_1\frac{a}{2}\right)-\epsilon_{11}\delta_2 A_{\phi,2}\cosh\left(-\delta_2\frac{a}{2}\right)\\
&+e_{14}iq_x A_{y,1}\cos\left(-\delta_1\frac{a}{2}\right)+e_{14}iq_x A_{y,2}\cosh\left(-\delta_2\frac{a}{2}\right)\\
&=-\epsilon_0\phi_-|\mathbf{q}_\parallel|e^{-|\mathbf{q}_\parallel|a/2}.
\end{aligned} \tag{62}$$

Evidently, combining Eqs. (41), (50), (57), and (61) leads to a 4×4 matrix equation in the coefficients $A_{y,1}$, $A_{y,2}$, $A_{\phi,1}$, $A_{\phi,2}$ whose determinental equation specifies the dispersion relation (q_x, ω_n) of acousto-optical type-I phonons and the associated coupled (u_x, u_z) phonon modes.

For type-II modes, Eqs. (53)–(54) still apply. Continuity of the transverse electric field at $z = \pm a/2$ yields

$$\phi_+ = \frac{A_{\phi,1}\cos\left(\delta_1\frac{a}{2}\right)+A_{\phi,2}\cosh\left(\delta_2\frac{a}{2}\right)}{e^{-|\mathbf{q}_\parallel|a/2}}, \tag{63}$$

$$\phi_- = \phi_+. \tag{64}$$

Imposing continuity in the normal electric displacement at $z = \pm a/2$ gives one constraint (type-II modes)

$$\begin{aligned}
&\epsilon_{11}\delta_1 A_{\phi,1}\sin\left(\delta_1\frac{a}{2}\right)-\epsilon_{11}\delta_2 A_{\phi,2}\sinh\left(\delta_2\frac{a}{2}\right)\\
&+e_{14}iq_x A_{y,1}\sin\left(\delta_1\frac{a}{2}\right)+e_{14}iq_x A_{y,2}\sinh\left(\delta_2\frac{a}{2}\right)\\
&=\epsilon_0\phi_+|\mathbf{q}_\parallel|e^{-|\mathbf{q}_\parallel|a/2},
\end{aligned} \tag{65}$$

Now, combining Eqs. (42), (51), (63), and (65) leads to a 4×4 matrix equation in the coefficients $A_{y,1}$, $A_{y,2}$, $A_{\phi,1}$, $A_{\phi,2}$ whose determinental equation specifies the dispersion relation (q_x, ω_n) of acousto-optical type-II phonons and the associated coupled (u_x, u_z) phonon modes.

Note that the permittivity ϵ_{11} is a function of frequency (refer to Eq. (99) in the next section) and this is central in the discussion of optical phonon modes. We also point to that for zincblende only one permittivity constant ϵ_{11} exists. Hence we shall in the following sections replace ϵ_{11} by ϵ.

Formally, the fact that we know (and assume) solutions have a specific symmetry in z, there is, in principle, no point in enforcing boundary conditions at both $z = a/2$ and $z = -a/2$.

2.3 Dispersion Relations for the Coupled (u_y, u_ϕ) Acousto-Optical Phonon Modes

It follows from Eq. (50) and the expressions (type I modes)

$$\begin{aligned} -2iq_x e_{14}\delta_1 A_{y,1} + \epsilon_{11}\left(q_x^2+\delta_1^2\right)A_{\phi,1} &= 0, \\ 2iq_x e_{14}\delta_2 A_{y,2} + \epsilon_{11}\left(q_x^2-\delta_2^2\right)A_{\phi,2} &= 0, \end{aligned}$$

that

$$\begin{aligned} &\left[-c_{44}\delta_1\epsilon_{11}\left(q_x^2+\delta_1^2\right) - 2e_{14}^2 q_x^2\delta_1\right]\sin\left(\delta_1\frac{a}{2}\right)A_{\phi,1} \\ &+\left[c_{44}\delta_2\epsilon_{11}\left(\delta_2^2-q_x^2\right) - 2e_{14}^2 q_x^2\delta_2\right]\sinh\left(\delta_2\frac{a}{2}\right)A_{\phi,2} = 0. \end{aligned} \quad (66)$$

It now further follows from Eq. (61)

$$\begin{aligned} &\left[-\epsilon_{11}\delta_1\cos\left(\delta_1\frac{a}{2}\right) + \frac{\epsilon_{11}\left(q_x^2+\delta_1^2\right)}{2\delta_1}\cos\left(\delta_1\frac{a}{2}\right) - \epsilon_0 q_x\sin\left(\delta_1\frac{a}{2}\right)\right]A_{\phi,1} \\ +&\left[-\epsilon_{11}\delta_2\cosh\left(\delta_2\frac{a}{2}\right) + \frac{\epsilon_{11}\left(\delta_2^2-q_x^2\right)}{2\delta_2}\cosh\left(\delta_2\frac{a}{2}\right) - \epsilon_0 q_x\sinh\left(\delta_2\frac{a}{2}\right)\right]A_{\phi,2} = 0. \end{aligned} \quad (67)$$

This set of two equations in two unknowns $A_{\phi,1}$, $A_{\phi,2}$ has solutions only if the determinant vanishes.

2.4 Proof that Piezoelectric Coupling Does Not Allow for Acousto-Optical Phonon Mode Excitation at the LO Frequency

We shall now prove an important result that distinguishes piezoelectric materials from non-piezoelectric materials. At the LO phonon frequency where the permittivity is zero coupled acousto-optical phonon modes cannot be excited. In contrast, in non-piezoelectric media confined optical phonon modes exist at the LO phonon frequency.

First, if $\epsilon_{11} = 0$, Eq. (26) shows that

$$2iq_x e_{14}\frac{\partial u_y}{\partial z} = 0, \tag{68}$$

and

$$u_y = A_y e^{iq_x x}. \tag{69}$$

Then Eq. (24) yields

$$\phi = (\phi_0 + \phi_1 z)\, e^{iq_x x}, \tag{70}$$

$$\left(\rho\omega^2 - q_x^2 c_{44}\right) A_y + 2iq_x e_{14}\frac{\partial \phi}{\partial x} = 0. \tag{71}$$

From Eq. (50):

$$c_{44}\frac{\partial u_y}{\partial z} + e_{14}\frac{\partial \phi}{\partial x}|_{z=\pm a/2} = 0, \tag{72}$$

it follows that

$$\left(\phi_0 + \phi_1\frac{a}{2}\right) iq_x e_{14} = 0, \tag{73}$$

$$\left(\phi_0 - \phi_1\frac{a}{2}\right) iq_x e_{14} = 0, \tag{74}$$

so

$$\phi_0 = \phi_1 = 0. \tag{75}$$

Continuity of the transverse electric field component guarantees further

$$\phi_+ = \phi_- = 0, \tag{76}$$

and finally from Eq. (59)

$$e_{14}\frac{\partial u_y}{\partial x}|_{z=\pm a/2} = iq_x e_{14} A_y = 0, \tag{77}$$

i.e.,

$$A_y = 0. \tag{78}$$

Hence, both the electric potential ϕ and the phonon u_y component must be zero simultaneously. In other words, it is not possible to excite coupled $u_y - u_\phi$ modes at the LO phonon frequency!

2.5 Proof that Confined Coupled Acousto-Optical $u_y - u_\phi$ Modes Cannot Exist in Piezoelectric Media Except at Certain Discrete q_x Wavenumber Values

It is well-known that confined LO phonon modes exist at the LO phonon frequency. We will now prove that confined coupled acousto-optical $u_y - u_\phi$ modes cannot exist in piezoelectric media except at certain discrete q_x wavenumber values.

A confined phonon mode must be characterized by an electric field that vanishes at the slab interfaces. If this is the case, coupled $u_y - u_\phi$ solutions can be sought quite generally either in the form (type I modes)

$$u_y = A_y \cos\left(\frac{m\pi}{a}z\right) e^{iq_x x}, \tag{79}$$

$$\phi = A_\phi \sin\left(\frac{m\pi}{a}z\right) e^{iq_x x}, \tag{80}$$

where $m = 2, 4, 6, ...$, or (type II modes)

$$u_y = A_y \sin\left(\frac{m\pi}{a}z\right) e^{iq_x x}, \tag{81}$$

$$\phi = A_\phi \cos\left(\frac{m\pi}{a}z\right) e^{iq_x x}, \tag{82}$$

where $m = 1, 3, 5, ...$ Let us restrict our analysis to the first case as the other case is a trivial extension. From Eq. (24) we have

$$\left(-q_x^2 c_{44} - \left(\frac{m\pi}{a}\right)^2 c_{44} + \rho\omega^2\right) A_y + 2iq_x e_{14}\left(\frac{m\pi}{a}\right) A_\phi = 0, \tag{83}$$

and from Eq. (26)

$$-2iq_x e_{14}\left(\frac{m\pi}{a}\right) A_y + \epsilon_{11}\left(q_x^2 + \left(\frac{m\pi}{a}\right)^2\right) A_\phi = 0. \tag{84}$$

The latter expression gives

$$A_\phi = \frac{2iq_x e_{14}\left(\frac{m\pi}{a}\right)}{\epsilon_{11}\left(q_x^2 + \left(\frac{m\pi}{a}\right)^2\right)} A_y. \tag{85}$$

and inserting Eq. (85) in Eq. (83) determines the possible mode frequencies

$$\omega_m = \sqrt{\frac{c_{44}}{\rho}} \left[q_x^2 + \left(\frac{m\pi}{a}\right)^2 + \frac{4q_x^2 e_{14}^2 \left(\frac{m\pi}{a}\right)^2}{\epsilon_{11} c_{44} \left(q_x^2 + \left(\frac{m\pi}{a}\right)^2\right)} \right]^{1/2} . \tag{86}$$

It follows immediately from Eqs. (79)–(80) that Eq. (50) is fulfilled:

$$T_4|_{z=a/2} = c_{44} \frac{\partial u_y}{\partial z} + e_{14} \frac{\partial \phi}{\partial x}|_{z=a/2} = 0, \tag{87}$$

and also that

$$\phi|_{z=a/2} = 0, \tag{88}$$

so from continuity of transverse electric field:

$$\phi_+ = 0. \tag{89}$$

We then need to check if the normal displacement vanishes at the slab interface, i.e.,

$$D_3|_{z=a/2} = -\epsilon_{11} \frac{\partial \phi}{\partial z} + e_{14} \frac{\partial u_y}{\partial x}|_{z=a/2} = 0. \tag{90}$$

Again by use of Eqs. (79)–(80), this condition is fulfilled if

$$A_\phi = \frac{i q_x e_{14}}{\epsilon_{11} \left(\frac{m\pi}{a}\right)} A_y. \tag{91}$$

Combining Eqs. (85) and (91) requires

$$q_x^2 \equiv q_{x,m}^2 = \left(\frac{m\pi}{a}\right)^2 . \tag{92}$$

Hence we have proven the conjecture that confined coupled $u_y - u_\phi$ phonon modes cannot exist in piezoelectric media except for certain discrete wavenumber values $\{q_{x,m}\}$ in contrast to the case for non-piezoelectric media!

3 Born-Huang Equations

The phenomenological Born-Huang equations, describing coupling between the electric-field induced polarization $\mathbf{P} = \chi\epsilon_0\mathbf{E}$ ($\mathbf{D} = \epsilon_0\mathbf{E} + \mathbf{P}$) and the ionic displacement $\mathbf{u}_{ion} = \mathbf{u}_+ - \mathbf{u}_-$, are

$$\frac{\partial^2 \mathbf{w}}{\partial t^2} = b_{11}\mathbf{w} + b_{12}\mathbf{E}, \tag{93}$$

$$\frac{\partial^2 \mathbf{P}}{\partial t^2} = b_{21}\mathbf{w} + b_{22}\mathbf{E}, \tag{94}$$

where

$$\mathbf{w} = \sqrt{\frac{m_r}{\Omega}}\mathbf{u}_{ion}, \tag{95}$$

$$m_r = \frac{m_+ m_-}{m_+ + m_-}, \tag{96}$$

and m_+, m_- are the ion masses, $\Omega = \frac{V}{N}$ is the unit cell volume, i.e., N is the number of unit cells in the volume V. Combining the latter equations gives

$$\epsilon(\omega) = \epsilon_0 + \left(b_{22} - \frac{b_{21}b_{12}}{b_{11} - \omega^2}\right). \tag{97}$$

Further, conservation of energy gives

$$b_{12} = b_{21}. \tag{98}$$

An alternative expression for the permittivity is

$$\epsilon(\omega) = \epsilon(\infty) + \frac{\epsilon(0) - \epsilon(\infty)}{1 - \frac{\omega^2}{\omega_{TO}^2}}. \tag{99}$$

and it follows that

$$\epsilon(\infty) \equiv \lim_{\omega\to\infty} \epsilon(\omega) = \epsilon(0) + b_{22}, \tag{100}$$

$$\epsilon(0) \equiv \lim_{\omega\to 0} \epsilon(\omega) = \epsilon_0 + \left(b_{22} - \frac{b_{21}b_{12}}{b_{11}}\right), \tag{101}$$

and

$$b_{11} = -\omega_{TO}^2, \tag{102}$$

$$b_{22} = \epsilon(\infty) - \epsilon_0, \tag{103}$$

$$b_{12} = b_{21} = \sqrt{\epsilon(0) - \epsilon(\infty)}\omega_{TO}. \tag{104}$$

It also follows that once the electric field is determined through the coupled scheme above, the ionic displacement can be found from Eq. (93):

$$\mathbf{w} = -\frac{b_{12}}{b_{11} + \omega^2}\mathbf{E}. \tag{105}$$

Hence, the present formalism determines the lattice displacement $\mathbf{u}$ *and* the ionic displacement $\mathbf{u}_{ion}$ simultaneously.

4 Normalization of Phonon Modes

Since the phonon modes generally consist of coupled ion and lattice displacements we need to formulate appropriate normalization conditions. These are

$$\int d\mathbf{r}\left[\sqrt{\frac{m_r}{\Omega}}\mathbf{u}^*_{ion}(\mathbf{r},t)\right]\left[\sqrt{\frac{m_r}{\Omega}}\mathbf{u}_{ion}(\mathbf{r},t)\right] + \int d\mathbf{r}\left[\sqrt{\rho}\mathbf{u}^*(\mathbf{r},t)\right]\left[\sqrt{\rho}\mathbf{u}(\mathbf{r},t)\right] = \frac{\hbar}{2\omega_n}. \tag{106}$$

Upon carrying out normalization of the coupled phonon field, acousto-optical phonon modes are determined.

5 Deformation Potential Coupling and Deformation Hamiltonian

The deformation potential coupling between carriers and acoustic phonons stems from the change in energies of the electron and hole band edges due to lattice deformations. The associated Hamiltonian is

$$H_{def} = D_{c,v}\nabla \cdot \mathbf{u}, \tag{107}$$

where $D_{c,v}$ are the deformation potentials of the conduction, valence bands. Note that shear deformations in principle also contribute and are important for the valence bands but we shall neglect them here for the sake of simplicity. Rewriting the deformation potential Hamiltonian in terms of the normalized phonon modes gives

$$H_{def} = D_{c,v}\sum_{\mathbf{q}_\| n}\sqrt{\frac{2a\hbar}{V\rho\omega_n(\mathbf{q}_\|)}}\left(a_{n,q} + a^\dagger_{n,-q}\right)\nabla \cdot \left[\mathbf{w}_n(\mathbf{q}_\|, z)e^{i\mathbf{q}_\|\cdot\mathbf{r}_\|}\right], \tag{108}$$

where $V = 2aA$ is the slab volume with A the slab cross-sectional area, and

$$\mathbf{w}_n(\mathbf{q}_\|, z) = \sqrt{\rho}\mathbf{u}_n(\mathbf{q}_\|, z). \tag{109}$$

6 Discussions

Consider first the influence of anisotropy on the phonon dispersion for the coupled $u_x - u_z$ lattice displacement components in a cubic piezoelectric GaAs slab. We use the following physical properties of GaAs: $c_{11} = 11.88 \times 10^{10}$ Pa, $c_{12} = 5.38 \times 10^{10}$ Pa, $c_{44} = 5.94 \times 10^{10}$ Pa, $\rho = 5.318 \times 10^3$ kg/m^3, $e_{14} = 0.154$ C/m^2, $\epsilon(0) = 12.9\epsilon_0$, $\hbar\omega_{LO} = 33.2$ meV, $\hbar\omega_{TO} = 36.1$ meV, $\epsilon_\infty = \epsilon(0)\frac{\omega_{TO}^2}{\omega_{LO}^2}$.

Since all cubic materials are anisotropic, the isotropic relation

$$c_{44} = \frac{c_{11} - c_{12}}{2} \tag{110}$$

is an approximation. In fact, the degree of anisotropy defined as

$$\alpha = \left| \frac{\frac{c_{11}-c_{12}}{2} - c_{44}}{c_{44}} \right|, \tag{111}$$

is large in GaAs and equal to 45%. Hence, substantial deviations in the phonon modes and dispersion relations due to anisotropy must be expected for GaAs. We point to that unit-cell inversion asymmetry and piezoelectricity do not influence the $u_x - u_z$ phonon modes and dispersion.

In Fig. 1 the coupled type-I $u_x - u_z$ dispersion curves are shown for a 3 nm GaAs slab. It is clear that accounting for the full anisotropy changes substantially both energies and shape of the dispersion curves.

Next, let us consider the dispersion curves for coupled $u_y - \phi$ acousto-optical phonon modes.

In Fig. 2 the $u_y - \phi$ type-I dispersion curves are shown for a 2 nm GaAs slab. It is evident that the coupled acousto-optical phonon modes are strongly influenced by anisotropy, as in the case with the coupled acoustic $u_x - u_z$ phonons. Moreover, a close inspection of the dispersion curves around the LO phonon energy of 36.1 meV reveals that when piezoelectricity is included, and for any strictly positive $q_x a$ value, crossing of the forbidden LO phonon energy does not take place in agreement with the proof given above.

We stress that that interface phonons are not captured in the plots as they correspond to imaginary q_x values but the procedure for finding them is the same as for non-pizoelectric and isotropic media [17].

A combined treatment of coupled acoustic and optical phonons in zincblende media, by virtue of piezoelectricity, was presented that allows determination of the lattice displacement vector and the internal ionic displacement vector simultaneously. It was demonstrated that phonons generally exist in pairs due to couplings between

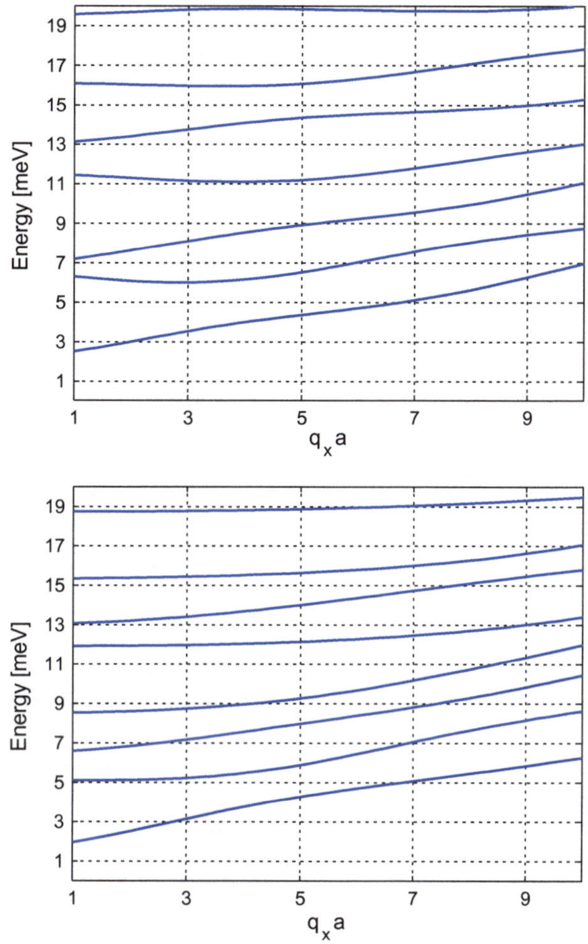

Fig. 1 Type-I dispersion relations for the coupled $u_x - u_z$ acoustic phonon modes of a 3 nm GaAs slab. (Upper plot) GaAs and (lower plot) GaAs but using the isotropic assumption: $c_{44} = \frac{c_{11}-c_{12}}{2} = 3.25 \times 10^{10}$ Pa. The first axis is $q_x a$ and the second axis the phonon band energy in meV

either the lattice displacement component perpendicular to the phonon wavevector (u_y coordinate) and the electric potential ϕ or the two other lattice displacement components (u_x and u_z). Three new conjectures were derived for piezoelectric cubic slabs: (1) isolated optical phonon modes cannot exist unless the in-plane phonon wavenumber vanishes ($q_x = 0$); (2) we have shown that confined acousto-optical phonon modes only exist in piezoelectric cubic slabs for a discrete set of in-plane wavenumbers; (3) at the LO phonon frequency coupled acousto-optical phonons cannot exist. We point to that the present method can be applied directly to other piezoelectric zincblende materials and is easily generalized to piezoelectric materials of a different crystal symmetry.

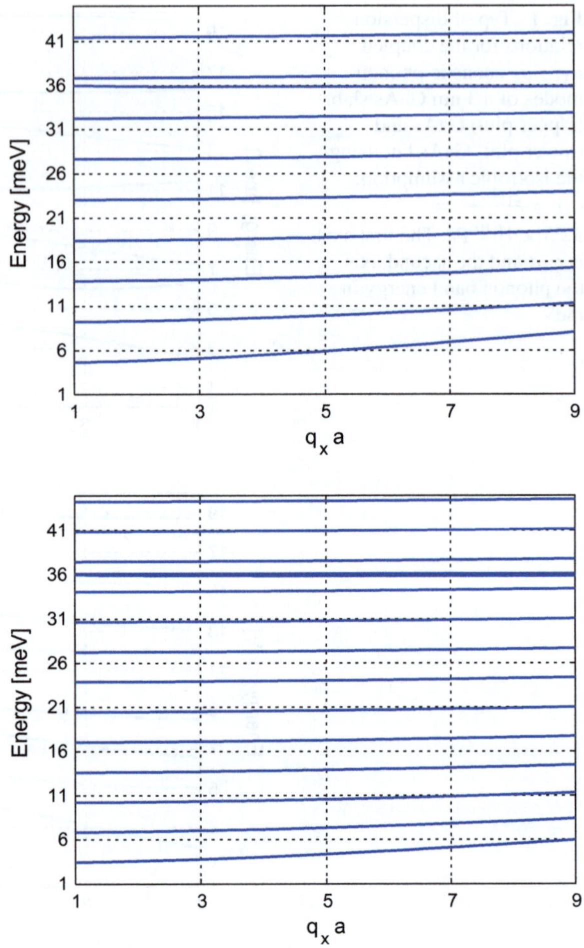

Fig. 2 Type-I dispersion relations for the coupled $u_y - \phi$ acousto-optical phonon modes of a 3 nm GaAs slab. (Upper plot) GaAs including piezoelectricity and anisotropy and (lower plot) GaAs but using the isotropic non-piezoelectric assumption: $c_{44} = \frac{c_{11}-c_{12}}{2} = 3.25 \times 10^{10}$ Pa and $e_{14} = 0$

Acknowledgements This research work partly emerged from a research stay at the Beijing Institute of Nanoenergy and Nanosystems, Chinese Academy of Sciences (BINN). MW gratefully acknowledges financial support from BINN.

References

1. J.H. Quilter, A.J. Brash, F. Liu, M. Glässl, A.M. Barth, V.M. Axt, A.J. Ramsay, M.S. Skolnick, A.M. Fox, Phys. Rev. Lett. **114**, 137401 (2015)
2. A.E. Siegman, *Lasers* (University Science Books, Oxford, 1986)
3. M. Glässl, A.M. Barth, V.M. Axt, Phys. Rev. Lett. **110**, 147401 (2013)
4. M.J. Gullans, Y.-Y. Liu, J. Stehlik, J.R. Petta, J.M. Taylor, Phys. Rev. Lett. **114**, 196802 (2015)
5. M. Sargent, M. Scully, W. Lamb, *Laser Physics* (Perseus Books Group, New York, 1978)
6. Y.Y. Liu, J. Stehlik, C. Eichler, M.J. Gullans, J.M. Taylor, J.R. Petta, Science **347**, 285 (2015)

7. P. Kaer, T.R. Nielsen, P. Lodahl, A.-P. Jauho, J. Mork, Phys. Rev. Lett. **104**, 157401 (2010)
8. A. Aufféves, J.M. Gerard, J.P. Poizat, Phys. Rev. A **79**, 053838 (2009)
9. T. Yoshie, A. Scherer, J. Heindrickson, G. Khitrova, H.M. Gibbs, G. Rupper, C. Ell, O.B. Shchekin, D.G. Deppe, Nature (London) **432**, 200 (2004)
10. H.J. Carmichael, R.J. Brecha, M.G. Raizen, H.J. Kimble, P.R. Rice, Phys. Rev. A **40**, 5516 (1989)
11. E. Knill, R. Laflamme, G.J. Milburn, Nature **409**, 46 (2001)
12. M.W. Graham, S.-F. Shi, D.C. Ralph, J. Park, P.L. McEuen, Nat. Phys. **9**, 103 (2009)
13. F.D. Natterer, Y. Zhao, J. Wyrick, Y.-H. Chan, W.-Y. Ruan, M.-Y. Chou, K. Watanabe, T. Taniguchi, N.B. Zhitenev, J.A. Stroscio, Phys. Rev. Lett. **114**, 245502 (2015)
14. D.N. Basov, M.M. Fogler, A. Lanzara, F. Wang, Y. Zhang, Rev. Mod. Phys. **86**, 959 (2014)
15. B.A. Auld, *Acoustic Fields and Waves in Solids, Part 1* (Krieger Publ. Co. 1990)
16. M. Willatzen, Z.L. Wang, Phys. Rev. B **92**, 224101 (2015)
17. J.J. Licari, R. Evrard, Phys. Rev. B **15**, 2254 (1977)

Part II
Modeling Biological Phenomena from Nano- to Macro-scales

Stochastic Models of Tumor Induced Angiogenesis

Luis L. Bonilla, M. Carretero and F. Terragni

Abstract Angiogenesis is a complex multiscale process by which diffusing vessel endothelial growth factors induce sprouting of blood vessels that carry oxygen and nutrients to hypoxic tissue. There is strong coupling between the kinetic parameters of the relevant branching—growth—anastomosis stochastic processes of the capillary network, at the microscale, and the family of interacting underlying biochemical fields, at the macroscale. A hybrid mesoscale tip cell model involves stochastic branching, fusion (anastomosis) and extension of active vessel tip cells with reaction-diffusion growth factor fields. Anastomosis prevents indefinite proliferation of active vessel tips, precludes a self-averaging stochastic process and ensures that a deterministic description of the density of active tips holds only for ensemble averages over replicas of the stochastic process. Evolution of active tips from a primary vessel to a tumor adopts the form of an advancing soliton that can be characterized by ordinary differential equations for its position, velocity and a size parameter. A short review of other angiogenesis models and possible implications of our work is also given.

1 Introduction

The growth of blood vessels out of a primary vessel or *angiogenesis* is a complex multiscale process responsible for organ growth and regeneration, tissue repair, wound healing and many other natural operations in living beings [1–5]. Angiogenesis is triggered by lack of oxygen (hypoxia) experienced by cells in some tissue. Such cells segregate growth factors that diffuse and reach a nearby primary blood vessel.

L. L. Bonilla (✉) · M. Carretero · F. Terragni
Department of Materials Science and Engineering, G. Millan Institute,
Universidad Carlos III de Madrid, Madrid, Spain
e-mail: bonilla@ing.uc3m.es

M. Carretero
e-mail: manuel.carretero@uc3m.es

F. Terragni
e-mail: filippo.terragni@uc3m.es

L. L. Bonilla et al. (eds.), *Coupled Mathematical Models for Physical and Biological Nanoscale Systems and Their Applications*, Springer Proceedings in Mathematics & Statistics 232, https://doi.org/10.1007/978-3-319-76599-0_6

In response, the vessel wall opens and issues endothelial cells that move towards the hypoxic region, build capillaries and bring blood, oxygen and nutrients to it. Once blood and oxygen have reached the hypoxic region, segregation of growth factors stops, anti-angiogenic substances may be segregated and a regular vessel network may have been put in place, after pruning capillaries with insufficient blood flow. In normal functioning, angiogenic and anti-angiogenic activities balance. Imbalance may result in many diseases including cancer [6]. In fact, after a tumor installed in tissue reaches some 2 mm size, it needs additional nutrients and oxygen to continue growing. Its hypoxic cells segregate growth factors and induce angiogenesis. Unlike normal cells, cancerous ones continue issuing growth factors and attracting blood vessels, which also supply them with a handy transportation system to reach other organs in the body.

Tumor-induced angiogenesis research started with J. Folkman's pioneering work in 1971 [6]. In addition to vast experimental research [7], models and theory [8] substantially contribute to understanding angiogenesis and developing therapies. In angiogenesis, events happening in cellular and subcellular scales unchain endothelial cell motion and proliferation and build millimeter scale blood sprouts and networks thereof [2–5]. Models range from very simple to extraordinarily complex and often try to illuminate some particular mechanism; see the review [8]. Realistic microscopic models involve postulating mechanisms and a large number of parameters that cannot be directly estimated from experiments, but they often yield qualitative predictions that can be tested. An important challenge is to extract mesoscopic and macroscopic descriptions of angiogenesis from the diverse microscopic models.

During angiogenesis, the relevant branching, growth and anastomosis (vessel fusion) stochastic processes of the capillary network at the microscale are strongly coupled to the interacting underlying biochemical and mechanical fields at the macroscale. In Sect. 2, we consider a hybrid mesoscale tip cell model that involves stochastic branching, anastomosis and extension of active vessel tip cells with reaction-diffusion growth factor fields [9]. Numerical simulations of the model show that anastomosis prevents indefinite proliferation of active vessel tips [10]. Then fluctuations about the mean of the density of active tips are not small and the stochastic process is not self-averaging. However, as shown in Sect. 3, it is possible to obtain a deterministic description of the density of active tips for ensemble averages over replicas of the stochastic process. The deterministic description consists of an integro-partial differential equation for the density of active vessel tips coupled to a reaction-diffusion equation for the growth factor [9, 10]. As shown in Sect. 4, the evolution of active tips from a primary vessel to a tumor adopts the form of an advancing soliton-like wave that can be characterized by ordinary differential equations for its position, velocity and a size parameter [11, 12]. These results may pave the way to assess optimal control of angiogenesis and therapies based on it.

What are the implications of our work? As described in Sect. 5, there are other models related to ours in which the vessel extension is described by random walks [13, 14], and our methodology may be used to extract deterministic descriptions for the density of active tips amenable to analysis. We could also seek to extend microscopic cellular Potts models (described in Sect. 6) to mesoscales and study them using our methods. The role of blood flow in remodeling vascular networks is briefly considered in Sect. 7. Further remarks are included in our conclusions in Sect. 8.

2 Langevin Tip Cell Models

Tip cell models assume that the tip cells are motile and non-proliferating whereas stalk cells build the blood vessel following the trajectories of the former. Assuming that the tip cells form point particles, their trajectories constitute the blood vessels advancing toward the tumor. In 1991, Stokes and Lauffenburger considered the capillary sprouts as particles of unit mass subject to chemotactic, friction and white noise forces [15, 16]. The distribution of tumor angiogenic factors (TAF) issuing from a small circular tumor is a known stationary non-uniform function. Associated to each sprout, its cell density satisfies a rate equation that takes into account proliferation, elongation, redistribution of cells from the parent vessel, branching and anastomosis. They did not consider the depletion effect that advancing sprouts would have on the TAF concentration. Later tip cell models combined a continuum description of fields influencing cell motion (chemotaxis, haptotaxis, ...) with random walk motion of individual sprouts that experience branching and anastomosis. Capasso and Morale [17] used ideas from these approaches to propose a hybrid model of Langevin-Ito stochastic equations for the sprouts undergoing chemotaxis, haptotaxis, branching and anastomosis coupled to reaction-diffusion equations for the continuum fields. In this model, the evolution of the continuum fields is influenced by the growing capillary network through smoothed (or mollified) versions thereof [18]. Capasso and Morale also attempted to derive a continuum equation for the density of moving tip cells from the stochastic equations but could not account for branching and anastomosis [17]. In what follows, we present a simplified hybrid model that ignores haptotaxis and derive a deterministic description for the density of active tips [9, 10, 19]. As in the Capasso-Morale model, the influence of haptotaxis can be included by adding reaction-diffusion equations for fibronectin and matrix-degrading enzymes [20]. The influence of blood circulation through the newly created blood vessels and secondary branching therefrom can be modeled as in [21].

We shall consider a slab geometry as indicated in Fig. 1, which is the result of a numerical simulation of the stochastic model. The extension of the ith capillary sprout with position $\mathbf{X}^i(t)$ and velocity $\mathbf{v}^i(t)$ is given by the nondimensional Langevin-Ito stochastic equation

$$\begin{aligned} d\mathbf{X}^i(t) &= \mathbf{v}^i(t)\,dt \\ d\mathbf{v}^i(t) &= \beta\left[-\mathbf{v}^i(t) + \mathbf{F}\left(C(t,\mathbf{X}^i(t))\right)\right]dt + \sqrt{\beta}\,d\mathbf{W}^i(t) \end{aligned} \tag{1}$$

for $t > T^i$ (T^i is the random birth time of the ith tip). Here $C(t,\mathbf{x})$ is the TAF concentration. At time T^i, the velocity of the newly created tip is selected out of a

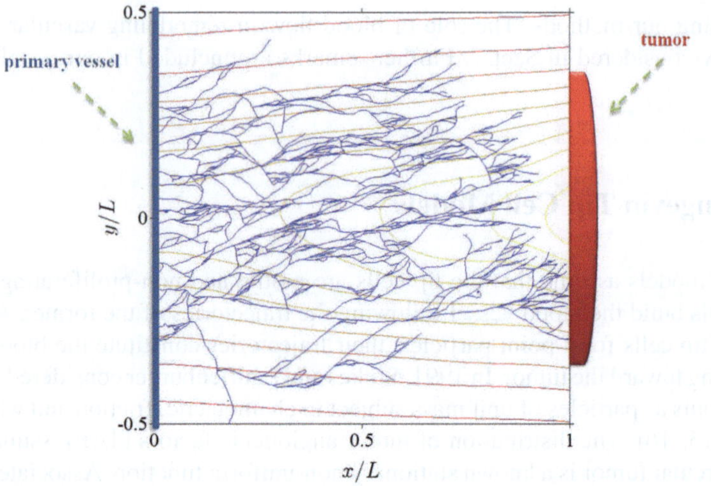

Fig. 1 Network of blood vessels simulated by the stochastic model of tumor induced angiogenesis. The level curves of the density of the tumor angiogenic factor (vessel endothelial growth factor) are also depicted [11]

normal distribution with mean $\mathbf{v}_0$ and variance σ_v^2, while the probability that a tip branches from one of the existing ones during an infinitesimal time interval $(t, t + dt]$ is proportional to

$$\sum_{i=1}^{N(t,\omega)} \alpha(C(t, \mathbf{X}^i(t)))dt. \tag{2}$$

Here $N(t, \omega)$ is the number of tips at time t for a realization ω of the stochastic process and

$$\alpha(C) = \frac{A\,C}{C+1}, \tag{3}$$

where A is a positive constant. We ignore secondary angiogenesis from newly formed capillaries [21]. The tip i disappears at a later random time Θ^i, either by reaching the tumor or by anastomosis, i.e., by meeting another capillary. At time t, anastomosis for the ith tip occurs at a point $\mathbf{x}$ such that $\mathbf{X}^i(t) = \mathbf{x}$ and $\mathbf{X}^j(s) = \mathbf{x}$ for another tip that was at $\mathbf{x}$ previously, at time $s < t$. In (1), $\mathbf{W}^i(t)$ are i.i.d. Brownian motions, and β (friction coefficient) is a positive parameter [9, 10, 12]. The chemotactic force $\mathbf{F}$ controlling tip cell migration in response to the TAF released by tumor cells is

$$\mathbf{F}(C) = \frac{\delta_1}{1+\Gamma_1 C}\nabla_x C\,, \tag{4}$$

where δ_1, and Γ_1 are positive parameters. The TAF diffuses and is consumed by advancing vessel tips according to [10]

$$\frac{\partial C}{\partial t}(t,\mathbf{x}) = \kappa_c \Delta_x C(t,\mathbf{x}) - \chi_c C(t,\mathbf{x}) \left| \sum_{i=1}^{N(t,\omega)} \mathbf{v}^i(t)\, \delta_{\sigma_x}(\mathbf{x} - \mathbf{X}^i(t)) \right|. \tag{5}$$

Here κ_c and χ_c are positive parameters, while δ_{σ_x} is a regularized delta function (e.g., a Gaussian with standard deviation σ_x). We are assuming that extending the vessel consumes TAF. As the vessel extends a length $|\mathbf{v}^i(t)|\,dt$ during the time interval between t and $t + dt$, the consumption should be proportional to $|\mathbf{v}^i(t)|$. The difference between the sum of the vector lengths and that indicated in (5) is negligible for the parameters and geometry considered in this paper (it amounts to having a coefficient 1.28 times larger than χ_c in the previous equation). Having the length of the flux vector as in (5) is convenient. Initial and boundary conditions for the TAF field C have been proposed in [9, 10].

The concentration of all vessels per unit volume in the physical space, at time t (i.e., the vessel network $\mathbf{X}(t,\omega)$) is [10]

$$\delta(\mathbf{x} - \mathbf{X}(t,\omega)) = \int_0^t \sum_{i=1}^{N(s,\omega)} \delta_{\sigma_x}(\mathbf{x} - \mathbf{X}^i(s,\omega))\, ds. \tag{6}$$

3 Deterministic Description

We shall see that we can understand the results of numerical simulations of the stochastic process described in the previous section by first finding a deterministic description of the density of active tips. The latter evolves in the form of a slowly varying soliton-like wave that we can analyze. Without performing numerical simulations of the stochastic process, we could guess that such a deterministic description could hold whenever the number of active tips arising from branching becomes very large. In such a case, we could use the law of large numbers to achieve such a description. This was the point of view adopted in the papers [9, 17]. However, anastomosis kills off so many active vessel tips that their number hardly grows to a hundred. Then we need a different point of view in order to derive a deterministic description. The alternative is the Gibbsian idea of considering an ensemble of replicas of the original stochastic process and carrying out arithmetic averages over the number of replicas.

We can find a deterministic description of the stochastic model for the densities of active vessel tips and the vessel tip flux, defined as ensemble averages over a sufficient number $\mathcal{N}$ of replicas (realizations) ω of the stochastic process:

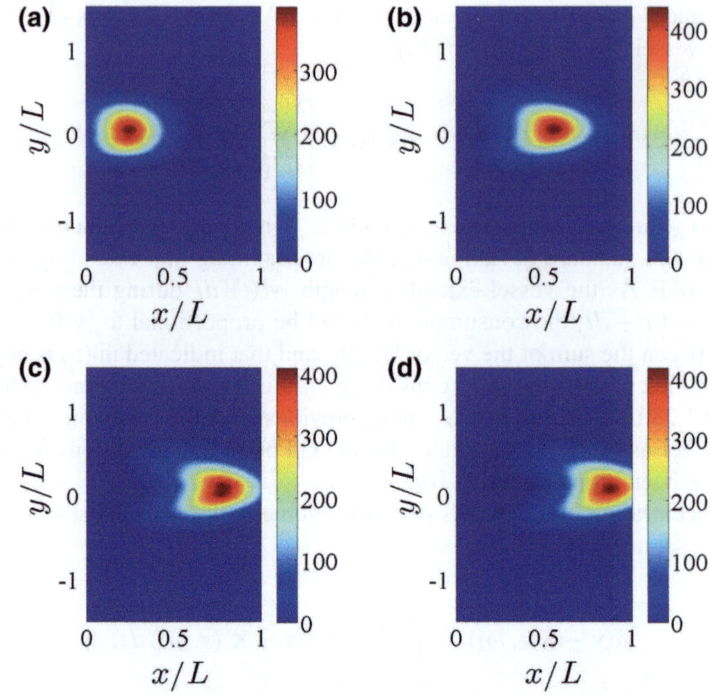

Fig. 2 Marginal density of active vessel tips resulting from an average over 400 replicas of the stochastic process according to Eq. (8) at four different times: **a** 12 h, **b** 24 h, **c** 32 h, and **d** 36 h. At these times, the numbers of active tips are (**a**) 56, (**b**) 69, (**c**) 72, and (**d**) 66, [10]

$$p_{\mathcal{N}}(t, \mathbf{x}, \mathbf{v}) = \frac{1}{\mathcal{N}} \sum_{\omega=1}^{\mathcal{N}} \sum_{i=1}^{N(t,\omega)} \delta_{\sigma_x}(\mathbf{x} - \mathbf{X}^i(t, \omega))\, \delta_{\sigma_v}(\mathbf{v} - \mathbf{v}^i(t, \omega)), \tag{7}$$

$$\tilde{p}_{\mathcal{N}}(t, \mathbf{x}) = \frac{1}{\mathcal{N}} \sum_{\omega=1}^{\mathcal{N}} \sum_{i=1}^{N(t,\omega)} \delta_{\sigma_x}(\mathbf{x} - \mathbf{X}^i(t, \omega)), \tag{8}$$

$$\mathbf{j}_{\mathcal{N}}(t, \mathbf{x}) = \frac{1}{\mathcal{N}} \sum_{\omega=1}^{\mathcal{N}} \sum_{i=1}^{N(t,\omega)} \mathbf{v}^i(t, \omega) \delta_{\sigma_x}(\mathbf{x} - \mathbf{X}^i(t, \omega)). \tag{9}$$

As $\mathcal{N} \to \infty$, these ensemble averages tend to the tip density $p(t, \mathbf{x}, \mathbf{v})$, the marginal tip density $\tilde{p}(t, \mathbf{x})$, and the tip flux $\mathbf{j}(t, \mathbf{x})$, respectively.

Figures 2 and 3 show the outcomes of typical simulations of ensemble averaged marginal densities: The two-dimensional lump shown in Fig. 2 is created at the primary vessel at $x = 0$ and marches to the tumor at $x = 1$. Its profile along the x axis is the soliton-like wave shown in Fig. 3.

Reference [10] shows that the angiogenesis model has a deterministic description based on the following equation for the density of vessel tips, $p(t, \mathbf{x}, \mathbf{v})$,

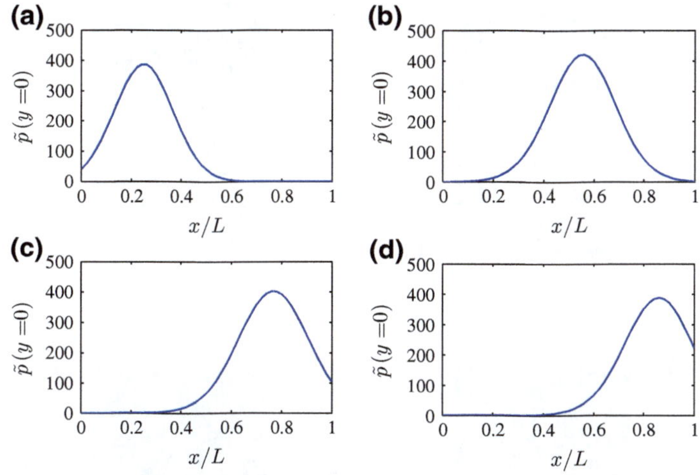

Fig. 3 Marginal density of active vessel tips at the x axis resulting from an average over 400 replicas of the stochastic process as in Fig. 2. The primary vessel at $x = 0$ issues a pulse that marches toward the tumor at $x = 1$, [10]

$$\frac{\partial p}{\partial t}(t, \mathbf{x}, \mathbf{v}) = \alpha(C(t, \mathbf{x}))\, p(t, \mathbf{x}, \mathbf{v})\delta_{\sigma_v}(\mathbf{v} - \mathbf{v}_0) - \Gamma\, p(t, \mathbf{x}, \mathbf{v}) \int_0^t \tilde{p}(s, \mathbf{x})\, ds$$

$$-\mathbf{v} \cdot \nabla_x p(t, \mathbf{x}, \mathbf{v}) - \beta \nabla_v \cdot [(\mathbf{F}(C(t, \mathbf{x})) - \mathbf{v}) p(t, \mathbf{x}, \mathbf{v})] + \frac{\beta}{2} \Delta_v p(t, \mathbf{x}, \mathbf{v}), \quad (10)$$

$$\tilde{p}(t, \mathbf{x}) = \int p(t, \mathbf{x}, \mathbf{v}')\, d\mathbf{v}'. \quad (11)$$

The two first terms on the right hand side of (10) correspond to vessel tip branching—from Eqs. (2) and (3)—and anastomosis, respectively. While the branching term follows from (2) and (3) in a straightforward manner, deducing the anastomosis integral term is the real breakthrough from past work achieved in [9]. The anastomosis coefficient, Γ, has to be fitted by comparison of the numerical solution of the deterministic equations and ensemble averages of the stochastic description, [10]. The other terms on the right hand side of (10) are in the Fokker-Planck equation that corresponds to the Langevin equation (1) in the usual manner [22]. While the branching term follows directly from the stochastic branching process, anastomosis occurs when a moving vessel tip at time $t > 0$ encounters a preexisting vessel whose tip was at the same place at an earlier time $s < t$. At time t, a moving vessel tip can reach an area $d\mathbf{x}$ about $\mathbf{x}$ that is either unoccupied or occupied by another vessel. In the latter case, it anastomoses. The occupation time density of the area $d\mathbf{x}$ about $\mathbf{x}$ is proportional to $\int_0^t \tilde{p}(s, \mathbf{x})\, ds$—the ensemble average of the vessel network density (6). Then the rate of anastomosis should be proportional to $p(t, \mathbf{x}, \mathbf{v})$ times this occupation time density [10]. Equation (5) becomes

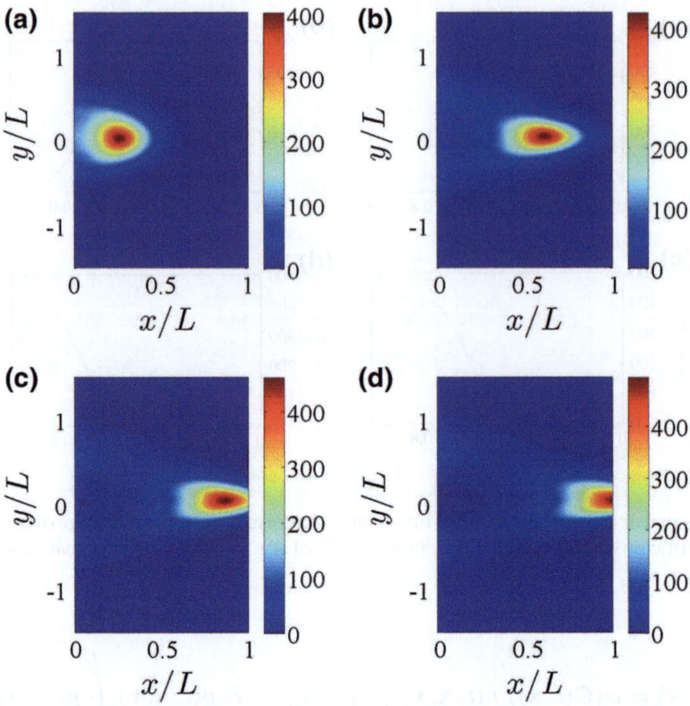

Fig. 4 Marginal density of active vessel tips resulting from a numerical simulation of the deterministic equations with appropriate boundary conditions for the same times as in Fig. 2 [9, 10]. Better agreement between both descriptions requires fine tuning of the boundary conditions

$$\frac{\partial C}{\partial t}(t, \mathbf{x}) = \kappa_c \Delta_x C(t, \mathbf{x}) - \chi_c C(t, \mathbf{x})\, |\mathbf{j}(t, \mathbf{x})|, \tag{12}$$

where $\mathbf{j}(t, \mathbf{x})$ is the current density (flux) vector at any point $\mathbf{x}$ and any time $t \geq 0$,

$$\mathbf{j}(t, \mathbf{x}) = \int \mathbf{v}' p(t, \mathbf{x}, \mathbf{v}')\, d\mathbf{v}'. \tag{13}$$

Carpio and collaborators have shown that the deterministic system of Eqs. (10)–(12) together with appropriate boundary and initial conditions has a unique solution that depends smoothly on parameters [23, 24].

Figure 4 shows that the outcome of a numerical simulation of the deterministic description is similar to that of the stochastic process.

4 Soliton and Collective Coordinates

In the overdamped limit of negligible inertia in (1), we obtain the simpler Langevin-Ito equation: $d\mathbf{X}^i(t) \approx \mathbf{F}(C(t, \mathbf{X}^i(t)))\, dt + \beta^{-1/2} d\mathbf{W}^i(t)$ [11]. By using the Chapman-Enskog perturbation method whose details are explained in [12], it is then possible to derive the following reduced equation for the marginal tip density,

$$\frac{\partial \tilde{p}}{\partial t} + \nabla_x \cdot (\mathbf{F}\tilde{p}) - \frac{1}{2\beta}\Delta_x \tilde{p} = \mu\, \tilde{p} - \Gamma \tilde{p} \int_0^t \tilde{p}(s, \mathbf{x})\, ds, \tag{14}$$

$$\mu = \frac{\alpha}{\pi}\left[1 + \frac{\alpha}{2\pi\beta(1+\sigma_v^2)} \ln\left(1 + \frac{1}{\sigma_v^2}\right)\right]. \tag{15}$$

The drift terms in Eq. (14) are those corresponding to the simpler Langevin-Ito equation for $\mathbf{X}^i(t)$ that results in the overdamped limit. The birth and death terms are obtained by integration of the corresponding ones on right hand side of (10) over velocity. However, the perturbation procedure changes the coefficient $\alpha(C)$ to the related function $\mu(C)$ in (15) [12]. Equation (14) has the following soliton-like solution for constant $\mathbf{F} = (F_x, F_y)$, μ, and zero diffusion, $1/\beta = 0$:

$$\tilde{p}_s = \frac{(2K\Gamma + \mu^2)c}{2\Gamma(c - F_x)} \text{sech}^2\left[\frac{\sqrt{2K\Gamma + \mu^2}}{2(c - F_x)}(x - X(t))\right], \quad \dot{X} \equiv \frac{dX}{dt} = c, \tag{16}$$

where K is a constant. In fact [11], consider $\tilde{p}_s = \partial P(x - ct)/\partial t = -c\, P'(\xi)$, $\xi = x - ct$, which, inserted in (14) with $1/\beta = 0$, yields

$$(F_x - c)P'' = \mu P' - \Gamma P P' \Longrightarrow (c - F_x)P' = \frac{\Gamma}{2}P^2 - K - \mu P.$$

Setting $P = \nu \tanh(\lambda\xi) + \mu/\Gamma$, we find $\nu^2 = (\mu^2 + 2K\Gamma)/\Gamma^2$ and $2\nu\lambda(c - F_x)/\Gamma = -\nu^2$, thereby obtaining

$$P = \frac{\mu}{\Gamma} - \frac{\sqrt{2K\Gamma + \mu^2}}{\Gamma}\tanh\left[\frac{\sqrt{2K\Gamma + \mu^2}}{2(c - F_x)}(\xi - \xi_0)\right].$$

Here ξ_0 is a constant of integration. Thus $\tilde{p}_s = \partial P/\partial t = -cP'$ is given by (16).

Numerical simulations on a slab geometry show that the marginal tip density evolves toward (16) after an initial stage [11, 12]. *It is an open problem to prove this stability result even for a one-dimensional version of Equation* (14) *on the whole real line and having constant values of* **F** *and* μ.

A small diffusion and slowly varying continuum field C produce a moving soliton whose shape and speed are slowly changing. We can find them by deducing evolution equations for the *collective coordinates* K, c, and X [11, 12]. Then the marginal density profile at $y = 0$ can be reconstructed from (16) with spatially averaged F_x

and μ [12]. Note that $\tilde{p}_s$ is a function of $\xi = x - X$ and also of $\mathbf{x}$ and t through $C(t, \mathbf{x})$,

$$\tilde{p}_s = \tilde{p}_s\left(\xi; K, c, \mu(C), F_x\left(C, \frac{\partial C}{\partial x}\right)\right). \quad (17)$$

We assume that the time and space variations of C, which appear when $\tilde{p}_s$ is differentiated with respect to t or x, produce terms that are small compared to $\partial \tilde{p}_s/\partial \xi$. As explained in [12], we shall consider that $\mu(C)$ is approximately constant, ignore $\partial C/\partial t$ because the TAF concentration varies slowly (the dimensionless coefficients κ_c and χ_c appearing in the TAF equation (12) are very small according to Table 2 of [12]) and ignore $\partial^2 \tilde{p}_s/\partial i \partial j$, where $i, j = K, F_x$. We now insert (16) into (14), thereby obtaining

$$\left(F_x - \dot{X}\right)\frac{\partial \tilde{p}_s}{\partial \xi} + \frac{\partial \tilde{p}_s}{\partial K}\dot{K} + \frac{\partial \tilde{p}_s}{\partial c}\dot{c} - \frac{1}{2\beta}\left(\frac{\partial^2 \tilde{p}_s}{\partial \xi^2} + 2\frac{\partial^2 \tilde{p}_s}{\partial \xi \partial F_x}\frac{\partial F_x}{\partial x} + \frac{\partial \tilde{p}_s}{\partial F_x}\Delta_x F_x\right)$$
$$+\tilde{p}_s \nabla_x \cdot \mathbf{F} + \frac{\partial \tilde{p}_s}{\partial F_x}\left(\frac{\partial F_x}{\partial t} + \mathbf{F}\cdot\nabla_x F_x\right) = \mu \tilde{p}_s - \Gamma \tilde{p}_s \int_0^t \tilde{p}_s dt. \quad (18)$$

Equation. (14) with $1/\beta = 0$ and constant $\mathbf{F}$ and μ has the soliton solution (16). Using this fact, we can eliminate the first term on the left hand side of (18) and also the right hand side thereof. Equation (18) then becomes

$$\frac{\partial \tilde{p}_s}{\partial K}\dot{K} + \frac{\partial \tilde{p}_s}{\partial c}\dot{c} = \mathscr{A}, \quad (19)$$

$$\mathscr{A} = \frac{1}{2\beta}\frac{\partial^2 \tilde{p}_s}{\partial \xi^2} - \tilde{p}_s \nabla_x \cdot \mathbf{F} - \frac{\partial \tilde{p}_s}{\partial F_x}\left(\mathbf{F}\cdot\nabla_x F_x - \frac{1}{2\beta}\Delta_x F_x\right) + \frac{1}{\beta}\frac{\partial^2 \tilde{p}_s}{\partial \xi \partial F_x}\frac{\partial F_x}{\partial x}. \quad (20)$$

We now find collective coordinate equations (CCEs) for K and c. As the lump-like angiton moves on the x axis, we set $y = 0$ to capture the location of its maximum. On the x axis, the profile of the angiton is the soliton (16). We first multiply (19) by $\partial \tilde{p}_s/\partial K$ and integrate over x. We consider a fully formed soliton far from primary vessel and tumor. As it decays exponentially for $|\xi| \gg 1$, the soliton is considered to be localized on some finite interval $(-\mathscr{L}/2, \mathscr{L}/2)$. The coefficients in the soliton formula (16) and the coefficients in (20) depend on the TAF concentration at $y = 0$, therefore they are functions of x and time and get integrated over x. The TAF concentration varies slowly on the support of the soliton, and therefore we can approximate the integrals over x by [12]

$$\int_{\mathscr{I}} F(\tilde{p}_s(\xi; x, t), x)dx \approx \frac{1}{\mathscr{L}}\int_{\mathscr{I}}\left(\int_{-\mathscr{L}/2}^{\mathscr{L}/2} F(\tilde{p}_s(\xi; x, t), x)d\xi\right)dx. \quad (21)$$

The interval $\mathscr{I}$ over which we integrate should be large enough to contain most of the soliton, of extension $\mathscr{L}$. Thus the CCEs hold only after the initial soliton formation stage. Near the primary vessel and near the tumor, the boundary conditions affect the soliton and we should exclude intervals near them from $\mathscr{I}$. We shall specify the integration interval $\mathscr{I}$ below. Acting similarly, we multiply (19) by $\partial\tilde{p}_s/\partial c$ and integrate over x. From the two resulting formulas, we then find $\dot{K}$ and $\dot{c}$ as fractions. The factors $1/\mathscr{L}$ cancel out from their numerators and denominators. As the soliton tails decay exponentially to zero, we can set $\mathscr{L} \to \infty$ and obtain the following CCEs [12]

$$\dot{K} = \frac{\int_{-\infty}^{\infty} \frac{\partial\tilde{p}_s}{\partial K}\mathscr{A}\,d\xi \int_{-\infty}^{\infty}\left(\frac{\partial\tilde{p}_s}{\partial c}\right)^2 d\xi - \int_{-\infty}^{\infty}\frac{\partial\tilde{p}_s}{\partial c}\mathscr{A}\,d\xi \int_{-\infty}^{\infty}\frac{\partial\tilde{p}_s}{\partial K}\frac{\partial\tilde{p}_s}{\partial c}d\xi}{\int_{-\infty}^{\infty}\left(\frac{\partial\tilde{p}_s}{\partial K}\right)^2 d\xi \int_{-\infty}^{\infty}\left(\frac{\partial\tilde{p}_s}{\partial c}\right)^2 d\xi - \left(\int_{-\infty}^{\infty}\frac{\partial\tilde{p}_s}{\partial c}\frac{\partial\tilde{p}_s}{\partial K}d\xi\right)^2}, \tag{22}$$

$$\dot{c} = \frac{\int_{-\infty}^{\infty} \frac{\partial\tilde{p}_s}{\partial c}\mathscr{A}\,d\xi \int_{-\infty}^{\infty}\left(\frac{\partial\tilde{p}_s}{\partial K}\right)^2 d\xi - \int_{-\infty}^{\infty}\frac{\partial\tilde{p}_s}{\partial K}\mathscr{A}\,d\xi \int_{-\infty}^{\infty}\frac{\partial\tilde{p}_s}{\partial K}\frac{\partial\tilde{p}_s}{\partial c}d\xi}{\int_{-\infty}^{\infty}\left(\frac{\partial\tilde{p}_s}{\partial K}\right)^2 d\xi \int_{-\infty}^{\infty}\left(\frac{\partial\tilde{p}_s}{\partial c}\right)^2 d\xi - \left(\int_{-\infty}^{\infty}\frac{\partial\tilde{p}_s}{\partial c}\frac{\partial\tilde{p}_s}{\partial K}d\xi\right)^2}. \tag{23}$$

In these equations, all terms varying slowly in space have been averaged over the interval $\mathscr{I}$. The last term in (20) is odd in ξ and does not contribute to the integrals in (22) and (23) whereas all other terms in (20) are even in ξ and do contribute. The integrals appearing in (22) and (23) are calculated in [12]. The resulting CCEs are

$$\dot{K} = \frac{(2K\Gamma + \overline{\mu}^2)^2}{4\Gamma\beta(c-\overline{F_x})^2}\,\frac{\frac{4\pi^2}{75}+\frac{1}{5}+\left(\frac{2\overline{F_x}}{5c}-\frac{2\pi^2}{75}-\frac{9}{10}\right)\frac{\overline{F_x}}{c}}{\left(1-\frac{4\pi^2}{15}\right)\left(1-\frac{\overline{F_x}}{2c}\right)^2}$$
$$-\frac{2K\Gamma+\overline{\mu}^2}{\Gamma c\left(2-\frac{\overline{F_x}}{c}\right)}\left(c\overline{\nabla_x\cdot\mathbf{F}}+\overline{\mathbf{F}\cdot\nabla_x F_x}-\frac{\overline{\Delta_x F_x}}{2\beta}\right), \tag{24}$$

$$\dot{c} = -\frac{7(2K\Gamma+\overline{\mu}^2)}{20\beta(c-\overline{F_x})}\,\frac{1-\frac{4\pi^2}{105}}{\left(1-\frac{4\pi^2}{15}\right)\left(1-\frac{\overline{F_x}}{2c}\right)}+\frac{\overline{\mathbf{F}\cdot\nabla_x F_x}-(c-\overline{F_x})\overline{\nabla_x\cdot\mathbf{F}}-\frac{\overline{\Delta_x F_x}}{2\beta}}{2-\frac{\overline{F_x}}{c}}, \tag{25}$$

$$\overline{g(x,y)} = \frac{1}{\mathscr{I}}\int_{\mathscr{I}} g(x,0)\,dx, \tag{26}$$

in which the functions of $C(t,x,y)$ have been averaged over the interval $\mathscr{I}$ after setting $y = 0$. We expect the CCEs (24) and (25) to describe the mean behavior of the soliton whenever it is far from primary vessel and tumor.

Both deterministic or stochastic simulations show that the soliton is formed after some time $t_0 = 0.2$ (10 h) following angiogenesis initiation. To find the soliton evolution afterwards, we need to solve the CCEs (24) and (25), in which the spatial averages depend on an interval $x \in \mathscr{I}$, which should exclude regions affected by boundaries. We calculate the spatially averaged coefficients in (24) and (25) by: (i) approximating all differentials by second order finite differences, (ii) setting $y = 0$, and (iii) averaging the coefficients from $x = 0$–0.6 by taking the arithmetic mean of their values at all

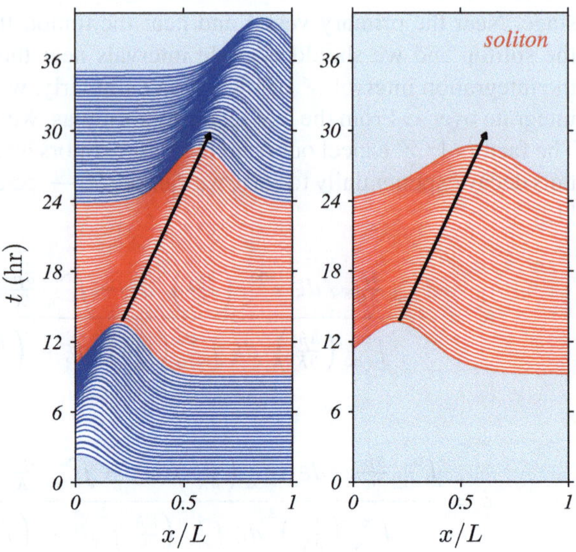

Fig. 5 Comparison of the marginal tip density profile $\tilde{p}(t, x, 0)$ (obtained from the stochastic description averaged over 400 replicas) to that of the moving soliton, [11]

grid points in the interval $\mathscr{I} = (0, 0.6]$. For $x > 0.6$, the boundary condition at $x = 1$ influences the outcome and therefore we leave values for $x > 0.6$ out of the averaging [12]. The initial conditions for the CCEs are set as follows. $X(t_0) = X_0$ is the location of the marginal tip density maximum, $\tilde{p}(t_0, x = X_0, 0)$. We find $X_0 = 0.2$ from the stochastic description. We set $c(t_0) = c_0 = X_0/t_0$. $K(t_0) = K_0$ is determined so that the maximum marginal tip density at $t = t_0$ coincides with the soliton peak. This yields $K_0 = 39$. Solving the CCEs (24) and (25) with these initial conditions and using (16), we obtain the curves depicted in Fig. 5.

5 Random Walk Tip Cell Models

These models describe the extension of blood vessels by random walks biased by chemotaxis or haptotaxis instead of using Langevin equations. The first such model, due to Anderson and Chaplain [13], is based on a reaction-diffusion description of angiogenesis. They consider a continuity equation for the density of endothelial cells (ECs) n (with zero-flux boundary conditions) coupled to equations for the TAF and fibronectin densities, C and f, respectively. In nondimensional form, these equations are [13]:

$$\frac{\partial n}{\partial t} = D\Delta n - \nabla \cdot \left(\frac{\chi}{1+\alpha C} n \nabla C \right) - \nabla \cdot (\rho n \nabla f), \tag{27}$$

$$\frac{\partial f}{\partial t} = \beta n - \gamma n f, \tag{28}$$

$$\frac{\partial C}{\partial t} = -\eta n C. \tag{29}$$

Here all parameters are positive. The three terms on the right hand side of (27) correspond to diffusion of ECs, chemotaxis and haptotaxis, respectively. Note that chemotaxis has the same form in this equation as in (10) with p replaced by n. Haptotaxis follows the gradient of fibronectin in the extracellular matrix. Note that proliferation and death of ECs are not contemplated by (27). In the next step, these equations are solved by an explicit Euler method in time and finite differences. The resulting equation for $n(t, x, y) \approx n^q_{l,m}$,

$$n^{q+1}_{l,m} = n^q_{l,m} W_0 + n^q_{l+1,m} W_1 + n^q_{l-1,m} W_2 + n^q_{l,m+1} W_3 + n^q_{l,m-1} W_4, \tag{30}$$

has the same form as a master equation for a random walk [22], except that the "transition probabilities" W_0 (staying), W_1 (moving to the left), W_2 (moving to the right), W_3 (moving downwards), and W_4 (moving upwards) are not normalized. However, this is easily fixed by defining

$$\mathscr{W}_i = \frac{W_i}{\sum_{j=0}^4 W_j}, \quad i = 0, 1, \ldots, 4, \tag{31}$$

as new transition probabilities. The random walk associated to these transition probabilities represents extension of vessel tips and replaces the Langevin equation (1). Branching and anastomosis are introduced as in the Langevin tip cell model, except that the tips have to wait some *maturity* time after branching before they are allowed to branch again. It should be straightforward to find equations for the density of active vessel tips by using the theory described in previous sections.

The Anderson-Chaplain idea is easy to implement starting from continuum models of angiogenesis (and therefore it can be immediately generalized by including more taxis mechanisms, influence of antiangiogenic factors [25], etc), but it has the drawback of having to rely on the finite difference grid or lattice. A few years later, Plank and Sleeman proposed non-lattice models independent of the grid [14] using biased circular random walk models previously introduced by Hill and Häder for swimming microorganisms [26]. If $\theta(t)$ is a continuous random walk on the unit circle biased by chemo and haptotaxis [14], the trajectory of the corresponding tip cell is

$$\frac{d\mathbf{x}}{dt} = v_0 (\cos\theta(t), \sin\theta(t)). \tag{32}$$

Thus the tip cells have the same speed v_0, directions given by $\theta(t)$ and their trajectories do not have to follow points on a lattice. While branching and anastomosis are modeled as in Sect. 2, the extensions of vessel tips are described by (32) and the biased circular random walk instead of Langevin equations. The master equation for the circular random walk is [14]

$$\frac{dP_n}{dt} = \hat{\tau}^+_{n-1} P_{n-1} + \hat{\tau}^-_{n+1} P_{n+1} - (\hat{\tau}^+_n + \hat{\tau}^-_n) P_n, \tag{33}$$

$$\hat{\tau}^\pm_n = 2\lambda \frac{\tau\left(n\delta \pm \frac{\delta}{2}\right)}{\tau\left(n\delta + \frac{\delta}{2}\right) + \tau\left(n\delta - \frac{\delta}{2}\right)}. \tag{34}$$

As $\delta \to 0$ and $n \to \infty$ so that $n\delta = \theta$, the master equation (33) becomes the Fokker-Planck equation [14]

$$\frac{\partial P}{\partial t}(t, \theta) = D \frac{\partial}{\partial \theta}\left[P(t, \theta) \frac{\partial}{\partial \theta}\left(\ln \frac{P(t, \theta)}{\tau(\theta)}\right)\right], \tag{35}$$

with $D = \lambda\delta^2$ for $P(t, \theta) = P(t, n\delta) = P_n(t)$. Chemo and haptotaxis are included in the model through the transition probability

$$\tau(\theta) = \frac{\exp[d_C \cos(\theta - \theta_C) + d_f \cos(\theta - \theta_f)]}{\int_{-\pi}^{\pi} \exp[d_C \cos(s - \theta_C) + d_f \cos(s - \theta_f)]\, ds}, \tag{36}$$

$$\tan\theta_C = \frac{\nabla C}{|\nabla C|}, \quad \tan\theta_f = \frac{\nabla f}{|\nabla f|}. \tag{37}$$

Here $\tau(\theta)$ is the stationary probability density of the Fokker-Planck equation (35). Comparisons between numerical simulations of the Anderson-Chaplain and Plank-Sleeman models are carried out in [14].

The random walk models of this Section get their input from continuum equations for EC, TAF and fibronectin densities, but the moving vessel tips characterized by the random walks do not affect the continuum fields. Their outcomes are numerical simulations of the stochastic processes, without further elaboration. In contrast to this somewhat artificial setting, the Langevin tip cell model of Sect. 2 is a hybrid model in which active vessel tips and continuum fields are fully coupled. Furthermore, we can derive an equivalent deterministic description from the Langevin tip cell model and analyze it in terms of a soliton-like attractor. This latter elaboration has also been carried out for a Langevin tip cell model that includes chemotaxis and haptotaxis [20]. It may also be worked out for the random walk models in an appropriate limit, as their governing master equation then becomes a Fokker-Planck equation (corresponding to a Langevin-Ito equation) [22].

6 Cellular Potts Models

In all the previous models, the cells are treated as point particles. For a more precise view of haptotaxis, i.e., the motion of ECs over the extracellular matrix (ECM), we need to consider adhesion and deformation of the cells. This requires a more microscopic view than that offered by tip cell models or by more complicated models that distinguish between tip and stalk ECs and add extra dynamics for them [8].

Often times, ECs and ECM are modeled by a cellular Potts model (CPM) with Monte Carlo dynamics coupled to continuum fields (elastic fields, TAF, ...) [27]. Space in these models consists of a lattice whose cells (lattice sites) may be in finitely many different states, denoted by type τ and representing ECs, matrix fibers, tissue cells and interstitial fluid. To account for individual entities (ECs, fibers, etc), each entity is further associated with a unique identifying number, denoted by σ, that is assigned to every lattice site occupied by it. At every Monte Carlo time step, the cell surface (represented by connected lattice vertices) is updated according to a set of cell behavior rules (e.g., target cell shape and size) that are translated in an energy change. Typically, we select randomly a cell $\mathbf{x}$, assign its type, $\tau(\mathbf{x})$, to a randomly chosen neighbor $\mathbf{x}'$, and update accordingly the total energy of the system, H. Using the Metropolis algorithm, a given update is accepted with probability one if the change in the total energy of the system, ΔH, is reduced and it is accepted with probability $e^{-\beta\Delta H}$ otherwise ($1/\beta$ is the Monte Carlo temperature). The energy in [27] is

$$H = \sum_{\text{sites}} J_{\tau,\tau'}(1-\delta_{\sigma\sigma'}) + \sum_{\text{cells}} \gamma_\tau (a_\sigma - A_\sigma)^2 - \sum_{\text{cells}}\sum_{\text{sites}} \mu_\sigma C(t,\mathbf{x}). \quad (38)$$

The first term in Eq. (38) is the contribution to total energy resulting from cell-cell and cell-medium adhesion. The second term allows deformation of cells with volume a_σ about a target volume (area in 2D space) A_σ, depending on the Potts parameters γ_τ. The target volume is twice that of the initial volume and it corresponds to the volume at which a cell undergoes mitosis, thereby creating a new cell. Thus cell proliferation is contemplated in this CPM. A variation of the last term in (38) is

$$\Delta H_{\text{chem}} = -\mu_\sigma[C(t,\mathbf{x}) - C(t,\mathbf{x}')], \quad (39)$$

where $\mathbf{x}$ and $\mathbf{x}'$ are two randomly picked neighboring lattice cells, $\mu_\sigma > 0$ is the chemical potential, and Eq. (39) represents chemotaxis favoring motion directed along the TAF gradient. The TAF concentration satisfies a reaction-diffusion equation [27]. The parameters appearing in the model are chosen in such a way that the progression of blood vessels occurs in the time scale observed in experiments [27].

Under this framework, each entity (ECs, ECM, ...) has a finite volume, a deformable shape and competes for space. ECs proliferate. Intercellular interactions occur only at the cells surface and have a cell-type-dependent surface (or adhesion) energy $J_{\tau,\tau'}$, which is a measure of the coupling strength between the entities

τ and τ'. Other CPMs include an ECM strain-dependent term that favors cell extension in the direction of principal strain (durotaxis). The force exerted by the ECs on the ECM is calculated by finite elements [28]. In more complicated models, each cell contains agents that signal to other cells and adhesion is modeled by a CPM [29].

As in the case of random walk tip cell models, there is a connection between CPM and a deterministic formulation for a density. In [30], Alber et al. have written a discrete time master equation for the probability density $P(t, \mathbf{r}, \mathbf{L})$ that a cell with its center of mass at $\mathbf{r}$ occupy a rectangle with sides $\mathbf{L} = (l_x, l_y)$ at time t. It is based on a CPM with energy given by (38), but with a target perimeter instead of the target area. The corresponding term in the energy is

$$H_{\text{perim}} = \sum_{\text{cells}} [\gamma_x (l_x - L_x)^2 + \gamma_y (l_y - L_y)^2]. \tag{40}$$

Here cells are always rectangles and do not proliferate nor die. Assuming that cells contain many lattice sites, they change little at each Monte Carlo step. Assuming further that cell-cell interactions are always binary, the authors derive a Fokker-Planck equation for $P(t, \mathbf{r}, \mathbf{L})$. These formulations would have to be extended to CPMs that include cell proliferation and be connected to mesoscopic angiogenesis models: from cell densities to densities of active vessel tip cells.

7 Blood Flow and Vascular Network

Once a vascular network is being created, blood flows through the capillaries, anastomosis enhances flow in some of them and secondary angiogenesis may start in new vessels. Pries and coworkers have modeled blood flow in a vascular network and the response thereof to changing conditions such as pressure differences and wall stresses [31, 32]. This response may remodel the vascular network by changing the radii of certain capillaries, and altering the distribution of blood flow [31, 32]. McDougall, Anderson and Chaplain [33] have used this formulation to add secondary branching from new capillaries induced by wall shear stress to the original random walk tip cell model [13]. Blood flows according to Poiseuille's law, mass is conserved, there are empirical expressions for blood viscosity and for the wall shear stresses, and radii of capillaries adapt to local conditions. Secondary vessel branching may occur after the new vessel has reached a certain level of maturation and before a basal lamina has formed about it [21, 33]. During such a time interval, the probability of secondary branching increases with both the local TAF concentration and the magnitude of the shear stress affecting the vessel wall. McDougall et al's. model can be used to figure out how drugs could be transported through the blood vessels and eventually reach a tumor [21, 33]. In dense vessel networks, secondary branching may have little effect on the number of active tips at a given time, as anastomosis could eliminate secondary branches quickly. Thus we may ignore secondary branching when considering the density of active tips in such networks. Of course we cannot ignore it when describing blood flow and network remodeling.

One missing feature of angiogenesis models that take blood flow into account seems to be pruning. It is known that capillaries with insufficient blood circulation may atrophy and disappear. Pruning such blood vessels is an important mechanism to achieve a hierarchical vascular network such as that observed in retinal vascularization during development [3, 4]. Global optimization and adaptation in developing networks has been recently shown to lead to highly optimized transport vascular systems [34, 35]. It would be interesting to adapt these studies to angiogenesis.

8 Conclusions

Angiogenesis is a complex multiscale process by which diffusing vessel endothelial growth factors induce sprouting of blood vessels that carry oxygen and nutrients to hypoxic tissue. Cancerous tumor cells profit from this process to prosper, grow and eventually migrate to other organs. Mathematical models contemplate different aspects of angiogenesis. Here we have reviewed recent work on a simple tip cell model that encompasses vessel extension driven by chemotaxis and described by Langevin equations, stochastic tip branching and vessel fusion (anastomosis). From the stochastic description, we have derived a deterministic integropartial differential equation for the density of active tip cells coupled with a reaction-diffusion equation for the growth factor. The associated initial-boundary value problem is well posed. It is important to note that anastomosis prevents proliferation of active tips and therefore the deterministic description is based on ensemble averages over replicas of the stochastic process. Numerical simulations of both (deterministic and stochastic) descriptions show that the density of active tips adopts the shape of a two-dimensional soliton-like wave (angiton) after a formation stage. We have found an analytical formula for the one-dimensional projection of the soliton and ordinary differential equations for variables that provide its velocity, position and size. These equations also characterize the advance of the vessel network for single replicas. Much more work needs to be carried out to solve mathematical issues arising from our results, both from analysis of the deterministic description and from establishing more precise conditions for its validity. The description of the soliton should be extended to the true two-dimensional soliton (angiton) that appears in the numerical simulations and to the case of a more general geometry than that of the slab. Fluctuations cannot be ignored in the case of ensemble averages, and future work predicting the evolution of a real vessel network should include confidence bands about averages. Anti-angiogenic treatments need to be improved [1, 2], and, in this respect, having better models and theories about their solutions should help. Therapies are related to optimal control of angiogenesis and they require accurate mathematical models, validated by comparison with real data (inverse problems—statistics of random geometric structures).

We have also related the specific model we study to other tip cells models in the literature that describe vessel extension by reinforced random walks instead of stochastic differential equations. Our methodology may be adapted to these other models as Langevin equations arise from reinforced random walks in appropriate limits. All these models describe mesoscales in which cells are just point particles, thereby ignoring their shapes and a microscopic description thereof. Other models consider the evolution of individual endothelial cells of variable shape and extension through cellular Potts models, but the continuation of these models toward the mesoscale has barely begun. Extending the analysis carried out for our mesoscopic stochastic tip cell model to microscopic models is a challenge for the future. Blood circulation through the angiogenic network favors certain vessels, others that do not have enough perfusion shrink and disappear and secondary branching may occur. Future work could delve deeper in the topics of vessel remodeling, pruning, formation of optimal vascular networks and transport of medicals through them.

Apart from the specific application to angiogenesis, we have presented in this paper methodological contributions for a sound mathematical modeling of stochastic vessel networks: (a) the use of stochastic distributions, and their mean densities, describing the vessels, which are random objects of Hausdorff dimension one, cf (6); (b) reduction of vessel distributions to integrals over time of active tip distributions, which are random objects of zero Hausdorff dimension, cf (8); (c) characterization of the attractor of the density of active tips as a soliton whose position, velocity and size are given as solutions of ordinary differential equations, cf (16), (22) and (23). In our system, which is strongly out of equilibrium, this attractor plays a similar role to the stable stationary equilibrium distribution of many physical systems.

Acknowledgements The authors thank V. Capasso and D. Morale from the Department of Mathematics of Università degli Studi di Milano, Milan, Italy, and B. Birnir from the Department of Mathematics of University of California at Santa Barbara, USA, for fruitful discussions and contributions. We also thank A. Lasanta from Universidad Carlos III de Madrid for useful comments on the manuscript. This work has been supported by the Ministerio de Economía y Competitividad grants MTM2014-56948-C2-2-P and MTM2017-84446-C2-2-R.

References

1. P.F. Carmeliet, Angiogenesis in life, disease and medicine. Nature **438**, 932–936 (2005)
2. P. Carmeliet, R.K. Jain, Molecular mechanisms and clinical applications of angiogenesis. Nature **473**, 298–307 (2011)
3. R.F. Gariano, T.W. Gardner, Retinal angiogenesis in development and disease. Nature **438**, 960–966 (2005)
4. M. Fruttiger, Development of the retinal vasculature. Angiogenesis **10**, 77–88 (2007)
5. P. Carmeliet, M. Tessier-Lavigne, Common mechanisms of nerve and blood vessel wiring. Nature **436**, 193–200 (2005)
6. J. Folkman, Tumor angiogenesis: therapeutic implications. New Engl. J. Med. **285**(21), 1182–1186 (1971)
7. J. Folkman, Angiogenesis. Annu. Rev. Med. **57**, 1–18 (2006)

8. T. Heck, M.M. Vaeyens, H. Van Oosterwyck, Computational models of sprouting angiogenesis and cell migration: Towards multiscale mechanochemical models of angiogenesis. Math. Model. Nat. Phen. **10**, 108–141 (2015)
9. L.L. Bonilla, V. Capasso, M. Alvaro, M. Carretero, Hybrid modeling of tumor-induced angiogenesis. Phys. Rev. E **90**, 062716 (2014)
10. F. Terragni, M. Carretero, V. Capasso, L.L. Bonilla, Stochastic model of tumor-induced angiogenesis: ensemble averages and deterministic equations. Phys. Rev. E **93**, 022413 (2016)
11. L.L. Bonilla, M. Carretero, F. Terragni, B. Birnir, Soliton driven angiogenesis. Sci. Rep. **6**, 31296 (2016)
12. L.L. Bonilla, M. Carretero, F. Terragni, Solitonlike attractor for blood vessel tip density in angiogenesis. Phys. Rev. E **94**, 062415 (2016)
13. A.R.A. Anderson, M.A.J. Chaplain, Continuous and discrete mathematical models of tumor-induced angiogenesis. Bull. Math. Biol. **60**, 857–900 (1998)
14. M.J. Plank, B.D. Sleeman, Lattice and Non-Lattice Models of Tumour Angiogenesis. Bull. Math. Biol. **66**, 1785–1819 (2004)
15. C.L. Stokes, D.A. Lauffenburger, Analysis of the roles of microvessel endothelial cell random motility and chemotaxis in angiogenesis. J. Theoret. Biol. **152**, 377–403 (1991)
16. C.L. Stokes, D.A. Lauffenburger, S.K. Williams, Migration of individual microvessel endothelial cells: stochastic model and parameter measurement. J. Cell Sci. **99**, 419–430 (1991)
17. V. Capasso, D. Morale, Stochastic modelling of tumour-induced angiogenesis. J. Math. Biol. **58**, 219–233 (2009)
18. S. Sun, M.F. Wheeler, M. Obeyesekere, C.W. Patrick Jr., Multiscale angiogenesis modeling using mixed finite element methods. Multiscale Mod. Simul. **4**(4), 1137–1167 (2005)
19. L.L. Bonilla, V. Capasso, M. Alvaro, M. Carretero, F. Terragni, On the mathematical modelling of tumour induced driven angiogenesis. Math. Biosci. Eng. **14**, 45–66 (2017)
20. L.L. Bonilla, M. Carretero, F. Terragni, Ensemble averages, soliton dynamics and influence of haptotaxis in a model of tumor-induced angiogenesis. Entropy **19**, 209 (2017)
21. A. Stéphanou, S.R. McDougall, A.R.A. Anderson, M.A.J. Chaplain, Mathematical modelling of the influence of blood rheological properties upon adaptative tumour-induced angiogenesis. Math. Comput. Model. **44**, 96–123 (2006)
22. C.W. Gardiner, Stochastic methods, in *A handbook for the natural and social sciences*, 4th edn. (Springer, Berlin, 2010)
23. A. Carpio, G. Duro, Well posedness of an angiogenesis related integrodifferential diffusion model. Appl. Math. Model. **40**, 5560–5575 (2016)
24. A. Carpio, G. Duro, M. Negreanu, Constructing solutions for a kinetic model of angiogenesis in annular domains. Appl. Math. Model. **45**, 303–322 (2017)
25. H.A. Levine, S. Pamuk, B.D. Sleeman, M. Nilsen-Hamilton, Mathematical modeling of the capillary formation and development in tumor angiogenesis: Penetration into the stroma. Bull. Math. Biol. **63**, 801–863 (2001)
26. N.A. Hill, D.P. Häder, A biased random walk model for the trajectories of swimming microorganisms. J. Theor. Biol. **186**, 503–526 (1997)
27. A.L. Bauer, T.L. Jackson, Y. Jiang, A Cell-Based Model Exhibiting Branching and Anastomosis during Tumor-Induced Angiogenesis. Biophys. J. **92**, 3105–3121 (2007)
28. R.F.M. Van Oers, E.G. Rens, D.J. La Valley, C.A. Reinhart-King, R.M.H. Merks, Mechanical cell-matrix feedback explains pairwise and collective endothelial cell behavior in vitro. PLoS Comput. Biol. **10**(8), el003774 (2014)
29. K. Bentley, C.A. Franco, A. Philippides, R. Blanco, M. Dierkes, V. Gebala, F. Stanchi, M. Jones, I.M. Aspalter, G. Cagna, S. Weström, L. Claesson-Welsh, D. Vestweber, H. Gerhardt, The role of differential VE-cadherin dynamics in cell rearrangement during angiogenesis. Nat. Cell Biol. **16**(4), 309–321 (2014)
30. N. Alber, N. Chen, P.M. Lushnikov, S.A. Newman, Continuous macroscopic limit of a discrete stochastic model for interaction of living cells. Phys. Rev. Lett. **99**, 168102 (2007)
31. A.R. Pries, T.W. Secomb, P. Gaehtgens, Structural adaptation and stability of microvascular netwoks: theory and simulation. Am. J. Physiol. Heart Circ. Physiol. **275**(44), H349–H360 (1998)

32. A.R. Pries, T.W. Secomb, Control of blood vessel structure: insights from theoretical models. Am. J. Physiol. Heart Circ. Physiol. **288**(3), H1010–H1015 (2005)
33. S.R. McDougall, A.R.A. Anderson, M.A.J. Chaplain, Mathematical modelling of dynamic adaptive tumour-induced angiogenesis: clinical implications and therapeutic targeting strategies. J. Theor. Biol. **241**, 564–589 (2006)
34. H. Ronellenfitsch, E. Katifori, Global optimization, local adaptation, and the role of growth in distribution networks. Phys. Rev. Lett. **117**, 138301 (2016)
35. H. Ronellenfitsch, J. Lasser, D.C. Daly, E. Katifori, Topological phenotypes constitute a new dimension in the phenotypic space of leaf venation networks. PLOS Comput. Biol. **11**(12), e1004680 (2016)

Biofilm Mechanics and Patterns

A. Carpio, E. Cebrián, D. R. Espeso and P. Vidal

Abstract From multicellular tissues to bacterial colonies, three dimensional cellular structures arise through the interaction of cellular activities and mechanical forces. Simple bacterial communities provide model systems for analyzing such interaction. Biofilms are bacterial aggregates attached to wet surfaces and encased in a self-produced polymeric matrix. Biofilms in flows form filamentary structures that contrast with the wrinkled layers observed on air/solid interfaces. We are able to reproduce both types of shapes through elastic rod and plate models that incorporate information from the biomass production and differentiation processes, such as growth rates, growth tensors or inner stresses, as well as constraints imposed by the interaction with environment.

Keywords Biofilms · Discrete rods · Helices · Von Karman plates · Wrinkles

1 Biofilm Shapes

Understanding how cellular systems evolve to adopt different shapes is an intriguing question which has motivated many theories. Here, we try to unravel this process in simple living beings: bacterial communities called biofilms. Environmental conditions seem to play a key role inducing changes. Biofilms growing in flows often form

A. Carpio (✉) · P. Vidal
Universidad Complutense de Madrid, Madrid, Spain
e-mail: ana_carpio@mat.ucm.es

P. Vidal
e-mail: pervidal@ucm.es

E. Cebrián
Universidad de Burgos, Burgos, Spain
e-mail: elenac@ubu.es

D. R. Espeso
Centro Nacional de Biotecnología, CSIC, Spain
e-mail: david.rodriguez@cnb.csic.es

L. L. Bonilla et al. (eds.), *Coupled Mathematical Models for Physical and Biological Nanoscale Systems and Their Applications*, Springer Proceedings in Mathematics & Statistics 232, https://doi.org/10.1007/978-3-319-76599-0_7

filaments constrained by the surrounding geometry. They may cross the current in corner flows [9] or wrap around tube walls forming helices [11]. Instead, biofilms spreading on semisolid agar surfaces exhibit different types of wrinkles [5, 20].

To understand the development of a biofilm one must take into account its nature. In a biofilm, bacteria are glued together and to a surface by a self-produced polymeric substance: the EPS matrix. Once the biofilm is formed, it can be seen as a biomaterial whose properties are controlled by the cellular activity [5, 20]. We discuss here how the material properties of a biofilm influence its shape in different environments.

2 Filamentary Structures in Flows

When spreading in flows, biofilms elongate along the current forming threads. The shape of the thread adapts to geometrical constraints, seeking to minimize adequate energies. Its time evolution until an equilibrium shape is reached can be described by discrete rod models. We tackle here two different experimental frameworks: biofilms in networks of cylindrical tubes and biofilms in corner flows.

2.1 *Helical Biofilms*

Consider the typical flow circuits used in medical systems, see Fig. 1. Injecting inside bacteria of the *Pseudomonas* genus, tubes fill with helical biofilms which wrap around the walls, see Fig. 2. Even if the Reynolds number is fairly small ($Re \sim 1$), the presence of connectors and junctions produces diameter variations, locally narrowing the passage. Vortices form past the stenoses. Vortical motion drives bacteria to the walls creating biofilm nucleation sites [11]. The biofilm then elongates following the streamlines until it undergoes a helical instability. Figure 3 shows the initial stages of the helical instability.

Elastic energies for filaments admit both straight and helical minimizers [7]. The presence of constraints that forbid the straight equilibrium prompt the appearance of helical structures [11]. A coiling effect is usually active at some end due to the presence of constrictions. When the biofilm hits a surface, deceleration results in

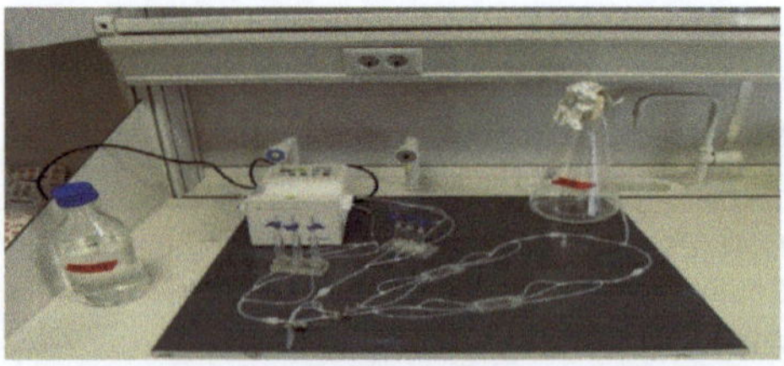

Fig. 1 Flow circuit. The fluid mixture flows from an initial reservoir to a pump (or a drip mechanism) that drives the liquid through a network of tubes, which may merge or split

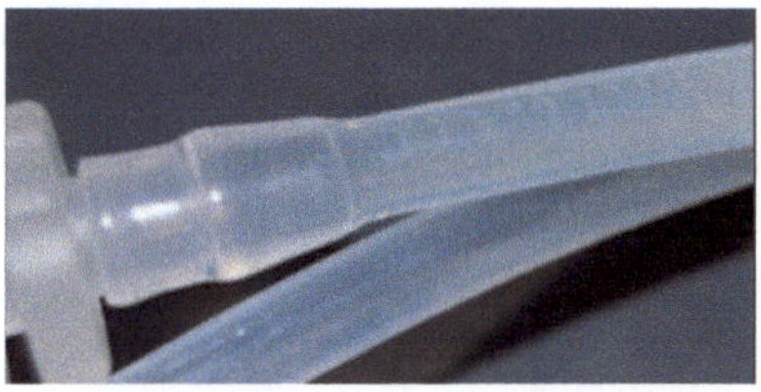

Fig. 2 Experimental image showing a biofilm helix wrapped around the tube wall past the junction. No biofilm is formed in the unperturbed branch. Bacteria were injected upstream the junction

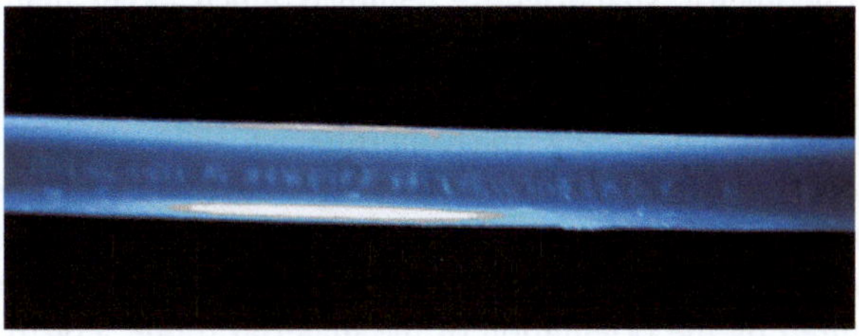

Fig. 3 Experimental image showing the onset of a helical instability along a biofilm thread in a 2 mm diameter tube

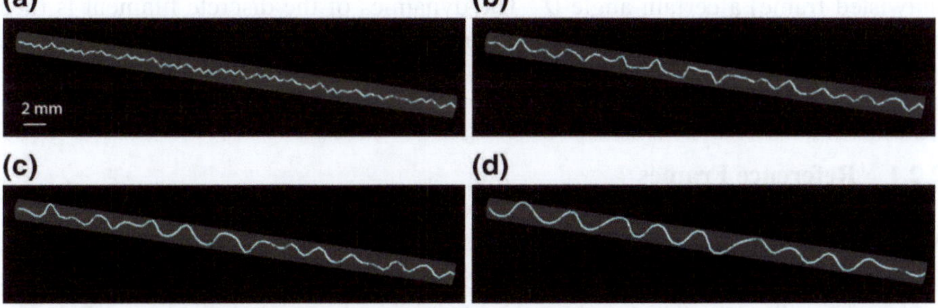

Fig. 4 Snapshots illustrating the in silico development and coarsening of a helical instability

coiling. Also, the presence of vortices at the stenosis may induce a helical beating of the thread. Additionally, biomass production causes a continuous length increase of the thread between constrictions. Helical shapes allow to allocate the excess length. Uninterrupted biomass production fosters the coarsening of the helical instability until it reaches the tube walls. More biomass is then allocated by narrowing the helix pitch. Notice that a filament of length L_f wraps around a tube of radius r_t and length L_t forming k steps of pitch L_t/k when $L_f^2 \sim L_t^2 + 4\pi^2 r_t^2 k^2$.

Figure 4 illustrates this dynamic process. A filament subject to twist develops a helical instability that coarsens until the helix reaches the tube wall thanks to a continuous length increase. These simulations are performed using the discrete rod model described in Sect. 2.2 [11].

2.2 *Discrete Rod Framework*

A filament is a geometric shape whose length is much larger than the rest of its dimensions. Any deformation of its cross section is expected to be small compared with variations of the total length. This fact motivates the description of a biofilm thread as a unidimensional curve γ (the centerline, which characterizes its position), plus a reference system at each point $\{\mathbf{t}, \mathbf{m}_1, \mathbf{m}_2\}$ (the material frame, which measures the twist). With this description, the movement of the thread can be fully captured: stretching and bending are computed by deforming the centerline, whereas twisting is captured by the orientation of the material frame. For dynamic simulations we use a discrete rod model [3, 15]. The filament is discretized using a sequence of nodes $\mathbf{x}^i$, $i = 0, \dots, n+1$, along the curve γ, and a reference system at each one, see Fig. 5. This frame is obtained at each location twisting the Bishop frame (a fixed untwisted frame) a certain angle θ^i. The dynamics of the discrete filament is then governed by equations for the angles θ^i, and for the node positions $\mathbf{x}^i$. We detail the procedure next.

2.2.1 Reference Frames

We assign to each point a local orthonormal frame (the material frame) $\{\mathbf{t}^i, \mathbf{m}_1^i, \mathbf{m}_2^i\}$, $i = 0, \dots, n$, describing the centerline orientation as follows. Let us denote by $\mathbf{e}^i = \mathbf{x}^{i+1} - \mathbf{x}^i$, $i = 0, \dots, n$, the straight segments joining the points $\{\mathbf{x}^0, \mathbf{x}^1, ..., \mathbf{x}^{n+1}\}$. The unit tangent vector associated to each edge is then $\mathbf{t}^i = \frac{\mathbf{e}^i}{\|\mathbf{e}^i\|}$, where $\|\,\|$ denotes the euclidean norm. Assuming the Bishop frame $\{\mathbf{t}^i, \mathbf{u}^i, \mathbf{v}^i\}$ known, the vectors $\mathbf{m}_1^i, \mathbf{m}_2^i$ are obtained rotating $\mathbf{u}^i, \mathbf{v}^i$ an angle θ^i in the plane orthogonal to $\mathbf{t}^i$:

$$\mathbf{m}_1^i = \cos(\theta^i)\mathbf{u}^i + \sin(\theta^i)\mathbf{v}^i, \quad \mathbf{m}_2^i = -\sin(\theta^i)\mathbf{u}^i + \cos(\theta^i)\mathbf{v}^i. \tag{1}$$

To define a Bishop frame we choose $\mathbf{u}^0 \perp \mathbf{t}^0$ and set $\mathbf{v}^0 = \mathbf{t}^0 \times \mathbf{u}^0$. The frames at the remaining edges are constructed by parallel transport [3]. We set

$$\mathbf{u}^i = P_i(\mathbf{u}^{i-1}), \quad \mathbf{v}^i = \mathbf{t}^i \times \mathbf{u}^i, \tag{2}$$

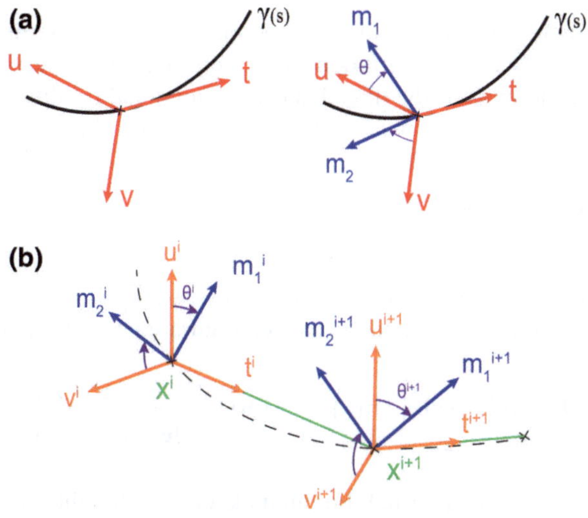

Fig. 5 **a** Continuous description of a filament [15]. The Bishop frame $\{\mathbf{t}, \mathbf{u}, \mathbf{v}\}$ defines the rest orientation at any point of the filament centerline $\gamma(s)$, parametrized by the arc length in R^3. Rotating $\{\mathbf{u}(s), \mathbf{v}(s)\}$ an angle $\theta(s)$ around $\mathbf{t}(s)$ we obtain the material frame characterizing the local orientation $\{\mathbf{t}(s), \mathbf{m}_1(s), \mathbf{m}_2(s)\}$. **b** Discrete description of a filament [3]. The centerline is discretized as a set of points $\{\mathbf{x}^0, \mathbf{x}^1, \ldots, \mathbf{x}^{n+1}\}$ and segments $\mathbf{e}^i = \mathbf{x}^{i+1} - \mathbf{x}^i$. Setting $\mathbf{t}^i = \frac{\mathbf{e}^i}{\|\mathbf{e}^i\|}$, the unit tangent vector per edge, a local orthonormal material frame $\{\mathbf{t}^i, \mathbf{m}_1^i, \mathbf{m}_2^i\}$ is assigned to each point

where P_i are rotation matrices about the curvature binormal defined by:

$$P_i(\mathbf{t}^{i-1}) = \mathbf{t}^i, \quad P_i(\mathbf{t}^{i-1} \times \mathbf{t}^i) = \mathbf{t}^{i-1} \times \mathbf{t}^i.$$

If $\mathbf{t}^{i-1} = \mathbf{t}^i$, P_i is the identity. The condition $\mathbf{u}^0 \perp \mathbf{t}^0$ must be maintained during the simulation. This is guaranteed when $\mathbf{t}^0$ is clamped. Otherwise, it can be reestablished by parallel transport in time (instead of space).

2.2.2 Equations for the Angles

The equations for the angles follow from energy arguments. When the undeformed configuration of the filament is straight and its elastic response is isotropic, the elastic energy due to torsion and bending takes the form [3]:

$$E = \sum_{i=1}^{n} \beta \frac{(\theta^i - \theta^{i-1})^2}{\overline{\ell}^i} + \sum_{i=1}^{n} \frac{\alpha}{2\overline{\ell}^i} \sum_{j=i-1}^{i} \|\mathbf{w}_i^j - \overline{\mathbf{w}}_i^j\|^2, \tag{3}$$

where α and β are the bending and torsion moduli, respectively. We may set $\alpha = E_Y I$ and $\beta = JG$, being E_Y the Young modulus of the thread, I the second moment of area, G the shear modulus of the thread and J the torsional rigidity constant. If we consider a thread composed of an isotropic elastic material, the Young modulus and shear modulus are related by the Poisson coefficient ν as $G = \frac{E_Y}{2(1+\nu)}$. For a filled cylinder we have $J = I$, hence $\beta = \frac{E_Y}{2(1+\nu)} I$.

In formula (3), $\overline{\ell}^i$ is the length of the segments $\overline{\mathbf{e}}^i = \overline{x}^{i+1} - \overline{x}^i$ in a reference undeformed configuration $\{\overline{\mathbf{x}}^0, \overline{\mathbf{x}}^1, ..., \overline{\mathbf{x}}^{n+1}\}$. The vectors $\mathbf{w}_i^j$, $\overline{\mathbf{w}}_i^j$, $j = i-1, i$, are material curvatures in the deformed and undeformed configurations, respectively:

$$\mathbf{w}_i^j = \left((\kappa\mathbf{b})_i \cdot \mathbf{m}_2^j, -(\kappa\mathbf{b})_i \cdot \mathbf{m}_1^j\right)^t, \quad (\kappa\mathbf{b})_i = \frac{2\mathbf{e}^{i-1} \times \mathbf{e}^i}{\|\overline{\mathbf{e}}^{i-1}\| \|\overline{\mathbf{e}}^i\| + \mathbf{e}^{i-1} \cdot \mathbf{e}^i}, \tag{4}$$

where $\kappa\mathbf{b}$ is the curvature binormal. For an undeformed straight shape $\overline{\mathbf{w}}_i^j = 0$. The general form of the elastic energy for anisotropic rods that adopt a nonstraight undeformed shape is given in [3].

The material frame is updated in a quasistatic way. Imposing

$$\frac{\partial E}{\partial \theta^i} = 0, \tag{5}$$

for all segments i not fixed by a boundary condition, this system of equations determines the angle configuration that minimizes the energy of the thread. Clamped ends are accounted for assigning the material frame for $i = 0$, $i = n$. No boundary condition corresponds to a stress free end.

2.2.3 Equations for the Positions

We keep track of the filament position displacing the nodes according to Newton's second law:

$$\mathcal{M}\frac{d^2\mathbf{x}}{dt^2} = -\frac{dE}{d\mathbf{x}} + \mathbf{F} = \mathbf{F}_T, \tag{6}$$

where $\mathbf{F}$ represents the external forces and $-\frac{dE}{d\mathbf{x}}$ the elastic forces. Explicit formulas for the elastic forces are given in [3]. $\mathcal{M}$ denotes the $3(n+2) \times 3(n+2)$ mass matrix, we set $\mathcal{M} = m\mathcal{I}$, where $\mathcal{I}$ is the identity matrix. $\mathbf{x} = (\mathbf{x}^0, \ldots, \mathbf{x}^{n+1})$ denotes the ordered sequence of 3D node coordinates. Notice that $\mathbf{x}^i = (x_1^i, x_2^i, x_3^i)$.

To approximate the solution of system (6), we apply a Verlet integrator to estimate the displacements and velocities, and then enforce an inextensibility constraint for each segment by using a manifold projection method [12]. This constraint facilitates a stable evolution from a numerical point of view.

For each new time $t_{k+1} = t_k + h$, the Verlet scheme provides the prediction:

$$\tilde{\mathbf{v}}_{k+1} = \mathbf{v}_k + h\mathcal{M}^{-1}\mathbf{F}_T(\mathbf{x}_k), \quad \tilde{\mathbf{x}}_{k+1} = \mathbf{x}_k + h\tilde{\mathbf{v}}_{k+1}, \tag{7}$$

starting from previous values $\mathbf{x}_k, \mathbf{v}_k$, where $\mathbf{v}_k = \dot{\mathbf{x}}_k$ and h is the time step. The projection method works as follows. We set $\mathbf{y}_0 = \tilde{\mathbf{x}}_{k+1}$. At each step j, we compute the next value $\mathbf{y}_{j+1} = \mathbf{y}_j + \delta\mathbf{y}_{j+1}$, where

$$\delta\mathbf{y}_{j+1} = -h^2\mathcal{M}^{-1}\nabla\mathbf{C}(\mathbf{y}_j)^t\,\delta\ell_{j+1}, \tag{8}$$

and $\mathbf{C}(\mathbf{y})$ defines the system of constraints. The vector $\delta\ell_{j+1}$ solves the linear system

$$h^2\nabla\mathbf{C}(\mathbf{y}_j)\mathcal{M}^{-1}\nabla\mathbf{C}(\mathbf{y}_j)^t\,\delta\ell_{j+1} = \mathbf{C}(\mathbf{y}_j). \tag{9}$$

The iteration stops if $\|\mathbf{C}(\mathbf{y}_{j+1})\| \leq \varepsilon$ for the desired tolerance $\varepsilon > 0$. The constraint enforcing velocity and position are then

$$\mathbf{v}_{k+1} = \frac{1}{h}(\mathbf{y}_{j+1} - \tilde{\mathbf{x}}_{k+1}), \quad \mathbf{x}_{k+1} = \mathbf{x}_k + h\mathbf{v}_{k+1}. \tag{10}$$

If inextensibility is the only constraint, $\mathbf{C}$ is defined by the system of equations $\|\mathbf{e}^i\|^2/\|\overline{\mathbf{e}}^i\| - \|\overline{\mathbf{e}}^i\| = 0$, for each edge i.

2.2.4 Coupling to the Flow

The force exerted by a fluid undergoing a given undisturbed flow on a long slender body is analyzed in [6]. The following asymptotic formula in terms of the ratio of the of the cross-sectional radius to the body length is given:

$$\frac{\mathbf{f}^t}{2\pi} = \mu_f\lambda\int_0^1\Bigg(\left[\frac{(\mathbf{U}-\mathbf{U}^*)^t}{\ln\kappa} + \frac{(\mathbf{U}-\mathbf{U}^*)^t\ln(2)}{(\ln\kappa)^2}\right]\cdot[\mathbf{t}\,\mathbf{t}^t - 2\mathcal{I}] + \frac{\frac{1}{2}(\mathbf{U}-\mathbf{U}^*)^t}{(\ln\kappa)^2}\cdot[3\mathbf{t}\,\mathbf{t}^t - 2\mathcal{I}]\Bigg)ds, \tag{11}$$

where $\mathbf{f}$ is the force acting on a filamentous body of length λ, $\mathbf{U}$ and $\mathbf{U}^*$ are the velocities of the unperturbed fluid and the thread, respectively, at the position $\gamma(s)$ and $\mathcal{I}$ is the 3×3 identity matrix. We denote by μ_f the viscosity of the fluid, $\gamma(s)$ the position of the thread centerline, s the arclength of the thread ($0 < s < 1$), $\mathbf{t}(s) = \frac{d\gamma(s)}{ds}$

the tangent vector at the position $\gamma(s)$ and $\kappa = \frac{r}{L}$ the ratio between cross-sectional radius r to characteristic thread length L.

This relation allows us to directly calculate the fluid force $\mathbf{f}^i$ acting on each node of the thread by using the difference of velocities between the fluid and the thread, the tangent vector at each node and the aspect ratio of the thread, an idea already exploited for filaments in 2D corner flows in [1]. Notice that, at each node $\mathbf{x}^i$, the thread velocity $\mathbf{U}^*(\mathbf{x}^i, t) = \mathbf{v}^i(t)$. In absence of other forces, $\mathbf{F} = (\mathbf{f}^0, \ldots, \mathbf{f}^{n+1})$ in Eq. (6).

Biofilm filaments live inside tubes of a certain shape. A simple way to incorporate this restriction is a penalty method. The idea is to include in the force term $\mathbf{F}$ in Eq. (6) additional forces $\mathbf{f}_w$ supported on the tube walls that point inside the tube and act on any node hitting the wall, sending it back inside. Alternatively, we might set the position equal to the effective maximum radius and reset the velocity equal to zero.

2.2.5 Increasing the Length

Increase in length of a filament can be due to the combined effect of different mechanisms: biomass production, biomass adhesion, elastic elongation, swelling... To reproduce an increase in length at a certain rate we enlarge the segments joining nodes in a controlled way, redefining the reference lengths at the same time. Directional mass addition may be represented adding nodes at an edge and redefining the reference configuration each time a node is added.

In practice, we alternate steps in which we solve the equations for the evolution of the discrete rod with steps in which we increase the length of the edges or the number of nodes, and reset the reference configuration before computing again the evolution of the enlarged filament.

2.2.6 Overall Procedure

Summarizing, to compute the evolution of a rod we proceed in the following steps:

- Initialization:
  - Define the Bishop frame at edge 0: $(\mathbf{t^0}, \mathbf{u^0}, \mathbf{v^0})$.
  - Set the position of the undeformed centerline: $\overline{\mathbf{x}}^0, \overline{\mathbf{x}}^1, ..., \overline{\mathbf{x}}^{n+1}$.
  - Select the initial position and velocity of the centerline: $(\mathbf{x}^0, \dot{\mathbf{x}}^0), (\mathbf{x}^1, \dot{\mathbf{x}}^1)$, ..., $(\mathbf{x}^{n+1}, \dot{\mathbf{x}}^{n+1})$.
  - Enforce the boundary conditions for the filament at the initial and final nodes.
  - Set the material curvatures using Eq. (4).
  - Set the material frame by means of Eqs. (1), (2) and (5).
- Iteration for each new time step:
  - Compute the elastic forces $-\frac{dE}{d\mathbf{x}}$ acting on the centerline, and possible additional forces $\mathbf{F}$ (see Sect. 2.2.4).

– Integrate Newton equations for the centerline (6) enforcing inextensibility and possible additional constraints (see Sect. 2.2.3).
– Update the Bishop frame using Eq. (2).
– Update the quasistatic material frame by means of Eqs. (1) and (5).
– Eventually, increase the length of the edges or the number of nodes, and reset the reference configuration (see Sect. 2.2.5).

The simulations shown in Fig. 4 start from a straight filament placed at the center of the tube. The initial node positions are randomly perturbed to ensure a slight initial excess length with respect to the tube length. A twist angle is imposed at the filament edges. The length of the filament is slowly increased as the helix develops to foster the coarsening process [11]. The unperturbed fluid velocity profile obeys a radial Hagen-Poiseuille distribution. However, the fluid force does not seem to play a role in helix formation, which is driven by elastic forces. It only causes slow downstream motion of the whole structure. Therefore, we may set it equal to zero to simplify the study of helix development. The force term **F** would only account for the presence of the walls in this case, constraining the helix radius. On the contrary, fluid forces are essential to produce the filaments crossing corner flows described in Sect. 2.3.

2.2.7 Nondimensional Equations

It is convenient to nondimesionalize Eqs. (3), (6) and (11) for numerical purposes. The change of variables $x = \lambda x'$, $t = Tt'$, $\mathbf{U} = U_0\mathbf{U}'$, $E = \frac{\alpha}{\lambda}E'$, $\mathbf{F} = \mu_f \lambda U_0 \mathbf{F}'$ yields:

$$\frac{d^2\mathbf{x}'}{dt'^2} = -\frac{T^2}{m\lambda^2}\frac{dE}{d\mathbf{x}'} + \frac{T^2}{m\lambda}\mathbf{F} = -\frac{\alpha T^2}{m\lambda^3}\frac{dE'}{d\mathbf{x}'} + \frac{\mu_f U_0 T^2}{m}\mathbf{F}', \tag{12}$$

where the force term includes the force exerted by the fluid **f** plus possible penalty forces $\mathbf{f}_w$ at the walls, that is, $\mathbf{F} = \mathbf{f} + \mathbf{f}_w$. In view of the definition (3) of the energy E, this change introduces the controlling parameters $\alpha' = \frac{\alpha T^2}{m\lambda^3}$, $\beta' = \frac{\beta T^2}{m\lambda^3}$ and $\delta' = \frac{\mu_f U_0 T^2}{m}$, U_0 being a characteristic velocity. Determining ranges of values of α', β' that lead to different types of filamentary structures we would obtain ranges for α, β whenever the density ρ and radius r of the filaments are experimentally quantified. Working with one dimensional filaments, we neglect the cross-sections. A three-dimensional cylindrical thread of density ρ, radius r, and length $L \gg r$ is approximated in this setting by a discrete rod with $n + 2$ nodes and $n + 1$ edges, with mass $m = \rho\pi r^2 L/(n + 1)$. From identities (3) and (12), the characteristic time associated to the elastic deformation of the thread can be estimated as $T_{elast} = \sqrt{\frac{m\lambda^3}{E_Y I}}$. If this value is chosen as characteristic time in our system and we use formula (11) we arrive at:

$$\frac{d^2\mathbf{x}'}{dt'^2} = -\frac{dE'}{d\mathbf{x}'} + \eta\mathbf{F}', \tag{13}$$

where $\eta = \frac{\mu_f U_0 \lambda}{\frac{E_Y I}{\lambda^2}}$ is the ratio of viscous forces to elastic forces, see [1].

2.3 Biofilm Threads in Corner Microflows

The situation described in Sect. 2.1 for helical biofilms is partly reminiscent of the observations made with *Pseudomonas Aeruginosa* in a laminar corner microflow [9]. Bacteria are driven to the wall by small secondary vortices past the corner, creating nucleation sites. Once a biofilm seed forms, threads made of bacteria joined by EPS matrix are issued. Initially, they align with the streamlines, like streamers. Unlike the case of helical biofilms, which are triggered by elastic forces and constrained by the tube walls, biofilm threads cross the streamlines of corner flows driven by the fluid force acting on them, as a result of the interaction fluid-structure. A two-dimensional model of an elastic filament in a corner flow shows that under certain conditions filaments cross the main stream and reach the opposite corner [1]. When the filament is long enough, it reaches the opposite corner adopting the equilibrium shape of an elastic rod in a corner flow. As in the case of helices, the final configuration seems to be a minimum of an elastic energy.

The 3D model summarized in Sect. 2.2 reproduces this behavior provided we include in the equations of motion (6) the force due to the fluid [6], given by (11). The pressure driven fluid velocity field $\mathbf{U}$ may be computed using finite element software such as COMSOL multiphysics in the unperturbed channel geometry, see Fig. 6a. We use this reference flow field during the whole simulation, ignoring perturbations due to the presence of the thin biofilm thread, as in [1, 6]. Figure 6b represents the initial biofilm configuration used in the simulations. Keeping the same parameters as in the experiments [1], we can study the influence of variables like the initial angle or

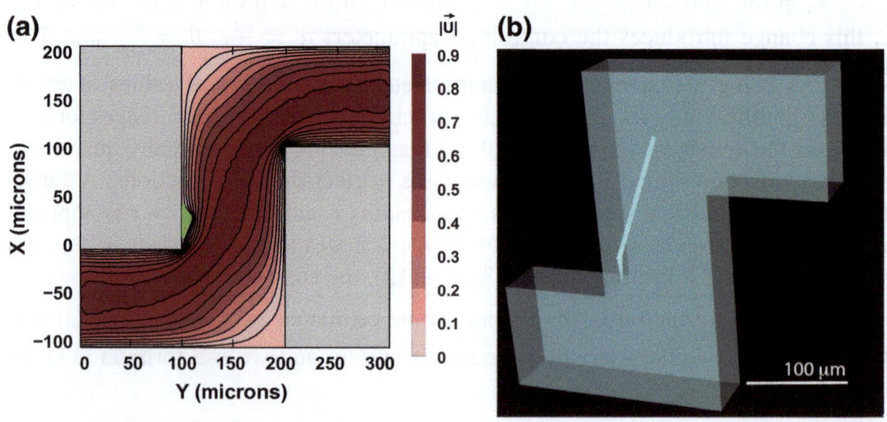

Fig. 6 **a** Normalized velocity modulus in the mid-plane of the there dimensional channel. **b** Initial filament configuration in the corner flow simulations

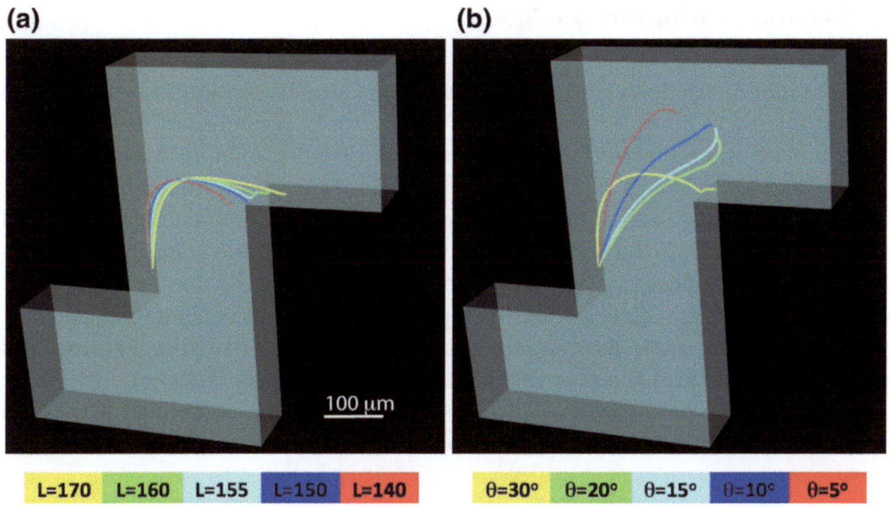

Fig. 7 Effect of the initial thread geometry on its dynamics during the simulation: **a** Length L and **b** angle θ. Parameter values are $\alpha' = 1$, $\beta' = 0.358$, $n = 100$, $\eta = 3000$, $\kappa = 0.05$, $\Delta t' = 10^{-3}$ time units. A final state joining opposite corners is reached depending on the initial orientation and length, although these constraints may change with the initial velocity of the thread

length on the filament evolution. Figure 7 shows some possible configurations. To obtain it, we integrated the evolution Eq. (13) setting the characteristic spatial length $\lambda = 1\mu$m and the characteristic time of the system equal to the elastic deformation time T_{elas}, which yields $\alpha' = 1$, $\beta' = \beta/\alpha$, $\delta' = \eta$.

As observed in [1], the parameter η regulates the effect on the fluid on the thread dynamics. For values of $\eta \leq 4000$, the contact of the thread with the opposite corner is only dependant of the filament initial position and length. A minimum length and angle are needed for the thread to reach the opposite corner. Within the range $4000 < \eta < 8000$ the fluid can substantially modify the trajectory of the thread. For $\eta > 8000$ the fluid strongly drags the thread in the direction of the stream, avoiding contact with the opposite corner. A steady position for the thread parallel to the streamlines was found for $\eta \sim 30,000$. These results are obtained setting the initial velocity of the thread equal to the fluid velocity at the node positions. The initial thread velocity affects the results. Setting it equal to zero, very low values of η would be required to cross the streamlines.

If we start from a short thread and implement directional mass increase adding nodes at an edge and redefining the reference configuration each time a node is added, then we can see the filament grow as it moves towards the opposite corner.

3 Wrinkled Biofilms on Agar

Whereas biofilms in flows tend to form filamentary structures, biofilms spreading on agar/air interfaces adopt wrinkled shapes [2, 5, 20]. Descriptions of their behavior may be made more precise than in the previous case due to an increasing amount of experimental evidence. Cell death has been shown to play a role on the onset of wrinkles in *Bacilus Subtilis* biofilms. The biofilm is formed by bacteria immersed in a polymer matrix, which gives the mixture a certain elastic cohesion. Dividing cells produce compression stresses. In addition, cells may die due to biochemical stress associated with high cell density, high waste and toxin concentration, and lack of resources. Dead areas allow to relieve that stress forming wrinkles. This explains the onset of wrinkles [2] but not the branching arrangements observed [10]. These arrangements can be understood incorporating information on cellular activity in mechanical models of biofilm expansion on a substrate, as we explain next.

3.1 Föppl-Von Karman Models

Let us consider a biofilm layer spreading on an agar substratum. We can reproduce wrinkle branching in the expanding biofilm resorting to Föppl-Von Karman descriptions of the interface biofilm/agar [8, 14, 17]:

$$\frac{\partial \xi}{\partial t} = \frac{1-2\nu_v}{2(1-\nu_v)}\frac{h_v}{\eta_v}\left[D(-\Delta^2\xi + \Delta C_M) + h\frac{\partial}{\partial x_\beta}\left(\sigma_{\alpha,\beta}(\mathbf{u})\frac{\partial \xi}{\partial x_\alpha}\right)\right] - \frac{\mu_v}{\eta_v}\xi, \quad (14)$$

$$\frac{\partial \mathbf{u}}{\partial t} = \frac{h_v h}{\eta_v}\nabla\cdot\sigma(\mathbf{u}) - \frac{\mu_v}{\eta_v}\mathbf{u}, \quad (15)$$

where h_v is the thickness of the viscoelastic substratum and μ_v, ν_v, η_v its rubbery modulus, Poisson ratio, and viscosity, respectively. The bending stiffness is $D = \frac{Eh^3}{12(1-\nu^2)}$, where E and ν represent the Young are Poisson moduli of the biofilm, whereas h is the film thickness. In these equations, ξ stands for the out of plane displacement and $\mathbf{u}$ for the in-plane displacement. α and β stand for x, y and summation over repeated indices is intended. Stresses $\boldsymbol{\sigma}$ and strains $\boldsymbol{\varepsilon}$ are defined in terms of in-plane displacements $\mathbf{u} = (u_x, u_y)$ [8, 16]:

$$\varepsilon_{\alpha,\beta} = \frac{1}{2}\left(\frac{\partial u_\alpha}{\partial x_\beta} + \frac{\partial u_\beta}{\partial x_\alpha} + \frac{\partial \xi}{\partial x_\alpha}\frac{\partial \xi}{\partial x_\beta}\right) + \varepsilon^0_{\alpha,\beta}, \quad (16)$$

$$\sigma_{xx} = \frac{E}{1-\nu^2}(\varepsilon_{xx} + \nu\varepsilon_{yy}), \quad \sigma_{xy} = \frac{E}{1+\nu}\varepsilon_{xy}, \quad \sigma_{yy} = \frac{E}{1-\nu^2}(\varepsilon_{yy} + \nu\varepsilon_{xx}). \quad (17)$$

The residual strains $\varepsilon^0_{\alpha,\beta}$ are expressed in terms of the growth tensor [8] as:

$$\varepsilon^0_{\alpha,\beta} = -\frac{1}{2}\left(\mathscr{G}_{\alpha\beta} + \mathscr{G}_{\beta\alpha} + \mathscr{G}_{z\alpha}\mathscr{G}_{z\beta}\right), \tag{18}$$

and should be computed from the cellular activity in the spreading biofilm configuration.

3.2 Bacterial Activity

Bacterial activity can be represented exploiting different agent based models. Cellular automata descriptions, for instance, provide a simple framework allowing for an easy transfer of information into macroscopic models. The biofilm is divided in cubic tiles, each of them containing a few cells. To simplify further, we may identify each tile with one cell. This approach has two advantages. First, we can use the same grid of tiles to discretize the equations for the relevant chemical concentrations and the displacements, and then solve them numerically. Second, we can calculate the growth tensors due to cell division, death, and other processes, and use them to estimate the residual stresses that enter the Föppl-Von Karman equations for the deformations. We have to decide for each cell which is its status. It may secrete chemicals, deactivate, divide creating a newborn cell that displaces the rest or die, being eventually reabsorbed by the rest. This may be done resorting to dynamic energy budget descriptions [4] or according to probabilities that depend on the relevant concentrations [10].

When there is an excess of oxygen, the concentration of nutrients c_n becomes the limiting concentration that restricts biofilm growth. The evolution of the concentration c_n in the biofilm/agar system is governed by

$$c_{n,t} - \mathrm{div}(D_n \nabla c_n) = r_n(c_n), \tag{19}$$

where $r_n(c_n) = -\tilde{k}_n \frac{c_n}{c_n+K_n}$, and $\tilde{k}_n$ is the uptake rate, equal to k_n at each alive cell location and zero otherwise. D_n and K_n denote the diffusion and half-saturation coefficients. No-flux boundary conditions are imposed at the interface with air. The evolution of the concentration of waste c_W in the biofilm/agar system is governed by

$$c_{w,t} - \mathrm{div}(D_w \nabla c_w) = r_w(c_w), \tag{20}$$

where $r_w(c_w) = \tilde{k}_w$, and $\tilde{k}_w$ is the waste production rate, equal to k_w at each alive cell location and zero otherwise. No-flux boundary conditions are imposed at the interface with air. The diffusion coefficients D_w and D_n may vary across the biofilm/agar system.

In the simplest cellular automata approach, tiles $\mathscr{C}$ occupied by alive cells are assumed to divide with probability [13]:

$$P_d(\mathscr{C}) = \frac{c_n(\mathscr{C})}{c_n(\mathscr{C}) + a_n}, \tag{21}$$

c_n being the limiting concentration and $a_n > 0$. Newborn cells inside the biofilm are reallocated by pushing existing cells in the direction of minimum mechanical resistance, that is, the shortest distance to the biofilm-air interface or to dead cells. Taking the concentration of waste c_w at a location as an indicator of death, a cell $\mathscr{C}$ is scheduled to die with probability:

$$P_w(\mathscr{C}) = \frac{c_w(\mathscr{C})}{c_w(\mathscr{C}) + a_w}, \tag{22}$$

c_w being the waste concentration and $a_w > 0$. Dead cells surrounded by enough alive neighbors may be reabsorbed by the rest, and its place occupied by a newborn cell. Otherwise, necrotic regions are created. This process may be further refined to account for cell differentiation into producers of different types of autoinducers [4, 10].

For a fixed distribution of cell types, the concentrations c_n and c_w relax fast to stationary values, which may be approximated by explicit finite difference schemes. Once the concentration values are calculated, we go through all the cells forming the film, creating new cells or killing existing ones with the selected probabilities. Then, a growth tensor may be defined at each tile by keeping track of all the new tiles created and the direction in which their predecessors where shifted. First we introduce a vector $\mathbf{w} = (w_1, w_2, w_3)a$, where a is the tile size. w_1 is evaluated at each location by adding ± 1 cumulatively for each tile shifted in the x direction in the positive or negative sense, respectively. w_2 and w_3 are calculated in a similar way, along the y and z directions, respectively. The resulting vector $\mathbf{w}$ is normalized to have norm a. Next, we compute $\nabla\mathbf{w}$ approximating the derivatives by finite differences. To estimate the growth tensor $\mathscr{G}(x, y)$ we average all the contributions from $\nabla\mathbf{w}(x, y, z)$ varying z.

3.3 Residual Strains

The residual strains defined in (18) can be computed using the growth tensor $\mathscr{G}$ introduced in Sect. 3.2. However, stochastic variations make it unsuitable to be inserted directly in the Föppl-Von Karman equations (14)–(15) because they cause numerical instability.

To smooth out the residual strains and visualize the underlying spatial variations, we average them over a number of runs of the step in which new cells are created or killed according to the selected probabilities, keeping the same initial configuration in all of them:

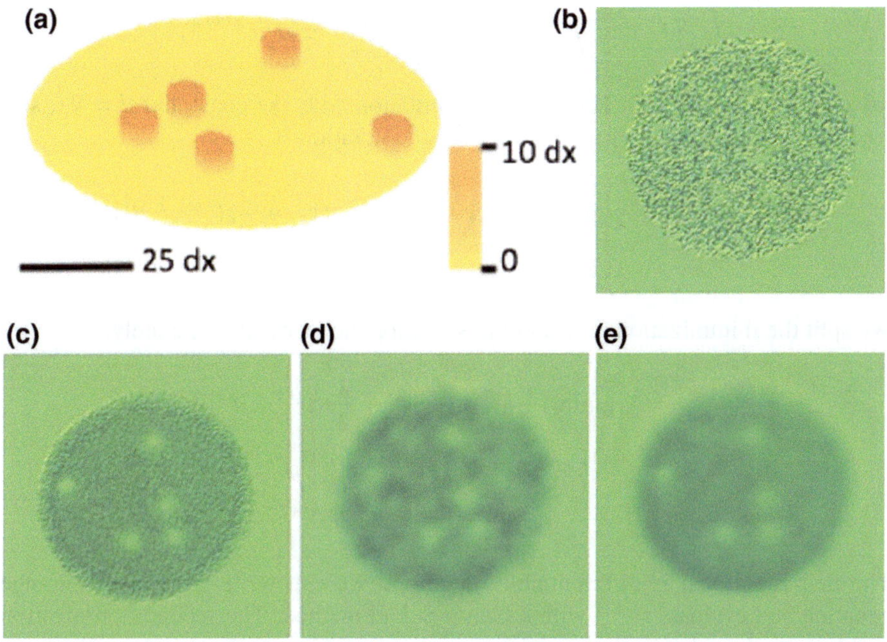

Fig. 8 **a** Biofilm containing regions where the cell density is higher. dx represents the spatial step. **b** $\varepsilon_{xx}^{0,av}$ component of the residual strain tensor due to growth with $N = 1$. **c** Averaged $\varepsilon_{xx}^{0,av}$ component with $N = 100$. **d** Filtered $\varepsilon_{xx}^{0,fil}$ component with $N = 1$. **e** Filtered $\varepsilon_{xx}^{0,fil}$ component with $N = 10$. The depressions correspond to the initial mounds and appear due to cell death caused by lack of resources. Strains are higher in the outer ring due to higher availability of resources, which results in higher division rates. The same scale of colors is used in all of them ranging from 3 (light yellow) to −3 (dark green)

$$\varepsilon^{0,av} = \frac{1}{N}\sum_{j=1}^{N}\varepsilon^{0,j}, \tag{23}$$

where $\varepsilon^{0,j}$ stands for the residual strain at trial j. Performing such ensemble averages for N large enough the averaged strains reproduce spatial variations reflecting cellular activity, see Fig. 8b, c. The resulting average becomes smoother as the number of runs N increases. However, the computational cost of this process is high.

Instead, we filter the residual fields using image processing techniques. This strategy yields smooth approximations with a clear spatial structure averaging just a few runs, see Fig. 8. The idea is to formulate a denoising problem: given an observed field $f^{obs} = f + n$, we seek the underlying smooth structure f obtained removing the noise n. To solve this problem we apply a split Bregman method to a ROF (Rudin, Osher, Fatemi) model of the denoising problem [19]. The ROF model consists in solving the variational problem: Find f minimizing

$$\int |\nabla f| + \frac{\mu}{2} \int |f - f^{obs}|^2 = TV(f) + \frac{\mu}{2} \|f - f^{obs}\|^2_{L^2},$$

for $\mu > 0$ large. The split Bregman reformulation adds the constraint $d = \nabla f$, sets $s(b,f,d) = \int |b + \nabla f - d|^2$ and introduces the iteration:

$$(f^{(k+1)}, d^{(k+1)}) = \text{Argmin}_{(f,d)}\{|d| + \frac{\mu}{2}\|f^{obs} - f\|^2_{L_2} + \frac{\lambda}{2} s(b^{(k)}, f, d)\},$$
$$b^{(k+1)} = b^{(k)} + \nabla f^{(k+1)} - d^{(k+1)}.$$

We split the minimization procedure to solve for each variable separately:

$$f^{(k+1)} = \text{Argmin}_f\{\frac{\mu}{2}\|f^{obs} - f\|^2_{L_2} + \frac{\lambda}{2} s(b^{(k)}, f, d^{(k)})\},$$
$$d^{(k+1)} = \text{Argmin}_d\{|d| + \frac{\lambda}{2} s(b^{(k)}, f^{(k+1)}, d)\},$$
$$b^{(k+1)} = b^{(k)} + \nabla f^{(k+1)} - d^{(k+1)}.$$

The first functional is differentiable, therefore, we can write the Euler-Lagrange equation and evaluate $f^{(k+1)}$ with a Gauss-Seidel method. The second optimization problem can be solved using shrinkage operators:

$$d^{(k+1)} = \text{shrink}(b^{(k)} + \nabla f^{(k+1)}, \frac{1}{\lambda}),$$
$$\text{shrink}(x, \gamma) = \frac{x}{|x|} \max(|x| - \gamma, 0).$$

The filtered fields reproduced in Fig. 8 have been produced setting $f^{obs} = \varepsilon^{0,av}_{xx}$ over the 2D grid in the plane XY, relabeling to transform it into a 1D vector, and using the algorithm:

- Initial guess $f^{(0)} = f^{obs}$, $d^{(0)} = 0$, $b^{(0)} = 0$.
- While $\|f^{(k)} - f^{(k-1)}\|_{L_2} > Tol$
  - $f^{(k+1)} = G^{(k)}$, where, for $j = 1, \ldots, M$,

$$G^{(k)}_j = \frac{\lambda}{\mu + 2\lambda}\left(f^{(k)}_{j+1} + f^{(k)}_{j-1} - (d^{(k)}_j - d^{(k)}_{j-1}) + (b^{(k)}_j - b^{(k)}_{j-1})\right) + \frac{\mu}{\mu + 2\lambda} f^{obs}_j,$$

    with $\nabla f^{(k+1)}_j = f^{(k+1)}_{j+1} - f^{(k+1)}_j$,
  - $d^{(k+1)} = \text{shrink}(b^{(k)} + \nabla f^{(k+1)}, \frac{1}{\lambda})$,
  - $b^{(k+1)} = b^{(k)} + \nabla f^{(k+1)} - d^{(k+1)}$.
- If $\|f^{(k)} - f^{(k-1)}\|_{L_2} \le Tol$, we set $f^{fil} = f^{(k)}$.

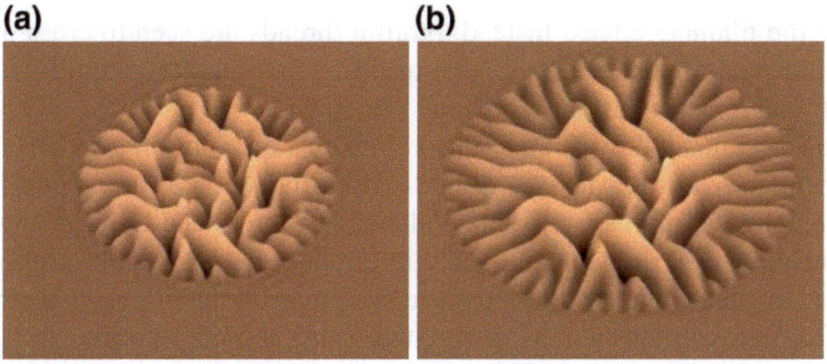

Fig. 9 In silico wrinkle coarsening and successive branching in a biofilm spreading on agar. The height of the wrinkles ranges from −5 to 5 dx and the final radius is about 250 dx, dx being the size of the tiles in the spatial discretization employed to evaluate the cellular activity

The resulting fields are smooth enough to be plugged in Eqs. (14)–(15) through (16)–(17) without causing numerical instability, allowing us to reproduce behaviors that resemble observed patterns.

Our simulations of biofilm behavior alternate steps in which we update the configuration of biofilm tiles, creating and killing cells, and then evaluate the resulting stresses, with steps in which the biofilm shape is deformed as determined by the Föppl-Von Karman equations, see [10] for details. Figure 9 shows wrinkles coarsening and opening up in radial branches. This phenomenon is associated to compression fronts expanding at certain speeds. Other observed arrangements, such as wrinkled coronas, that is, a corona of radial wrinkles surrounding a central core [5], can be reproduced varying the Young modulus as usual in corona instabilities: a swollen corona with diminished Young modulus around a harder core [10]. Two phase models have been proposed to describe swelling processes in [18]. However, an adequate way to handle water migration processes is still missing in our description.

4 Conclusions

Summarizing, we have shown that inserting in plate and rod models information from the cellular activity and the interaction with the environment we gain insight on biofilm shapes.

Coupling discrete rod models to the forces exerted by surrounding fluids and incorporating external constraints such as the presence of walls and constrictions, we are able to reproduce experimentally observed filamentary patterns. Helical biofilms in flows arise as elastic instabilities that coarsen as the length of the biofilm thread increases due to biomass production and finally wrap around tube walls reducing their pitch to accommodate more biomass. The main role of the flow in this case is to promote biofilm filament nucleation past constrictions and to provide mechanisms for twist at the filament edges. Instead, biofilm threads are seen to cross 3D corner microflows to join opposite corners as a result of the interaction fluid/structure, in agreement with experiments and previous 2D studies.

Whereas biofilms in flows tend to form filaments, biofilms on agar surfaces often spread forming wrinkled patterns. Successive radial wrinkle branching is associated to expanding compression fronts and can be reproduced by inserting residual stresses caused by cell division and death in Föppl-Von Karman descriptions of the out of plane displacements of the interface biofilm/agar. These residual stresses are obtained from growth tensors, computed here using the information on the cellular activity provided by a simple cellular automata model. To avoid instability caused by stochasticity and be able to visualize the spatial variations caused by the cellular activity, such residual stresses are smoothed out combining ensemble averages and denoising algorithms. More realistic models of cellular activity could be considered at the expense of increasing the computational cost. Additional processes affecting biofilm shapes such as water migration through the agar/biofilm system are missing in our description and should be the object of further consideration.

Acknowledgements This research has been supported by MINECO grants No. MTM2014-56948-C2 and MTM2017-84446-C2-1-R, and project C-ICT/3285 of the UE FP7.

References

1. N. Autrusson, L. Guglielmini, S. Lecuyer, R. Rusconi, H.A. Stone, The shape of an elastic filament in a two-dimensional corner flow. Phys. Fluids **23**, 063602 (2011)
2. M. Asally, M. Kittisopikul, P. Rué, Y. Du, Z. Hu, T. Çağatay, A.B. Robinson, H. Lu, J. Garcia-Ojalvo, G.M. Süel, Localized cell death focuses mechanical forces during 3D patterning in a biofilm. Proc. Nat. Acad. Sci. USA **109**, 18891–18896 (2012)
3. M. Bergou, M. Wardetzky, S. Robinson, B. Audoly, E. Grinspun, Discrete elastic rods. ACM Trans. Gr. **27**, 63 (2008)
4. B. Birnir, A. Carpio, E. Cebrián, P. Vidal, Dynamic energy budget approach to evaluate antibiotic effects on biofilms. Commun. Nonlinear Sci. Numer. Simulat. **54**, 70–83 (2018)
5. L. Chai, H. Vlamakis, R. Kolter, Extracellular signal regulation of cell differentiation in biofilms. MRS Bull. **36**, 374–379 (2011)
6. R.G. Cox, The motion of long slender bodies in a viscous fluid. Part 1. General theory. J. Fluid Mech. **44**, 791–810 (1970)
7. N. Chouaieb, A. Goriely, J.H. Maddocks, Helices. Proc. Nat. Acad. Sci. USA **103**, 9398–9406 (2006)
8. J. Dervaux, M.B. Amar, Morphogenesis of growing soft tissues. Phys. Rev. Lett. **101**, 068101 (2008)
9. K. Drescher, Y. Shen, B.L. Bassler, H.A. Stone, Biofilm streamers cause catastrophic disruption of flow with consequences for environmental and medical systems. Proc. Nat. Acad. Sci. USA **110**, 4345–4350 (2013)
10. D.R. Espeso, A. Carpio, B. Einarsson, Differential growth of wrinkled biofilms. Phys. Rev. E **91**, 022710 (2015)
11. D.R. Espeso, A. Carpio, E. Martinez-Garcia, V. de Lorenzo, Stenosis triggers spread of helical Pseudomonas biofilms in cylindrical flow systems. Sci. Rep. **6**, 27170 (2016)
12. R. Goldenthal, D. Harmon, R. Fattal, M. Bercovier, E. Grispun, Efficient simulation of inextensible cloth. ACM Trans. Gr. **26**, 49 (2007)

13. S.W. Hermanovicz, A simple 2D biofilm model yields a variety of morphological features. Math. Biosci. **69**, 1–14 (2001)
14. R. Huang, S.H. Im, Dynamics of wrinkle growth and coarsening in stressed thin films. Phys. Rev. E **74**, 026214 (2006)
15. J. Langer, D.A. Singer, Lagrangian aspects of the Kirchhoff elastic rod. SIAM Rev. **38**, 605–618 (1996)
16. T. Mora, A. Boudaoud, Buckling of swelling gels. Eur. Phys. J. E Soft Matter **20**, 119–124 (2006)
17. Y. Ni, D. Yang, L. He, Spontaneous wrinkle branching by gradient stiffness. Phys. Rev. E **86**, 031604 (2012)
18. A. Seminara, T.E. Angelini, J.N. Wilking, H. Vlamakis, S. Ebrahim, R. Kolter, D.A. Weitz, M.P. Brenner, Osmotic spreading of Bacillus subtilis biofilms driven by an extracellular matrix. Proc. Natl. Acad. Sci. USA **109**, 1116–1121 (2012)
19. A. Torres, A. Marquina, J.A. Font, J.M. Ibañez, Total variation-based methods for gravitational wave denoising. Phys. Rev. D **90**, 084029 (2014)
20. J.N. Wilking, T.E. Angelini, A. Seminara, M.P. Brenner, D.A. Weitz, Biofilms as complex fluids. MRS Bull. **36**, 385–391 (2011)

Modelling the Unfolding Pathway of Biomolecules: Theoretical Approach and Experimental Prospect

Carlos A. Plata and Antonio Prados

Abstract We analyse the unfolding pathway of biomolecules comprising several independent modules in pulling experiments. In a recently proposed model, a critical velocity v_c has been predicted, such that for pulling speeds $v > v_c$ it is the module at the pulled end that opens first, whereas for $v < v_c$ it is the weakest. Here, we introduce a variant of the model that is closer to the experimental setup, and discuss the robustness of the emergence of the critical velocity and its dependence on the model parameters. We also propose a possible experiment to test the theoretical predictions of the model, which seems feasible with state-of-art molecular engineering techniques.

Keywords Single-molecule experiments · Free energy · Unfolding pathway
Critical velocity · Force-extension curve · Modular proteins

1 Introduction

The development of the so-called single-molecule experiments in the last decades has made it possible to carry out research at the molecular level. Biophysics is, undoubtedly, one of the fields where these techniques have had a bigger impact, triggering a whole new area of investigation on the elastic properties of biomolecules. Recent accounts of the current development of this enticing field can be found in Refs. [1–4].

Atomic force microscopy (AFM) stands out because of its extensive use. In particular, the role played by AFM in the study of modular proteins is crucial [5–7]. Figure 1 shows a sketch of the experimental setup in a pulling experiment of a

C. A. Plata · A. Prados (✉)
Física Teórica, Universidad de Sevilla, Apartado de Correos 1065, 41080 Seville, EU, Spain
e-mail: cplata1@us.es

A. Prados
e-mail: prados@us.es

L. L. Bonilla et al. (eds.), *Coupled Mathematical Models for Physical and Biological Nanoscale Systems and Their Applications*, Springer Proceedings in Mathematics & Statistics 232, https://doi.org/10.1007/978-3-319-76599-0_8

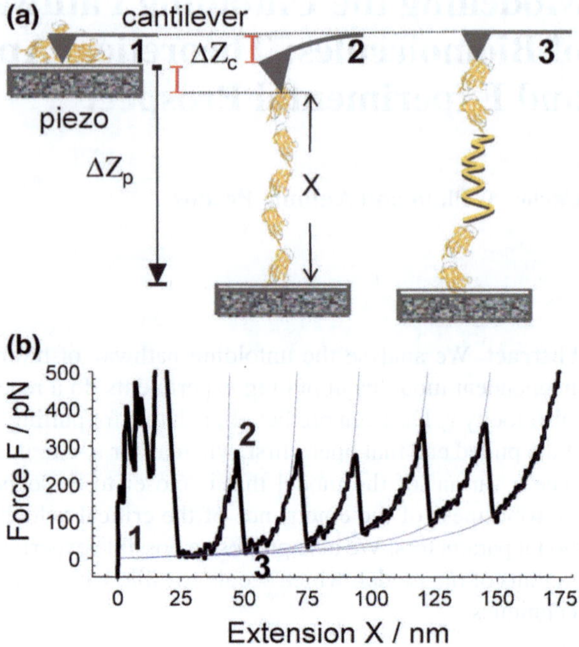

Fig. 1 **a** Sketch of the experimental setup in an AFM experiment with a modular protein. The position of the platform is shifted ΔZ_p from 1 to 2, producing an elongation of X over the molecule and bending the cantilever a magnitude ΔZ_c. From 2 to 3 the force is almost relaxed because of the unravelling of one of the modules. **b** Force-extension curve in a typical lenght-control AFM experiment with a polyprotein. Each rip in the force accounts for the unfolding of a module. Taken from [3]

molecule comprising several modules. The biomolecule is stretched between the platform and the tip of the cantilever. The spring constant of the cantilever is k_c, which is usually in the range of 100 pN/nm. Here, we consider the simplest situation, in which the total length of the system $\Delta Z_p = \Delta Z_c + X$ is controlled. The stretching of the molecule makes the cantilever bend by ΔZ_c, and then the force can be recorded as $F = k_c \Delta Z_c$.

The outcome of the above described experiment is a force-extension curve, similar to panel (b) in Fig. 1. This force-extension curve provides a fingerprint of the elastomechanical properties of the molecule under study. When molecules composed of several structural units, such as modular polyproteins, are pulled, a sawtooth pattern comes about in the force-extension curve [5–7]. At certain values of the length, there are almost vertical "force rips": each force rip marks the unfolding of one module. Interestingly, these force rips already appear when the molecule is quasi-statically pulled, a limit that can be explained by means of an equilibrium statistical mechanics description [8]. When the molecule is pulled at a finite rate, the appearance of these force rips can still be explained by the system partially sweeping the metastable region of the equilibrium branches [9].

The unfolding pathway is, roughly, the order and the way in which the structural units of the system unravel. It has been recently found out that the unfolding pathway depends on the pulling velocity and the pulling direction [7, 10–12]. Particularly, in [11], different unfolding pathways are observed in SMD simulations on the Maltose Binding Protein. The authors reported that for low pulling speeds the first unit to unfold is the least stable, whereas for high pulling speed it is the closest to the pulled end, regardless of their relative stability.

Very recently, a toy-like model has been proposed to qualitatively understand the above experimental framework [13]. Specifically, each module is represented by a nonlinear spring, characterised by a bistable free energy that depends on the module extension. Therein, the two basins represent its folded and unfolded states. The spatial structure of the system is retained in its simplest way: each module extends from the end point of the previous one to its own endpoint (which coincides with the start point of the next). Moreover, each module endpoint obeys an overdamped Langevin equation with forces stemming from the bistable free energies and white noise forces with amplitudes verifying the fluctuation-dissipation theorem.

In the above model, the unfolding pathway was found to depend on the pulling velocity. In the simplest non-trivial case, there is only one module that is different from the rest, which is also the furthest from the pulled end. In this situation, only one critical velocity v_c shows up: for pulling velocities $v_p < v_c$, it is the weakest module that opens first but for $v_p > v_c$ it is the module at the pulled end. In addition, analytical results were derived for this critical velocity by introducing some approximations: mainly two, (i) perfect length control and (ii) the deterministic approximation, that is, our neglecting of the stochastic forces. This was done by means of a perturbative solution of the deterministic equations in both the pulling velocity and an asymmetry parameter, which measures how different the potentials of the modules are.

The main aim of this work is twofold. First, we would like to refine the above theoretical framework, making it closer to the experimental setup in AFM. In particular, we would like to look into the effect of a more realistic modelling of the length-control device. Instead of considering perfect length-control, we consider a device with a finite value of the stiffness, both at the end of the one-dimensional chain (as originally depicted in Ref. [13]) and at the start point thereof, which is where it is usually situated in the AFM experiments, see Fig. 1. Second, we would like to discuss how our theory could be checked in a real experiment with modular proteins.

This chapter is structured as follows. First, we introduce the original model and discuss its most relevant results in Sect. 2. In Sect. 3, we study the role played by the location of the restoring spring and the finite value of its spring constant. We provide some details about the free energy modelling employed for each of the modules in Sect. 4. Section 5 is devoted to discuss a possible AFM experiment in order to test our theory. Finally, we wrap up the main conclusions which emerge from our work.

2 Revision of the Model and Previous Results

Here, we briefly review the model that was originally put forward in Ref. [13]. We consider a polyprotein comprising N modules. When the molecule is stretched, the simplest description is to portray it as a one-dimensional chain. We define the coordinates q_i, $(i = 0, \dots, N)$ in such a way that the i-th unit extends from q_{i-1} to q_i, the

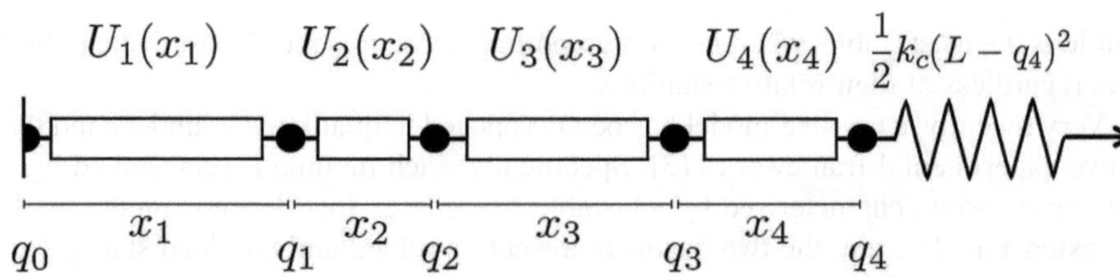

Fig. 2 Sketch of the model for a molecule comprising four units. Therein, the units are denoted by rectangles and have potentials $U_i(x_i)$, with x_i being the extension of the i-th unit's. The unit endpoints q_i are represented by the beads, and the extensions are thus $x_i = q_i - q_{i-1}$ (by definition, $q_0 = 0$). The spring stands for the length-controlling device attached to the pulled end q_4, whose contribution to the system free energy is assumed to be harmonic with stiffness k_c

extension of the i-th unit is $x_i = q_i - q_{i-1}$. Moreover, as shown in Fig. 2, the pulling device is assumed to be connected to the right (pulled) end of the chain.

We assume Langevin dynamics for the q_i coordinates ($q_0 = 0$),

$$\gamma \dot{q}_i = -\frac{\partial}{\partial q_i} A(q_0, \ldots, q_N) + \zeta_i \qquad i > 0, \tag{1}$$

in which the ζ_i are Gaussian white noise forces, such that

$$\langle \zeta_i(t) \rangle = 0, \qquad \langle \zeta_i(t) \zeta_j(t') \rangle = 2\gamma k_B T \delta_{ij} \delta(t - t'), \tag{2}$$

with k_B being the Boltzmann constant, and γ and T being the friction coefficient (assumed to be common for all the units) and the temperature of the fluid in which the protein is immersed, respectively. The global free energy function of the system is[1]

$$A(q_0, \ldots, q_N) = \sum_{i=1}^{N} U_i(q_i - q_{i-1}) + \frac{1}{2} k_c (L - q_N)^2. \tag{3}$$

In the previous equation, we have considered an elastic term due to the finite stiffness k_c of the controlling device, which is located at the pulling end as shown in Fig. 2. Finally, L stands for the desired length program and $U_i(x_i)$ is the single unit contribution, which only depends on the extension, to A. Consequently, the force exerted over the biomolecule is $k_c(L - q_N)$. We consider length-controlled experiments at constant pulling velocity, that is, $\dot{L} = v_p$.

When $k_c \to \infty$, the control over the length is perfect and $q_N \to L$ in such a way that $k_c(L - q_N) \to F$, being F a Lagrange multiplier. That is, the perfect length-controlled situation is the same that a force-controlled one but with F the force needed to main-

[1]In Ref. [13], this free energy was denoted by G. Here, we have preferred to employ A because the relevant potential in length-controlled situations is the Hemlholtz-like free energy, and G is usually the notation reserved for the Gibbs-like potential $G = A - FL$, which is the relevant one in force-controlled experiments [8, 9].

tain a total length equal to L. Note that there is no contribution to the free energy coming from the pulling device in this limit, since $k_c(L-q_N)^2/2 = F^2/(2k_c) \to 0$.

The approach in Ref. [13] tries to keep things as simple as possible. Then, the evolution equations for the extensions are written by assuming (i) perfect length control and (ii) the deterministic approximation, obtained by neglecting the noise terms. Note that the evolution equations in the latter approximation are sometimes called the macroscopic equations [14], which are

$$\gamma \dot{x}_1 = -U_1'(x_1) + U_2'(x_2), \tag{4a}$$
$$\gamma \dot{x}_i = -2U_i'(x_i) + U_{i+1}'(x_{i+1}) + U_{i-1}'(x_{i-1}), \quad 1 < i < N, \tag{4b}$$
$$\gamma \dot{x}_N = -2U_N'(x_N) + U_{N-1}'(x_{N-1}) + F, \tag{4c}$$
$$F = \gamma v_p + U_N'(x_N). \tag{4d}$$

So far, nothing has been said about the shape of the single-unit contributions U_i to the free energy. In order to maintain a general approach, we only request the functions $U_i(x) - Fx$ to become a double well for some interval of forces. Each well stands for the folded and the unfolded basins of each module. Now, we separate these functions in a main part common to all units and a separation from this main part weighted by an asymmetry parameter ξ,

$$U_i(x) = U(x) + \xi \delta U_i(x), \quad U_i'(x) = U'(x) + \xi \delta f_i(x). \tag{5}$$

We have done the same separation in the derivative, by defining $\delta f_i(x) = \delta U_i'(x)$.

It is possible to solve the system (4) by means of a perturbative expansion in the pulling velocity v_p and the asymmetry ξ. Indeed, if we retain only linear order terms in v_p and ξ, the corrections due to finite pulling rate and asymmetry are not coupled. This perturbative solution, when the system starts from an initial condition in which all the units are folded, is shown to be [13]

$$x_i = \ell + \frac{\xi \overline{\delta f}(\ell) - v_p \gamma \dfrac{N^2-1}{6N}}{U''(\ell)} + \frac{v_p \gamma \dfrac{i(i-1)}{2N} - \xi \delta f_i(\ell)}{U''(\ell)}, \tag{6}$$

where $\ell = L/N$ stands for the specific length per module and the over-bar means average over the units. Note that, if all the free energies are equal ($\xi = 0$) and we are not pulling ($v_p = 0$), the total length will be reasonably equally distributed among all the units. Moreover, it is worth emphasising that this solution is approximate, it diverges when $U''(\ell) \to 0$. This shows that this perturbative solution breaks down when the average length per module ℓ reaches the *stability threshold* ℓ_b, such that $U''(\ell_b) = 0$.

We are interested in a criterion that allows us to discern which unit is the first to unfold and we hope that our perturbative solution is good enough in this regard. Since the folded state ceases to exist when x reaches ℓ_b, it is reasonable to assume that the first module to unfold is precisely the one for which $x_i = \ell_b$ is attained for the shortest time. In Eq. (6), we can see that the finite pulling term favours the unfolding of modules that are nearer to the pulled end, whereas the asymmetry term favours the unfolding of the weaker units (those with the lowest values of δf_i).

We can compute the pulling velocities $v^i(j)$ for which each couple of modules (i,j), $j > i$, reach simultaneously the stability threshold. They are determined by the condition

$$x_i(\ell_c) = x_j(\ell_c) = \ell_b, \tag{7}$$

which gives both the value of ℓ_c (or time t_c) at which the stability threshold is reached and the relationship between v_p and ξ. Then, in a specific system with known δf_i's we can predict what are the critical velocities that separate regions inside which the first unit to unfold is different. In Ref. [13], some examples of the use of this theory are provided, which show a good agreement with simulations of the Langevin dynamics (1).

The simplest configuration in which a critical velocity arises is the following. Let us consider a chain of N units, all of them with the same contribution to the free energy, except the first one (the furthest from the pulled end). Therefore, $\delta f_i(x) = 0$, $i \neq 1$, $\delta f_1(x) \neq 0$. Moreover we will assume that $\delta f_1(\ell_b) < 0$, that is, the first module is weaker than the rest. For such a configuration, there appears only one critical velocity, which is given by [13]

$$\frac{\gamma v_c}{\xi} = -\frac{\delta f_1(\ell_b)}{N-1}. \tag{8}$$

For $v_p < v_c$, the first module to unfold is the weakest one, whereas for $v_p > v_c$ the unfolding starts from the pulled end. In Ref. [13], a more general situation is investigated but, here, we restrict ourselves to this configuration.

3 Relevance of the Stiffness

In a real AFM experiment, the stiffness is finite and, as a result, the control over the length is not perfect. Furthermore, the position that is externally controlled is, usually, that of the platform and the main elastic force stems from the bending of the tip of the cantilever, as depicted in Fig. 1. Thus, it seems more reasonable to model the pulling of the biomolecule in the way sketched in Fig. 3.

Some authors [15] have used other elastic reactions that reflect the attachment by means of flexible linkers among the platform and the pulled end q_N, and between consecutive modules. Here, we will consider a perfect absorption, in order to keep the model as simple as possible. In the next two subsections, we study the effect of the finite value of the stiffness k_c and the location of the spring, respectively, on the unfolding pathway.

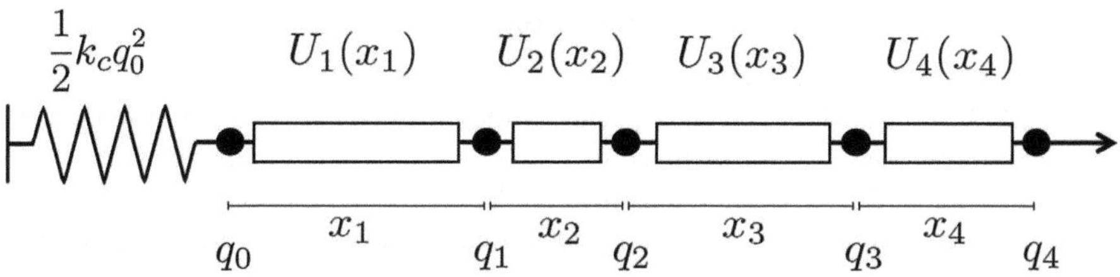

Fig. 3 Sketch of the model for a protein with four units. It is identical to Fig. 2, except for the position of the length-controlling device, which is now located at the fixed end

3.1 *Finite Stiffness*

Here, we still consider the model depicted in Fig. 2, that is, the spring is situated at the end of the chain, but with a finite value of the stiffness k_c. Also, we consider the macroscopic equations (zero noise), which are

$$\gamma \dot{x}_1 = -U_1'(x_1) + U_2'(x_2), \tag{9a}$$

$$\gamma \dot{x}_i = -2U_i'(x_i) + U_{i+1}'(x_{i+1}) + U_{i-1}'(x_{i-1}), \quad 1 < i < N, \tag{9b}$$

$$\gamma \dot{x}_N = -2U_N'(x_N) + U_{N-1}'(x_{N-1}) + k_c\left(L - \sum_{k=1}^{N} x_k\right). \tag{9c}$$

This system differs from that in Eq. (4) because in the last equation the Lagrange multiplier F is substituted by the harmonic force $k_c(L - \sum_{k=1}^{N} x_k)$. As in the previous case, this system is analytically solvable by means of a perturbative expansion in v_p and ξ. The approximate solution for the extension x_i is

$$x_i = \ell + \frac{\xi N k_c \overline{\delta f}(\ell) - v_p \gamma k_c \dfrac{[3U''(\ell) + k_c(N-1)]N(N+1)}{6[Nk_c + U''(\ell)]}}{U''(\ell)[Nk_c + U''(\ell)]} + \frac{v_p \gamma k_c i(i-1) - 2\xi[Nk_c + U''(\ell)]\delta f_i(\ell)}{2U''(\ell)[Nk_c + U''(\ell)]}. \tag{10}$$

Here $\ell \neq L/N$, it stems from the relation

$$U'(\ell) = k_c(L - N\ell). \tag{11}$$

We can see easily how we reobtain (6) taking the limit $k_c \to \infty$ in (10), as it should be. Although the solution is slightly different, it still breaks down when $U''(\ell)$ vanishes, that is, when $\ell \to \ell_b$. Therefore, to the lowest order, again we have to seek a solution of (7), with the extensions given by (10), and substitute $\ell_c \simeq \ell_b$ therein. This leads to the same critical velocities found for the infinite stiffness limit.

3.2 Location of the Elastic Reaction

As depicted in Fig. 1, in an AFM experiment the distance between the moving platform and the fixed cantilever is the controlled quantity. Then, the model sketched in Fig. 3 is closer to the experimental setup: the left end corresponds to the fixed cantilever, with q_0 standing for ΔZ_c, and the right end represents the moving platform. Thus, the free energy of this setup is given by

$$A(q_0, \dots, q_N) = \sum_{i=1}^{N} U_i(q_i - q_{i-1}) + \frac{1}{2} k_c q_0^2 . \tag{12}$$

From the free energy (12), we derive the Langevin equations by making use of Eq. (1). The macroscopic equations (zero noise) read

$$\gamma \dot{x}_1 = -2U_1'(x_1) + U_2'(x_2) + k_c \left(L - \sum_{k=1}^{N} x_k \right), \tag{13a}$$

$$\gamma \dot{x}_i = -2U_i'(x_i) + U_{i+1}'(x_{i+1}) + U_{i-1}'(x_{i-1}), \quad 1 < i < N, \tag{13b}$$

$$\gamma \dot{x}_N = -U_N'(x_N) + U_{N-1}'(x_{N-1}) + v_p. \tag{13c}$$

In the infinite stiffness limit, $k_c \to \infty$, the harmonic contribution tends to a new Lagrange multiplier F such that $\sum x_i = L$. Therefrom, it is obtained that $F = U_1'(x_1)$ and the resulting system is exactly equal to that in Eq. (4). This is logical: if the spring is totally stiff and then the control over the length is perfect, the two models are identical.[2]

The system (13) can be solved in an analogous way, by means of a perturbative expansion in the asymmetry ξ and the pulling velocity v_p. The result is

$$x_i = \ell + \frac{\xi N k_c \overline{\delta f}(\ell) - v_p \gamma k_c \dfrac{[3U''(\ell) + k_c(N-1)]N(N+1)}{6[Nk_c + U''(\ell)]}}{U''(\ell)[Nk_c + U''(\ell)]} + \frac{v_p \gamma k_c i \left(i - 1 + \dfrac{2U''(\ell)}{k_c} \right) - 2\xi[Nk_c + U''(\ell)]\delta f_i(\ell)}{2U''(\ell)[Nk_c + U''(\ell)]}. \tag{14}$$

Again, we can reobtain (6) taking the infinite stiffness limit in (14). Although the final solution for the extension is different from the previous one, when we look for

[2]It is worth emphasising that the two variants of the model, with the spring at either the fixed or moving end, have the same number of degrees of freedom. In Fig. 2, $q_0 = 0$ and our degrees of freedom are q_i, $i = 1, \dots, N$, whereas in Fig. 3 we have the dynamical constraint $q_N = L$ and the degrees of freedom are q_i, $i = 0, \dots, N-1$. In the limit as $k_c \to \infty$, we have both constraints, $q_0 = 0$ and $q_N = L$, in both models, making it obvious that they are identical.

the critical velocities and make the approximation $\ell_c \simeq \ell_b$ we get the same analytical results for them.

Our main conclusion is that the existence of a set of critical velocities, setting apart regions where the first unit to unfold is different, is not an artificial effect of the limit $k_c \to \infty$. Indeed, at the lowest order, all the versions of the studied model, independently of the value of the stiffness and the location of the spring, give the same critical velocities. This robustness is an appealing feature of the theory, and makes it reasonable to seek this phenomenology in real experiments.

4 Shape of the Bistable Potentials

Different shapes for the double-well potentials have been considered in the literature, mainly simple Landau-like quartic potentials to understand the basic mechanisms underlying the observed behaviours [8, 9, 11] and more complex potentials when trying to obtain a more detailed, closer to quantitative, description of the experiments [9, 16–18]. For the sake of concreteness, we restrict ourselves to the proposal made by Berkovich et al. [16, 17]. Therein, the free energy of a module is represented by the sum of a Morse potential, which mimics the enthalpic minimum of the folded state, and a worm-like-chain (WLC) term [19], which accounts for the entropic contribution to the elasticity of the unfolded state. Specifically, the free energy is written as

$$U(x) = U_0 \left[\left(1 - e^{-2b\frac{x-R_c}{R_c}} \right)^2 - 1 \right] + \frac{k_B T}{4P} L_c \left(\frac{1}{1-\frac{x}{L_c}} - 1 - \frac{x}{L_c} + \frac{2x^2}{L_c^2} \right). \quad (15)$$

This shape has shown to be useful for some pulling experiments with actual proteins as titin I27 or ubiquitin [16, 17]. Therein, each parameter has a physical interpretation. First, in the WLC part, we have: (i) the contour length L_c, which is the maximum length for a totally extended protein, and (ii) the persistence length P, which measures the characteristic length over which the chain is flexible. Both of them, L_c and P, can be measured in terms of the number of amino acids in the chain. Second, for the Morse contribution, we have: (iii) R_c, which gives the location of the enthalpic minimum and (iv) U_0 and b, which measure the depth and the width (in a non-trivial form) of the folded basin. The stability threshold ℓ_b cannot be provided as an explicit function of the parameters in Berkovich et al.'s potential. However, we can always estimate it numerically, solving $U''(\ell_b) = 0$ for a specific set of parameters.

As we anticipated in Sect. 2, here we will focus in a very specific configuration where only the first module is different from the rest. Consistently, we use $U(x)$ to represent the free energy of each of the identical modules, and $U_1(x)$ for that of the first one. In particular, we consider that the first unit has a slightly different contour length, $L_c + \Delta$. Therefore, we can linearise $U_1(x)$ around $U(x)$, using the natural asymmetry parameter $\xi = \Delta/L_c \ll 1$. Therefore,

$$U_1'(x; L_c + \Delta) \simeq U'(x; L_c) + \xi \delta f_1(x; L_c), \tag{16}$$

where

$$\delta f_1(x; L_c) \equiv L_c \frac{\partial U'(x; L_c)}{\partial L_c} = -\frac{k_B T}{2P} \left[\frac{\frac{x}{L_c}}{\left(1 - \frac{x}{L_c}\right)^3} + \frac{2x}{L_c} \right]. \tag{17}$$

This linearisation is useful for the direct application of our theory to some engineered systems, see the next section.

5 Experimental Prospect

In the experiments, the observation of the unfolding pathway is not trivial at all. The typical outcome of AFM experiments is a force-extension curve (FEC) in which the identification of the unfolding events is, in principle, not possible when the modules are identical. Thus, in order to test our theory, molecular engineering techniques that manipulate proteins adding some extra structures, such as coiled-coil [20] or Glycine [21] probes, come in handy. For instance, a polyprotein in which all the modules except one have the same contour length may be constructed in this way. A reasonable model for this situation is a chain with modules described by Berkovich et al. potentials with the same parameters for all the modules, with the exception of the contour length of one of them. According to our theory, a critical velocity emerges (8) and it may be observed because the unfolding of the unit that is different can be easily identified in the FEC, see below.

Let us consider an example of a possible real experiment for a polyprotein with $N = 10$ modules. We characterise the modules by Berkovich et al. potentials with parameters,

$$P = 0.4 \text{ nm}, \qquad L_c = 30 \text{ nm}, \qquad R_c = 4 \text{ nm}, \tag{18a}$$

$$b = 2, \qquad U_0 = 100 \text{ pN nm}, \qquad k_B T = 4.2 \text{ pN nm}, \tag{18b}$$

and the friction coefficient $\gamma = 0.0028 \text{ pN nm}^{-1}\text{s}$ [16]. We call this system M_{10}: since all the modules are equal in M_{10}, it is not a very interesting system from the point of view of our theory. Nevertheless, we can consider a mutant species M_{10}' that is identical to M_{10} except for the module located in the first position (the fixed end), which has an insertion adding Δ to its contour length. Our theory gives an estimate for the critical velocity v_c by using (8).

In Fig. 4, we compare the theoretical estimate for the critical velocity with the actual critical velocity obtained by integration of the dynamical system (13). Specifically, we have considered a system with spring constant $k_c = 100$ pN/nm. The numer-

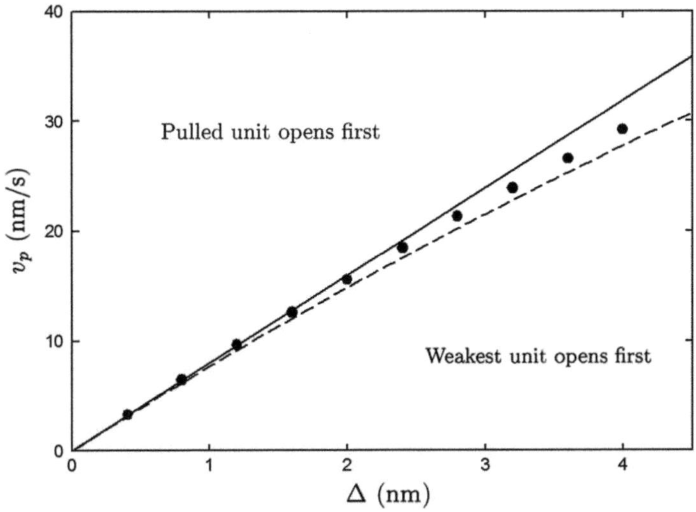

Fig. 4 Critical velocities for M'_{10} systems. The parameter Δ stands for the additional contour length of the first module. Numerical values (circles) are compared with two theoretical estimates: "complete" (dashed line) and linear (solid line)

ical strategy to determine v_c has been the following: starting from a completely folded state we let the system evolve obeying (13), with a "high" value of v_p (well above the critical velocity), up to the first unfolding. We tune v_p down until it is observed that the first module that unfolds is the weakest one: this marks the actual critical velocity. There are two theoretical lines: the solid line stems from the rigorous application of (8), with δf_1 given by (17), and v_c is a linear function of ξ, whereas the dashed line corresponds to the substitution in (8) of $\xi \delta f_1(x)$ by $U_1'(x; L_c + \Delta) - U'(x; L_c)$, without linearising in the asymmetry ξ. Note the good agreement between theory and numerics, specially in the "complete" theory where, for the range of plotted values, the relative error never exceeds 5%. Interestingly, the computed values of the critical velocity lie in the range of typical AFM pulling speeds, from 10 to 10^4 nm/s [22].

Below v_c, it is always the weakest unit that unfolds first. Above v_c, the unit that unfolds first is the pulled one. For the sake of concreteness, from now we consider an specific molecule M'_{10} fixing $\Delta = 2$ nm. Using the linear estimation (17) in (8), we get a critical velocity $v_c \simeq 16$ nm/s that, as stated above, is in the range of the typical pulling speeds in AFM experiments.

In Fig. 5, we plot the extension of each unit vs the total extension $q_N - q_0$ in our notation (X in Fig. 1), which is a good reaction coordinate [23]. We have numerically integrated Eq. (13) for two values of v_p: one below and one above v_c, namely $v_p = 10$ nm/s and $v_p = 22$ nm/s. The red trace stands for the weakest unit extension whereas the blue one corresponds to the pulled module. We can see that, for $v_p = 10$ nm/s $< v_c$, the first unit that unfolds is the weakest one, whereas for $v_p = 22$ nm/s $> v_c$ that

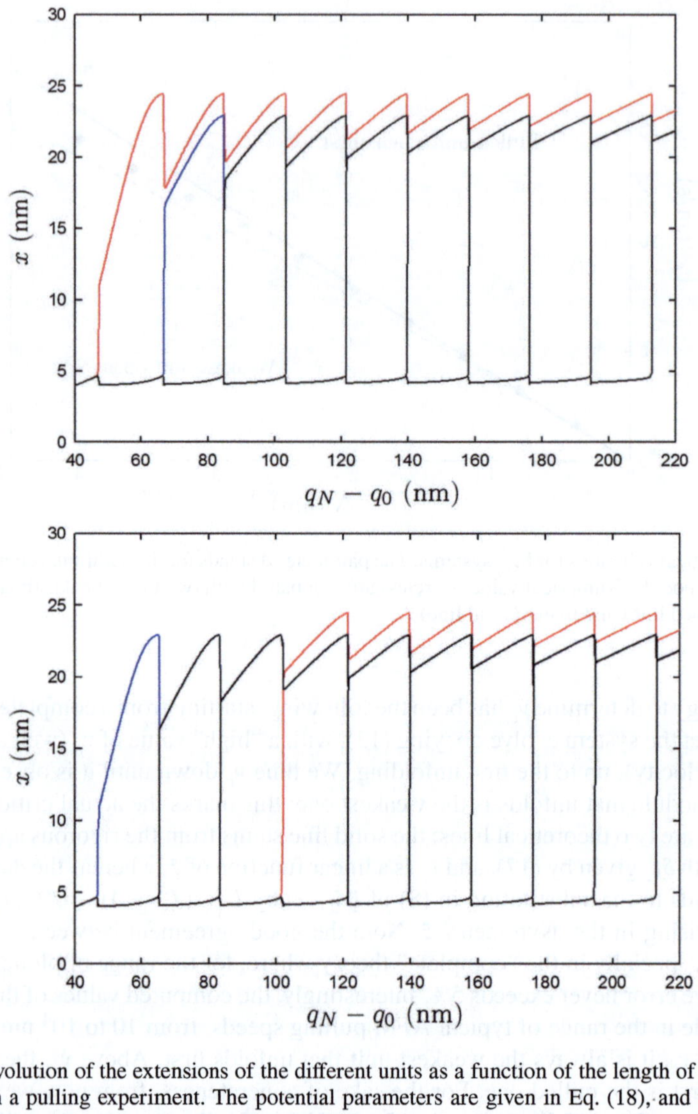

Fig. 5 Evolution of the extensions of the different units as a function of the length of the system $q_N - q_0$ in a pulling experiment. The potential parameters are given in Eq. (18), and the pulling speeds are $v_p = 10$ nm/s $< v_c$ (top) and $v_p = 22$ nm/s $> v_c$ (bottom). The stiffness is $k_c = 100$ pN/nm, which lies in the range of typical AFM values. The red line corresponds to the weakest unit and the blue line to the pulled one

is no longer the case. Specifically, the first unit that unfolds is the pulled one, and the weakest unfolds in the 4-th place.

The plots in Fig. 5 are the most useful in order to detect the unfolding pathway of the polyprotein. Unfortunately, they are not accessible in the real experiments, for which the typical output is the FEC. Thus, we have also plotted the FEC in order to bring to light the expected outcome of a real experiment. In Fig. 6, we show the FEC

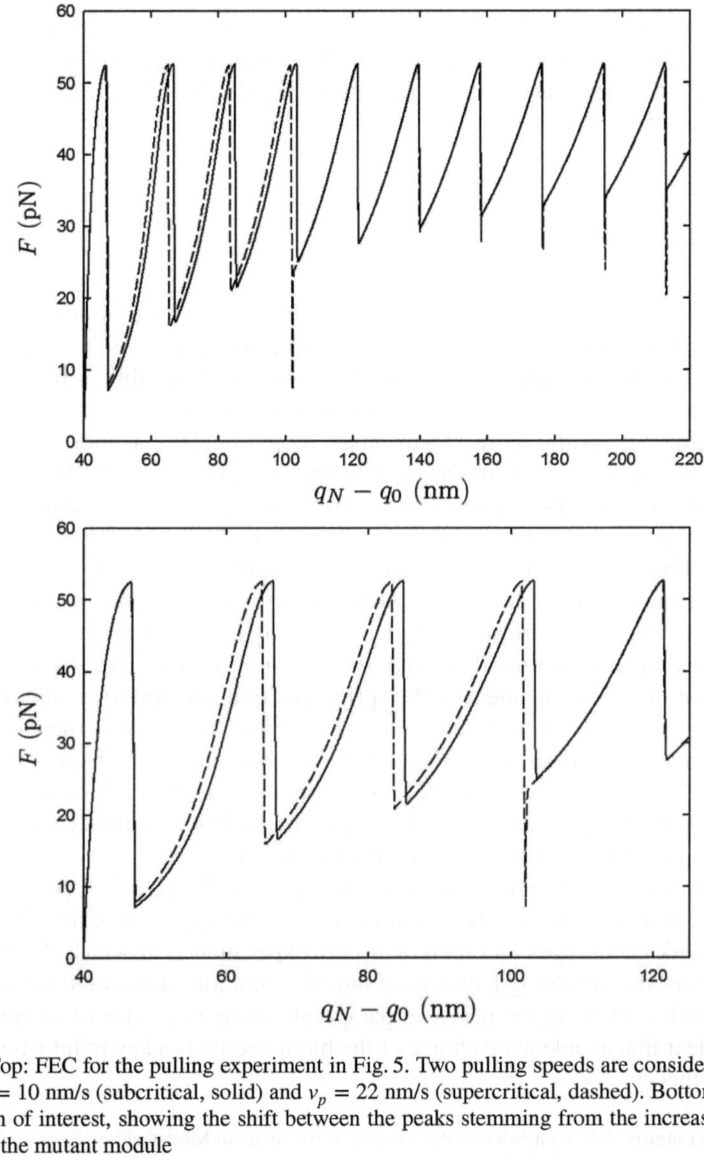

Fig. 6 Top: FEC for the pulling experiment in Fig. 5. Two pulling speeds are considered, specifically $v_p = 10$ nm/s (subcritical, solid) and $v_p = 22$ nm/s (supercritical, dashed). Bottom: zoom of the region of interest, showing the shift between the peaks stemming from the increased contour length of the mutant module

for the two considered velocities in the same graph (solid line for the lower speed and dashed line for the higher one).

The FECS in Fig. 6 are superimposed until the first force rip, which corresponds to the first unfolding event (that of the mutant module for the slower velocity and that of the pulled unit for the faster one). As the mutant unit has a longer contour length than the rest, a shift between the curves in the next three pikes is found, because the effective contour length of the polyprotein has an extra contribution of 2 nm. Reasonably, for the higher velocity, this shift disappears when the mutant module unfolds, and the curves are once again superimposed. This plot clearly shows how the existence of a critical velocity in a real experiment could be sought.

6 Conclusions

We have provided a useful theoretical framework in the context of modular proteins or, in general, of biomolecules comprising several units that unfold (almost) independently. Therein, according to our theory, it should be possible to find the emergence of a set of critical velocities which separate regions where the first module that unfolds is different. Although we focus on the biophysical application of the theory, it is worth highlighting that similar models are used in other fields. Many physical systems are also "modular", since they comprise several units [24], and thus a similar phenomenology may emerge. Some examples can be found in studies of plasticity [25, 26], lithium-ion batteries [27, 28] or ferromagnetic alloys [29].

The development of our theory has shown that the position and value of the elastic constant k_c of the length-controlled device is roughly irrelevant for the existence and value of the critical velocities. The derived expressions for the critical velocities are, to the lowest order, independent of the spring position and stiffness. Notwithstanding, our theory completely neglects the noise contributions and thus the units unfold when they reach their limit of stability, that is, at the force for which the folded basin disappears. This is expected to be relevant for biomolecules that follow the *maximum hysteresis path*, using the same terminology as in [9, 24], completely sweeping the metastable part of the intermediate branches of the FEC.

The numerical integration of the evolution equations for a realistic potential point out that our proposal for an experiment is, in principle, completely feasible. Therefore, our work encourages and motivates new experiments, in which the predicted features about the unfolding pathway of modular biomolecules could be observed. Finally, the discussion in the previous paragraph on the relevance of thermal noise makes it clear that an adequate choice of the biomolecule is a key point when trying to test our theory.

Acknowledgements We acknowledge the support of the Spanish Ministerio de Economía y Competitividad through Grant FIS2014-53808-P. Carlos A. Plata also acknowledges the support from the FPU Fellowship Programme of the Spanish Ministerio de Educación, Cultura y Deporte through Grant FPU14/00241. We also thank Prof. P. Marszalek's group for sharing with us all their knowledge and giving the opportunity to start exploring AFM experiments during a research stay of Carlos A. Plata at Duke University in summer 2016, funded by the Spanish FPU programme. Finally, we thank the book editors for their organisation of the meeting at the excellent BIRS facilities in Banff, Canada, in which a preliminary version of this work was presented in August 2016.

References

1. F. Ritort, J. Phys, Condens. Matter **18**, R531 (2006)
2. D. Kumar, M.S. Li, Phys. Rep. **486**, 1 (2010)
3. P.E. Marszalek, Y.F. Dufrêne, Chem. Soc. Rev. **41**, 3523 (2012)
4. T. Hoffmann, L. Dougan, Chem. Soc. Rev. **41**, 4781 (2012)
5. M. Carrión-Vázquez, A.F. Oberhauser, S.B. Fowler, P.E. Marszalek, S.E. Broedel, J. Clarke, J.M. Fernandez, Proc. Natl. Acad. Sci. USA **96**, 3694 (1999)
6. T.E. Fisher, P.E. Marszalek, J.M. Fernandez, Nat. Struct. Biol. **7**, 719 (2000)
7. C. Hyeon, R.I. Dima, D. Thirumalai, Structure **14**, 1633 (2006)
8. A. Prados, A. Carpio, L.L. Bonilla, Phys. Rev. E **88**, 012704 (2013)
9. L.L. Bonilla, A. Carpio, A. Prados, Phys. Rev. E **91**, 052712 (2015)
10. M.S. Li, S. Kouza, J. Chem. Phys. **130**, 145102 (2009)
11. C. Guardiani, D. Di Marino, A. Tramontano, M. Chinappi, F. Cecconi, J. Chem. Theory Comput. **10**, 3589 (2014)
12. M. Kouza, C.K. Hu, M.S. Li, A. Kolinski, J. Chem. Phys. **139**, 065103 (2013)
13. C.A. Plata, F. Cecconi, M. Chinappi, A. Prados, J. Stat. Mech. P08003 (2015)
14. N.G. van Kampen, *Stochastic Processes in Physics and Chemistry* (North-Holland, Amsterdam, 1992)
15. G. Hummer, A. Szabo, Biophys. J. **85**, 5 (2003)
16. R. Berkovich, S. Garcia-Manyes, M. Urbakh, J. Klafter, J.M. Fernandez, Biophys. J. **98**, 2692 (2010)
17. R. Berkovich, R.I. Hermans, I. Popa, G. Stirnemann, S. Garcia-Manyes, B.J. Bernes, J.M. Fernandez, Proc. Nat. Acad. Sci. **109**, 14416 (2012)
18. L.L. Bonilla, A. Carpio, A. Prados, EPL **108**, 28002 (2014)
19. J.F. Marko, E.D. Siggia, Macromolecules **28**, 8759 (1995)
20. Q. Li, Z.N. Scholl, P.E. Marszalek, Angew. Chem. Int. Ed. **53**, 13429 (2014)
21. I. Schwaiger, A. Kardinal, M. Schleicher, A.A. Noegel, M. Rief, Nat. Struct. Mol. Biol. **11**, 81 (2003)
22. J.L.R. Arrondo, A. Alonso, *Advanced Techniques in Biophysics* (Springer, 2006)
23. G. Arad-Haase, S.G. Chuartzman, S. Dagan, R. Nevo, M. Kouza, B.K. Mai, H.T. Nguyen, M.S. Li, Z. Reich, Biophys. J. **99**, 238 (2010)
24. I. Benichou, Y. Zhang, O.K. Dudko, S. Givli, J. Mech. Phys. Solids **95**, 44 (2016)
25. I. Mller, P. Villaggio, Arch. Ration. Mech. Anal. **65**, 25 (1977)
26. G. Puglisi, L. Truskinovsky, Contin. Mech. Thermodyn. **14**, 437 (2002)
27. W. Dreyer, J. Jamnik, C. Guhlke, R. Huth, J. Moskon, M. Gaberscek, Nat. Mater. **9**, 448 (2010)
28. W. Dreyer, C. Guhlke, M. Herrmann, Contin. Mech. Thermodyn. **23**, 211 (2011)
29. I. Benichou, E. Faran, D. Shilo, S. Givli, Appl. Phys. Lett. **102**, 011912 (2013)

The Geometry of Most Probable Trajectories in Noise-Driven Dynamical Systems

John C. Neu, Akhil Ghanta and Stephen W. Teitsworth

Abstract This paper presents a heuristic derivation of a geometric minimum action method that can be used to determine most-probable transition paths in noise-driven dynamical systems. Particular attention is focused on systems that violate detailed balance, and the role of the stochastic vorticity tensor is emphasized. The general method is explored through a detailed study of a two-dimensional quadratic shear flow which exhibits bifurcating most-probable transition pathways.

Keywords Bifurcating transition paths · Geometric action · Detailed balance violation

1 Introduction

In the final chapter of Feynman and Hibbs' classic text on path integrals [1], the scattering of a fast particle from nucleii in a slab of material is examined in order to answer the question: what is the most probable path of the particle from emitter to detector? The relative probabilities of noise sequences which produce paths with the required endpoints can be represented as a functional of the paths themselves. The variational characterization of the most probable path is closely analogous to least action paths in mechanics, or geodesic paths in geometry (a special case of the former). Independently, a rigorous theory of most probable paths in stochastic dynamical systems was developed by Wentzell and Freidlin [2, 3], and subsequently elaborated and explored by many others, see e.g. [4]. Additionally, a similar

J. C. Neu
Department of Biomedical Engineering, Duke University, Box 90281, Durham, NC 27708-0281, USA
e-mail: john.neu@duke.edu; neu@berkeley.edu

A. Ghanta · S. W. Teitsworth (✉)
Department of Physics, Duke University, Box 90305, Durham, NC 27708-0305, USA
e-mail: teitso@phy.duke.edu

A. Ghanta
e-mail: akhil.ghanta@duke.edu

L. L. Bonilla et al. (eds.), *Coupled Mathematical Models for Physical and Biological Nanoscale Systems and Their Applications*, Springer Proceedings in Mathematics & Statistics 232, https://doi.org/10.1007/978-3-319-76599-0_9

formalism for treating large noise-induced deviations has been applied more recently to spatially-extended hydrodynamic models [5, 6].

In applications to concrete problems, physicists [7, 8] often employ the Legendre transformation to convert the Lagrangian formulation of most probable paths into an equivalent Hamiltonian formulation. Here we retain the Lagrangian formulation, to see very directly some geometric aspects of most probable paths (see Sect. 2). A geometrical characterization of most probable paths was recently introduced by Heymann and Vanden-Eijnden [9–11], and we recapitulate aspects of their treatment in the language of classical mechanics at the level of a typical physics graduate course. A chief advantage is that there are very clear expressions of some common insights: in the small noise limit, almost all paths with fixed endpoints closely follow the most probable path in an almost deterministic manner. Furthermore, the speed along the most probable path is the deterministic speed. As the noise decreases, the transit time between endpoints remains fixed, but the expected time between successful transits becomes exponentially large.

In Sect. 3, we discuss the geometric meaning of *detailed balance*, a fundamental notion from statistical physics. The traditional meaning of detailed balance, or its breaking, is framed in terms of the probability current in the Fokker-Planck equation: "fence off" a bounded region of state space with the impermeable boundary condition, and allow the probability density to relax to a steady state within that region. For a detailed balance system, the probability current vanishes identically, and there is a potential energy function so that the equilibrium probability density is proportional to a Boltzmann factor with this potential. If detailed balance is broken, the steady probability current is divergence-free, as it must be, but does *not* vanish identically [12, 13]. We review how this traditional characterization of detailed balance restricts the velocity field and noise tensor of the stochastic dynamics, so that the *stochastic vorticity tensor* vanishes. Furthermore, there is a connection to the geometry of most probable paths: in detailed balance systems, the "backward" path associated with an interchange of the starting and destination endpoints is the reversal of the original "forward" path. If detailed balance is broken, this reversibility is broken. In this case, the path from starting to destination point and then back again to the starting point forms a loop, and the sense of rotation around the loop is determined by the vorticity [14, 15].

2 The Geometric Action

We begin by reviewing the statistics of noise driven trajectories, in the spirit that was formulated many years ago by Feynman and Hibbs [1]. The state space is $\boldsymbol{R}^N$, and trajectories are curves $\mathbf{x} = \mathbf{x}(t)$ in state space parametrized by time t. The $\mathbf{x}(t)$ satisfy a Langevin equation

$$\dot{\mathbf{x}} - \mathbf{u}(\mathbf{x}) = \sqrt{\varepsilon}\sigma\mathbf{w}(t). \tag{1}$$

Here, $\mathbf{u}(\mathbf{x})$ is a given flow vector field on state space, the components of $\mathbf{w}(t)$ are independent unit white noises, and σ is a noise tensor, assumed uniform and constant. $\varepsilon > 0$ is a gauge parameter, so we can formalize the small noise limit $\varepsilon > 0$. The Langevin equation (1) with uniform and constant noise tensor σ is physically appropriate when the noise represents external forcing as a function of time, by degrees of freedom not included in $\mathbf{x}$.

The trajectories we consider are solutions of (1) which pass through two given points $\mathbf{a}$ and $\mathbf{b}$ of state space in succession. There is no restriction on the time of flight T from $\mathbf{a}$ to $\mathbf{b}$. Due to autonomy, we can set the origin of time so $\mathbf{x}(0) = \mathbf{a}$. Then $\mathbf{x}(T) = \mathbf{b}$. The essential idea is that the relative probability of such a trajectory from $\mathbf{a}$ to $\mathbf{b}$ is the same as the relative probability of the white noise $\mathbf{w}(t)$ in $0 < t < T$ which produces it. The relative probability of a noise sequence $\mathbf{w}(t)$ in $0 < t < T$ is expressed formally as

$$e^{-\frac{1}{4}\int_0^T |\mathbf{w}(t)|^2 dt}. \tag{2}$$

For the intuition behind (2), model the sample space of white noises in $0 < t < T$ by piecewise constant functions. Any one component of $\mathbf{w}(t)$ is represented by $w(t) \equiv w_j$ in $(j-1)\Delta t < t < j\Delta t$. Here, the time increment is $\Delta t = \frac{T}{N}$ for some positive integer N. There are N time intervals of piecewise constant $w(t)$ in $0 < t < T$. The constant values of $w(t)$ in different time intervals are statistically independent, and the probability density of each w_j is proportional to the Gaussian

$$e^{-\frac{w_j{}^2}{2(\frac{2}{\Delta t})}} = e^{-\frac{1}{4}w_j{}^2\Delta t}, \tag{3}$$

with mean square $\frac{2}{\Delta t}$. Due to the independence of the w_j for different j, the probability density in the $\boldsymbol{R}^N$ of N-tuples $(w_1, \ldots w_N)$ which characterize whole noise sequences in $0 < t < T$ is the product of the Gaussians (3),

$$e^{-\frac{1}{4}(\sum_1^N w_j{}^2)\Delta t}. \tag{4}$$

We now see that the formal integral (2) is a shorthand reminder of this construction. The choice of mean square $\frac{2}{\Delta t}$ for each w_j gives the correct behavior of the Brownian motion,

$$B(t) := \int_0^t w(t')dt'. \tag{5}$$

At $t = n\Delta t$, we have

$$\langle B^2(t)\rangle = n(\frac{2}{\Delta t})(\Delta t)^2 = 2n\Delta t = 2t. \tag{6}$$

In this little review, we have presented (2) as the relative probability of noise sequences subject to no constraints. The noise sequences $\mathbf{w}(t)$ which actually produce a trajectory with starting point $\mathbf{a}$ and endpoint $\mathbf{b}$ are in a restricted class—i.e., they must obey the boundary conditions, and we assume that the relative probabilities within this restricted class are still characterized by (2).

The relative probability (2) is converted into a functional of the trajectory $\mathbf{x} = \mathbf{x}(t)$ in $0 < t < T$ simply by "solving" the stochastic differential equation (1) for $\mathbf{w}(t)$,

$$\mathbf{w}(t) = \frac{1}{\sqrt{\varepsilon}}\sigma^{-1}(\dot{\mathbf{x}} - \mathbf{u}), \tag{7}$$

and substituting this $\mathbf{w}(t)$ into (2). The relative probability is thereby expressed as

$$e^{-\frac{1}{\varepsilon}S[\mathbf{x}(t)]}, \tag{8}$$

where $S[\mathbf{x}(t)]$ is the *stochastic action functional*, given by

$$S[\mathbf{x}(t)] := \frac{1}{4}\int_0^T (\dot{\mathbf{x}} - \mathbf{u}) \cdot D^{-1}(\dot{\mathbf{x}} - \mathbf{u})dt. \tag{9}$$

Here,

$$D := \sigma\sigma^T \tag{10}$$

is the *diffusion tensor* associated with the noise tensor σ. In this whole train of thought, we are assuming that σ is invertible [11].

The global minimizer $\mathbf{x}(t)$ of the action (9) is called the *most probable trajectory*. The minimization of the action determines the *geometric curve* C that $\mathbf{x}(t)$ traces out the path from $\mathbf{a}$ to $\mathbf{b}$, and also the time sequence of positions along C. In particular, the flight time T is precisely determined by (21) below. The traditional analysis [8] begins with the Euler-Lagange equation of the action (9), with T fixed. The Lagangian is

$$L(\mathbf{x}, \mathbf{v} := \dot{\mathbf{x}}) := \frac{1}{4}(\mathbf{v} - \mathbf{u}) \cdot D^{-1}(\mathbf{v} - \mathbf{u}). \tag{11}$$

Due to autonomy, there is a conserved "energy"

$$h(\mathbf{x}, \mathbf{v}) = \mathbf{v} \cdot \nabla_{\mathbf{v}} L - L = \frac{1}{4}(\mathbf{v} \cdot D^{-1}\mathbf{v} - \mathbf{u} \cdot D^{-1}\mathbf{u}). \tag{12}$$

Some presentations carry out a Legendre transformation from Lagrangian to Hamiltonian dynamics [8, 16]. Here, we stay within the Lagrangian framework. Think of D^{-1} as the metric tensor of the $\boldsymbol{R}^N$ in which $\dot{\mathbf{x}}$ and $\mathbf{u}$ live. The associated inner product is

$$\mathbf{f} \bullet \mathbf{g} := D^{-1}\mathbf{f} \cdot \mathbf{g}, \tag{13}$$

and the squared "length" of $\mathbf{f}$ is

$$|\mathbf{f}|^2 := \mathbf{f} \bullet \mathbf{f}. \tag{14}$$

In geometric notation, the action (9) is

$$\begin{aligned} S &= \frac{1}{4}\int_0^T |\dot{\mathbf{x}} - \mathbf{u}|^2 dt \\ &= \frac{1}{4}\int_0^T (\dot{\mathbf{x}} \bullet \dot{\mathbf{x}} - 2\mathbf{u} \bullet \dot{\mathbf{x}} + \mathbf{u} \bullet \mathbf{u}) dt \\ &= \frac{1}{4}\int_0^T (|\dot{\mathbf{x}}|^2 - 2|\dot{\mathbf{x}}||\mathbf{u}| + |\mathbf{u}|^2) dt + \frac{1}{2}\int_0^T (|\mathbf{u}||\dot{\mathbf{x}}| - \mathbf{u} \bullet \dot{\mathbf{x}}) dt, \end{aligned}$$

or, finally,

$$S = \frac{1}{4}\int_0^T (|\dot{\mathbf{x}}| - |\mathbf{u}|)^2 dt + \frac{1}{2}\int_0^T (|\mathbf{u}||\dot{\mathbf{x}}| - \mathbf{u} \bullet \dot{\mathbf{x}}) dt. \tag{15}$$

The first integral on the right hand side achieves its minimum value of *zero* if the speed $|\dot{\mathbf{x}}(t)|$ of the trajectory matches the speed $|\mathbf{u}(\mathbf{x}(t))|$ of the deterministic flow,

$$|\dot{\mathbf{x}}(t)| = |\mathbf{u}(\mathbf{x}(t))|. \tag{16}$$

Given (16), the action (15) reduces to

$$S = \frac{1}{2}\int_0^T (|\mathbf{u}||\dot{\mathbf{x}}| - \mathbf{u} \bullet \dot{\mathbf{x}}) dt. \tag{17}$$

In (17), we recognize

$$\mathbf{dx} = \dot{\mathbf{x}} dt, \tag{18}$$

and

$$ds = |\mathbf{dx}| = |\dot{\mathbf{x}}| dt. \tag{19}$$

as increments of displacement and arclength in elapsed time dt, respectively. Hence, the action (9) reduces to a geometric line integral along the curve C connecting **a** to **b**,

$$S = \frac{1}{2}\int_C (|\mathbf{u}| ds - \mathbf{u} \bullet \mathbf{dx}). \tag{20}$$

In summary, the most probable trajectory $\mathbf{x}(t)$ from **a** to **b** proceeds with deterministic speed as in (16), and the geometric curve C traced out by $\mathbf{x}(t)$ minimizes the line integral (20).

We acknowledge some collateral insights: since the speed along C is the deterministic speed, the time of flight from **a** to **b** is

$$T = \int_C \frac{ds}{|\mathbf{u}|}. \tag{21}$$

A further consequence of (16) is that the most probable trajectory lives on the null surface of the energy (12) in $\mathbf{x}$, $\dot{\mathbf{x}}$ space. In geometric notation, (12) reads

$$h = \frac{1}{4}(|\dot{\mathbf{x}}|^2 - |\mathbf{u}|^2) \tag{22}$$

and (16) implies $h = 0$. In the important special case where the most probable paths connect *critical points*, we can deduce directly that $h \equiv 0$ and hence $|\dot{\mathbf{x}}| = |\mathbf{u}|$ because $\dot{\mathbf{x}}$ and $\mathbf{u}$ both vanish at critical points. But the null Hamiltonian character of most probable paths is more general. In the preceeding argument, we have seen that the speed along the least action flucatuational path between *any* two endpoints equals the deterministic speed. Hence the null Hamiltonian character of most probable paths applies to any pair of endpoints, be they critical or not.

The propagation along C at deterministic speed seems deeply peculiar when we reflect that it is an asymptotic result of the small noise limit $\varepsilon \to 0$. Notice that we can halve the noise amplitude, and half the noise still drives the *same* speed of propogation along C. Here is another collateral insight: A change in ε can be absorbed by scaling the diffusion tensor D. Look at the contribution to the time of flight T from a small segment $\mathbf{dx}$ of C. The increment of arclength is

$$ds = \sqrt{D^{-1}\dot{\mathbf{x}} \cdot \dot{\mathbf{x}}}dt, \tag{23}$$

the speed is

$$|\mathbf{u}| = \sqrt{D^{-1}\mathbf{u} \cdot \mathbf{u}}, \tag{24}$$

and it is clear that the scaling of D does not change the time increment $dt = \frac{ds}{|\mathbf{u}|}$.

A qualitative explanation might go like this: The noise induced sequence of kicks that drives the speed $|\mathbf{u}|$ must have some "winning" combination of strength and unidirectionality, and it must last throughout the flight time T. As the noise amplitude goes to zero, the time duration of a "winning streak" ($\mathbf{x}(t)$ goes from $\mathbf{a}$ to $\mathbf{b}$) remains near T, cf. (21), but the expected time *between* winning streaks becomes exponentially large.

3 Broken Detailed Balance and the Geometry of Least Action Paths

We begin with some details behind the overview of detailed balance in the introduction. Let $\rho(\mathbf{x}, t)$ be the ensemble probability density. The probability current generated by the stochastic dynamics (1) is

$$\mathbf{J} := \rho\mathbf{u} - D\nabla\rho. \tag{25}$$

Here, $\mathbf{u}(\mathbf{x})$ is the velocity field of the stochastic differential equation (1), and D is the diffusion tensor (10). Take a bounded region $\mathscr{R}$ of state space, and impose the impermeable boundary condition, $\mathbf{J} \cdot \mathbf{n} = 0$ on $\partial\mathscr{R}$. Then the probability density in $\mathscr{R}$ asymptotes to a time independent steady state $\rho = \rho(\mathbf{x})$. In a detailed balance system, the probability current is not only divergence-free, but vanishes identically, so

$$\rho\mathbf{u} = D\nabla\rho \tag{26}$$

throughout $\mathscr{R}$. It follows from (24) that

$$\nabla(\log\rho) = D^{-1}\mathbf{u} \tag{27}$$

or, equivalently, the *stochastic vorticity tensor* Ω, with components

$$\Omega_{ij} := \partial_j(D^{-1}\mathbf{u})_i - \partial_i(D^{-1}\mathbf{u})_j, \tag{28}$$

vanishes.

To see the role of vorticity in the geometry of most probable paths, we write down the variational differential equation of the geometric action (20). Let $\mathbf{x} = \mathbf{x}(s)$ be the parametric representation of a most probable path with respect to the arclength s. Then, the variational equation is

$$\frac{d}{ds}(|\mathbf{u}|D^{-1}\frac{d\mathbf{x}}{ds}) - \nabla|\mathbf{u}| = \Omega\frac{d\mathbf{x}}{ds}. \tag{29}$$

Evoking the close analogy to geometric optics [17], any curve traced out by a solution of (27) is called a *ray*. The LHS of (27) is invariant under orientation reversal ($s \to -s$), but the RHS changes sign if the vorticity is nonzero. Hence, the forward and backward paths between two endpoints in a region of nonzero vorticity are generally not reversals of each other. The addition of forward and backward paths makes a closed loop, whose orientation is determined by the vorticity tensor.

A simple local analysis of rays informs the structure of forward plus backward loops whose endpoints are close to each other. Let $C : \mathbf{x} = \mathbf{x}(s)$ be a curve parametrized by arclength s. Set the origin of arclength so as to mark a given point $\mathbf{X}$ on C. In the Taylor series of $\mathbf{x}(s)$ as $s \to 0$, we have

$$\mathbf{x}(s) = \mathbf{X} + \mathbf{x}'(0)s + \frac{1}{2}\mathbf{x}''(0)s^2 + O(s^3), \tag{30}$$

where the tangent $\mathbf{x}'(0)$ at $\mathbf{X}$ is a unit vector, $|\mathbf{x}'(0)| = 1$ (since the expansion is with respect to arc length), and the second derivative is orthogonal to the tangent, $\mathbf{x}'(0) \bullet \mathbf{x}''(0) = 0$. If $\mathbf{x}''(0) \neq 0$, we see that C near $\mathbf{X}$ is asymptotic to a parabola in the plane spanned by $\mathbf{x}'(0)$ and $\mathbf{x}''(0)$. Now look at two particular rays $C_+ : \mathbf{x}_+(s)$ and $C_- : \mathbf{x}_-(s)$ which meet at $\mathbf{X}$ at $s = 0$, with opposite and equal tangents, $\hat{\mathbf{t}} := \mathbf{x}'_+(0) = -\mathbf{x}'_-(0)$. From the ray equation (27), it follows that the vorticity induces a *difference* of second derivatives,

$$[\mathbf{x}''(0)] := \mathbf{x}''_+(0) - \mathbf{x}''_-(0) = \frac{2}{|\mathbf{u}|} D\Omega\hat{\mathbf{t}}. \tag{31}$$

To see this, we evaluate (27) at $s = 0$ to obtain $|\mathbf{u}|D^{-1}\mathbf{x}''(0) + (\mathbf{x}'(0) \cdot \nabla)|\mathbf{u}| D^{-1}\mathbf{x}'(0) - \nabla|\mathbf{u}| = \Omega\mathbf{x}'(0)$. Note that the second and third terms on the LHS of this equation are invariant under reversal, $(s \to -s)$. Thus, taking differences evaluated along C_+ and C_-, we can write $|\mathbf{u}|D^{-1}[\mathbf{x}''(0)] = \Omega[\mathbf{x}'(0)] = 2\Omega\hat{\mathbf{t}}$, from which (29) follows directly.

The antisymmetry of Ω implies that $[\mathbf{x}''(0)]$ is orthogonal to $\hat{\mathbf{t}}$, as it must be, since $\mathbf{x}''_+(0)$ and $\mathbf{x}''_-(0)$ are. For $\Omega\hat{\mathbf{t}} \neq 0$, we can visualize C_+ and C_- near $\mathbf{X}$ as two parabolas tangent to each other at $\mathbf{X}$, but bending in different planes, C_+ in the plane of $\hat{\mathbf{t}}$ and $\mathbf{x}''_+(0)$, and C_- in the plane of $\hat{\mathbf{t}}$ and $\mathbf{x}''_-(0)$ as shown in Fig. 1. Now take two endpoints $\mathbf{a}$ and $\mathbf{b}$ close to $\mathbf{X}$, and take $\hat{\mathbf{t}}$ to be parallel to $\mathbf{b} - \mathbf{a}$. By small translations of the parabolas in Fig. 1 in the $-\mathbf{x}''_+(0)$ and $-\mathbf{x}''_-(0)$ directions, we can asymptotically construct the forward plus backward loop with these endpoints. Pictorially, the difference $\mathbf{x}''_+(0) - \mathbf{x}''_-(0)$ corresponds to the "openness" of the constructed fluctuation loop as shown in Fig. 2. The dashed vector marks the displacement between the turning points of the $\mathbf{a}$ to $\mathbf{b}$ and $\mathbf{b}$ to $\mathbf{a}$ parabolas; it is proportional to $[\mathbf{x}''(0)] = \frac{2}{|\mathbf{u}|} D\Omega\hat{\mathbf{t}}$.

In two dimensions, we can develop further intuition. In that case, the vorticity tensor takes the form

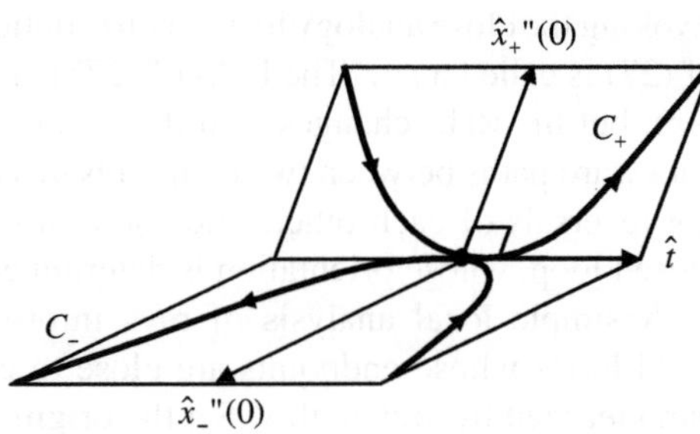

Fig. 1 Illustration of trajectories C_+ and C_- which just touch at position $\mathbf{X}$ with oppositely directed tangent vectors

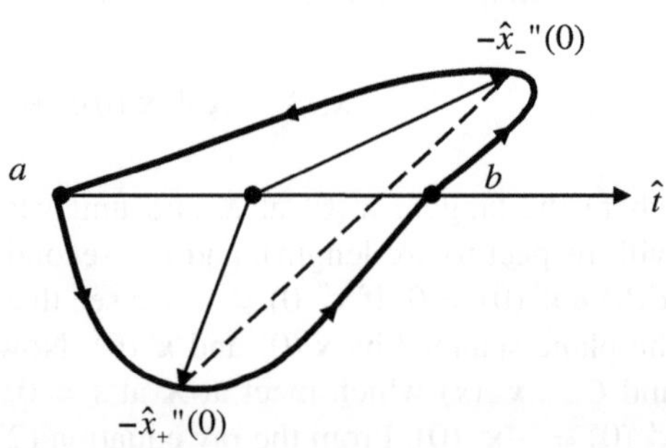

Fig. 2 Illustration of "forward" and "backward" loops for endpoints $\mathbf{a}$ and $\mathbf{b}$ in the neighborhood of $\mathbf{X}$, obtained by translating C_+ in the $-\mathbf{x}''_+(0)$ direction and C_- in the $-\mathbf{x}''_-(0)$ direction

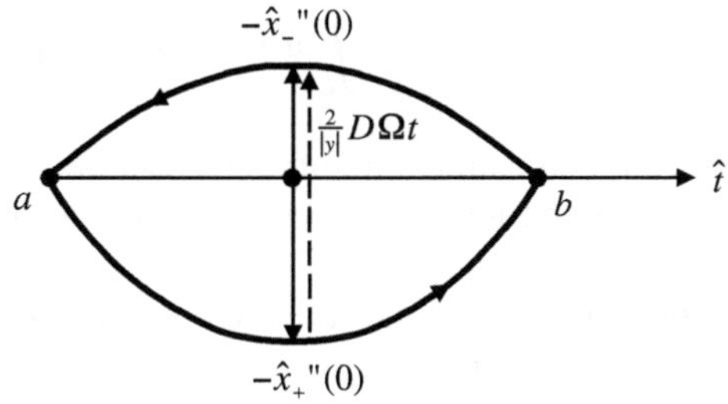

Fig. 3 Depiction of forward and backward loops in the two-dimensional case

$$\Omega = \begin{bmatrix} 0 & -\omega \\ \omega & 0 \end{bmatrix}, \tag{32}$$

where $\omega := \partial_1(D^{-1}\mathbf{u})_2 - \partial_2(D^{-1}\mathbf{u})_1$, is the *scalar vorticity*. For $\omega > 0$, (27) is the generator of a counterclockwise rotation. Figure 3 is the two dimensional version of Fig. 2, drawn assuming scalar diffusion (D proportional to the identity) and $\omega > 0$. The counterclockwise circulation around the loop agrees with the counterclockwise rotation associated with the vorticity. This is the choice you would make to reduce "headwinds" and thereby reduce the stochastic action.

4 Vorticity-Induced Bifurcations

The global effects of vorticity-induced bending of rays are most simply displayed in a class of two-dimensional examples we call *pure shear*: we have isotropic diffusion $D = I$, and the flows $\mathbf{u}$ on $\boldsymbol{R}^2$ are

$$\mathbf{u} = u(y)\hat{\mathbf{y}}, \tag{33}$$

where the *velocity profile* $u(y)$ is a given positive function. The x-component of the ray ODE (27) is

$$\frac{d}{ds}(u\frac{dx}{ds}) = u'(y)\frac{dy}{ds}, \tag{34}$$

which has the first integral

$$u(y)(1 - \frac{dx}{ds}) = \sigma = constant. \tag{35}$$

(33) is analogous to Snell's law in optics: that is, $\frac{dx}{ds} = \cos\theta$, where θ is the local angle of the ray with respect to $\hat{\mathbf{x}}$. Hence, (33) implies that θ changes with elevation y due to gradients of the velocity profile $u(y)$. In this sense, we have "refraction by shear." Due to the first integral (33), the geometric action of a ray C has the simple expression

$$S = \int_C u ds - u dx = \int_C u(1 - \frac{dx}{ds}) ds = \sigma L, \quad (36)$$

where L is the arclength of C. Deterministic paths are immediately recognized from the effective Snell's law (33) and (34): The geometric action vanishes if the constant σ is zero, and then $\frac{dx}{ds} \equiv 1$, which corresponds to horizontal straight lines oriented to the right. These lines are the obvious integral curves of the shear flow.

Next, we examine rays that undergo a net displacement in the negative x-direction, which is "against the wind." By the mean value theorem, such a ray must have a *turning point* $\mathbf{x}_*$, where $\frac{dx}{ds} = -1$. Evaluating (33) at the turning point, we have $\sigma = 2u_* := 2u(y_*)$, where y_* is the elevation of the turning point. Hence, (33) can be written as

$$u\frac{dx}{ds} = u - 2u_*, \quad (37)$$

and the geometric action becomes

$$S = 2u_* L. \quad (38)$$

We can determine a governing equation for the vertical component $y(s)$ of $\mathbf{x}(s)$, by evoking the geometric identity $(\frac{dx}{ds})^2 + (\frac{dy}{ds})^2 = 1$, and using the effective Snell's law (35):

$$u^2(\frac{dy}{ds})^2 = 4u_*(u(y) - u_*). \quad (39)$$

Since $u(y)$ is the deterministic speed along the ray, we recognize $u\frac{d}{ds}$ as the time derivative $\frac{d}{dt}$, and we write (35) and (37) respectively as

$$\dot{x} = u(y) - 2u_*, \quad (40)$$

$$\dot{y}^2 = 4u_*(u(y) - u_*). \quad (41)$$

The second equation is the first integral of the second order differential equation

$$\ddot{y} = 4u_* u'(y). \quad (42)$$

Here, $-u'(y)$ is the scalar vorticity of the shear flow, so (40) directly expresses "bending by vorticity."

We can construct rays explicitly for the instructive case of quadratic shear flow, with

$$u(y) = \frac{1}{2}(y^2 + \varepsilon^2), \quad (43)$$

where ε is a positive parameter. The minimum positive x-velocity happens along the x-axis. We expect that rays with a start to finish displacement in the negative x-direction bend toward the x-axis, to take advantage of the reduced headwind there.

Such a ray must have a turning point. By the x and t translation invariance of (38) and (39), we can use the turning point to mark the origins of x and t, so we have initial conditions $x(0) = 0$, $y(0) = y_*$, $\dot{y}(0) = 0$. The solutions of (38, 39) with these initial conditions are

$$y = y(t, y_*) = y_* \cosh \tau, \tag{44}$$

$$x = x(t, y_*) = \frac{{y_*}^2}{4\sqrt{{y_*}^2 + \varepsilon^2}}(\cosh \tau \sinh \tau - \tau) - \frac{\sqrt{{y_*}^2 + \varepsilon^2}}{2}\tau. \tag{45}$$

Here, τ is the scaled time

$$\tau := \sqrt{{y_*}^2 + \varepsilon^2}t. \tag{46}$$

For each value of y_* we obtain a corresponding ray represented by a parametric curve

$$\mathbf{x}(t) = x(t, y_*)\hat{\mathbf{x}} + y(t, y_*)\hat{\mathbf{y}}. \tag{47}$$

We present a pictorial narrative of these rays for the case $\varepsilon = 0.3$, and this shows how segments of rays produce net displacements in the negative x-direction. Figure 4 shows the ray with turning point elevation $y_* = 1$. The most striking feature is the teardrop-shaped loop with counterclockwise orientation. The "teardrop" persists for all $y_* > 0$. From (38), we see that the vertical tangents happen at the elevation y for which the shear velocity $u(y)$ is twice as high as at the elevation of the turning point. As $y_* \to 0$, the bottom of the teardrop between the vertical tangents flattens out into a long, left-oriented segment close to the x-axis. We see this in Fig. 5, which depicts the ray with $y_* = 0.01$.

A ray which interpolates between given starting and ending points $\mathbf{a}$ and $\mathbf{b}$ is called a *connector*. We examine the connectors with

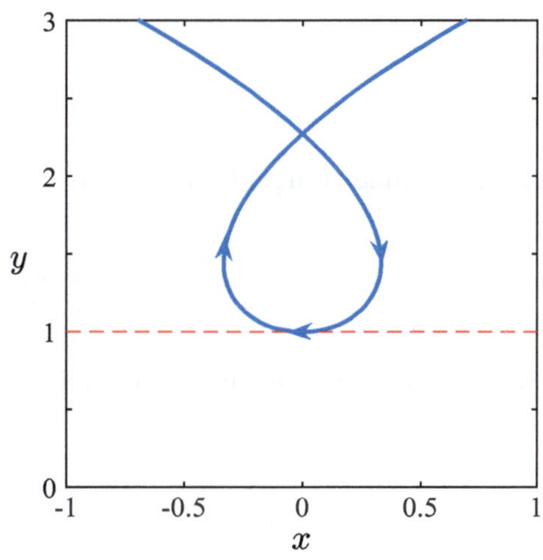

Fig. 4 Transition path ray for the case $y_* = 1$ and $\varepsilon = 0.3$

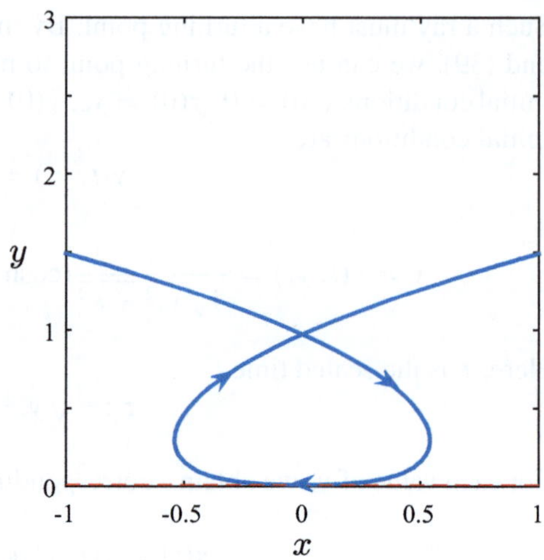

Fig. 5 Transition path ray for the case $y_* = 0.01$ and $\varepsilon = 0.3$

$$\mathbf{a} = \frac{l}{2}\hat{\mathbf{x}} + \hat{\mathbf{y}}, \, \mathbf{b} = -\frac{l}{2}\hat{\mathbf{x}} + \hat{\mathbf{y}}, \tag{48}$$

which makes a net displacement of length l in the negative x-direction. These connectors are segments of the "teardrop," symmetric about the x-axis. Given l, there are discrete choices for the turning point elevations y_* of possible connectors. Let $\tau = \tau_b > 0$ be the value of scaled time marking the ending point $\mathbf{b}$. By symmetry, $\tau = -\tau_b$ marks the starting point $\mathbf{a}$, and the connector passes through the turning point $y_*\hat{\mathbf{y}}$ at $\tau = 0$. At $\tau = \tau_b$, the elevation y is unity, so (42) implies

$$y_* \cosh \tau_b = 1. \tag{49}$$

Given $0 < y_* < 1$, we solve for τ_b,

$$\tau_b = \log \frac{1 + \sqrt{1 - y_*^2}}{y_*}. \tag{50}$$

The actual time of flight from $\mathbf{a}$ to $\mathbf{b}$ is

$$T = \frac{2\tau_b}{\sqrt{y_*^2 + \varepsilon^2}} + \frac{2}{\sqrt{y_*^2 + \varepsilon^2}} \log \frac{1 + \sqrt{1 - y_*^2}}{y_*} \tag{51}$$

At $\tau = -\tau_b$, $x = \frac{l}{2}$, and then (43) implies

$$\frac{l}{2} = \frac{y_*^2}{4\sqrt{y_*^2 + \varepsilon^2}}(\tau_b - \cosh \tau_b \sinh \tau_b) + \frac{\sqrt{y_*^2 + \varepsilon^2}}{2}\tau_b. \tag{52}$$

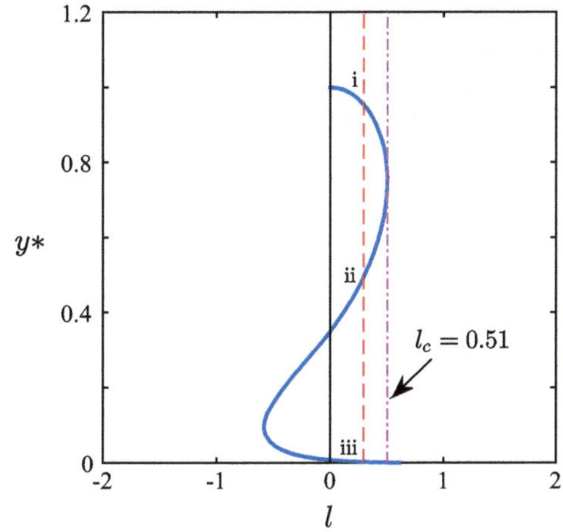

Fig. 6 Plot of Eq. (53) showing the allowed values of y_* as a function of l and $\varepsilon = 0.3$. The turning point at l_c corresponds to a saddle-node bifurcation in which two transition paths with endpoints **a** and **b** collide and annihilate

Substituting for τ_b from (48), this reduces to

$$l = (\sqrt{y_*^2 + \varepsilon^2} + \frac{y_*^2}{2\sqrt{y_*^2 + \varepsilon^2}}) \log \left[\frac{1 + \sqrt{1 - y_*^2}}{y_*}\right] - \frac{\sqrt{1 - y_*^2}}{2\sqrt{y_*^2 + \varepsilon^2}}. \tag{53}$$

The graph of y_* versus l in Fig. 6 based on (51) is plotted with $\varepsilon = 0.3$, but is qualitatively correct for $0 < \varepsilon << 1$. For connectors with a net displacement in the negative x-direction, the relevant portion of this graph has $l > 0$, marked by the solid curves. There are three branches of roots for y_* as functions of l, which we've labeled i, ii, and iii. Branches i and ii coelesce in a saddle-node bifurcation as $l \to l_c \simeq 0.51$. Furthermore, the branches i and ii have definite limits as $\varepsilon \to 0$, and the saddle-node bifurcation survives with $l_c \to 0.4852$. Figure 7 depicts the connecting pathways corresponding to branches i, ii and iii for $\varepsilon = 0.3$.

Which connector has the smallest geometric action? We can compute the arclength L of connectors, and then their geometric actions from $S = 2u_* L$ as in (36). Since $u = u(y)$ is the speed along a ray, the arclength of a connector C is

$$L = \int_C u dt = \int_C (\dot{x} + 2u_*) dt = -l + 2u_* T. \tag{54}$$

Substituting into (42) the time of flight T from (49), and l from (51), we obtain for the action

$$S = \frac{\sqrt{y_*^2 + \varepsilon^2}}{2}((y_*^2 + 2\varepsilon^2) \log \frac{1 + \sqrt{1 - y_*^2}}{y_*} + \sqrt{1 - y_*^2}). \tag{55}$$

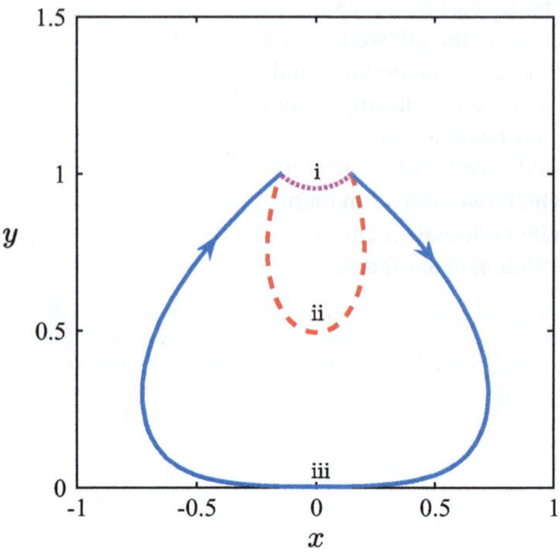

Fig. 7 Plot of the three connector solutions corresponding to the three branches of Fig. 6 for $l = 0.3$ and $\varepsilon = 0.3$

From (53), we find that the shorter branch i connector has the smallest geometric action. Apparently, the branch ii connector is a stationary point of geometric action, but not a minimum.

Branch iii in Fig. 7 and its corresponding connector persist for all $l > 0$. In the limit $\varepsilon \to 0$, branch iii of the y_* versus l relation is asymptotic to

$$y_* \sim 2 \exp\left[-\frac{1}{2\varepsilon^2} - \frac{l}{\varepsilon} \right]. \tag{56}$$

Figure 7 depicts the branch iii connector for $\varepsilon = 0.3$. The branch iii connector has a singular structure as $\varepsilon \to 0$. In particular, its arclength diverges like

$$L \sim \frac{1}{\varepsilon} + l. \tag{57}$$

Its geometric action is asymptotic to

$$S \sim \varepsilon. \tag{58}$$

In the limit $\varepsilon \to 0$ with l fixed, we see that the branch iii connector is the most probable path, beating out connector i, for $0 < l < l_c$. The "strategy" of of connector iii is clear: Descend from $y = 1$ to the x-axis, stay close to the x-axis for a long time, and finally ascend back up to $y = 1$. In this way, the resistance against the deterministic flow is minimized. The descent and ascent branches have little cost, because they are almost parallel to the deterministic flow. Most of the leftwards motion happens along the segment close to the x-axis, where the headwind is small.

We expect that as we close the gap between **a** and **b** by decreasing l, we'll eventually find $l = l_*$ so that the branch i connector becomes the most probable path for $0 < l < l_*$. By an elementary calculation, we find $l_* \sim \varepsilon$ as $\varepsilon \to 0$.

5 Conclusions and Future Directions

The breaking of detailed balance is "made visible" by the splitting of forward and backward most probable paths between two fixed endpoints. This is a first hint that detailed balance and its breaking can be inferred directly from recorded histories of a stochastic dynamical system, in a suitable space of observables. A first "obvious" proposal, while achievable in principle, may often be difficult to implement in practice: this involves collecting records of trajectories that connect two small regions around given endpoints, and observing directly the aforementioned splitting between forward and backward paths. At small noise levels, there are long waits to collect these trajectories, and then you might want to have many of them for averaging. It would be much better if detailed balance or its breaking can be detected by some simple processing of data from a few, or even one trajectory. Such a procedure in fact exists. It has been fully developed by the authors for linear stochastic dynamical systems, and numerically tested for a simple circuit example [15]. More recently, we've found that its nonlinear generalization is straightforward and that story will get its own paper [18].

Acknowledgements We gratefully acknowledge the assistance of Varun Gudapati with the preparation of Figs. 1, 2, and 3.

References

1. R.P. Feynman, A.R. Hibbs, *Quantum Mechanics and Path Integrals* (McGraw-Hill, New York, 1965)
2. A.D. Ventsel, M.I. Freidlin, Russ. Math. Surv. **1**, 1 (1970)
3. M.I. Freidlin, A.D. Wentzell, Random perturbations of dynamical systems, in *A Series of Comprehensive Studies in Mathematics* (Springer, 2012)
4. N. Berglund, B. Gentz, Noise-induced phenomena in slow-fast dynamical systems: a sample-paths approach, in *Probability and Its Applications* (Springer, 2006)
5. A. Prados, P.I. Lasanta, P.I. Hurtado, Phys. Rev. Lett. **107**, 140601 (2011)
6. L. Bertini, A. De Sole, D. Gabrielli, G. Jona-Lasinio, C. Landim, Rev. Mod. Phys. **87**, 593 (2015)
7. R.S. Maier, D.L. Stein, SIAM J. Appl. Math. **57**, 752 (1997)
8. D.G. Luchinsky, P.V.E. McClintock, M.I. Dykman, Rep. Prog. Phys. **61**, 905 (1998)
9. M. Heymann, E. Vanden-Eijnden, Phys. Rev. Lett. **100**, 140601 (2008)
10. M. Heymann, E. Vanden-Eijnden, Commun. Pure Appl. Math. **61**, 1052 (2008)
11. M. Heymann, *Minimum Action Curves in Degenerate Finsler Metrics*. Lecture Notes in Mathematics, vol. 2134 (Springer, 2015)
12. R.S. Maier, D.L. Stein, Phys. Rev. Lett. **71**, 1783 (1993)
13. R.K.P. Zia, B. Schmittmann, J. Stat. Mech.-Theory Exp P07012 (2007)
14. P.H. Dannenberg, J.C. Neu, S.W. Teitsworth, Phys. Rev. Lett. **113**, 020601 (2014)
15. A. Ghanta, J.C. Neu, S.W. Teitsworth, Phys. Rev. E **95**, 032128 (2017)
16. M.I. Dykman, E. Mori, J. Ross, P.M. Hunt, J. Chem. Phys. **100**, 5735 (1994)
17. M. Born, E. Wolf, Principles of Optics, 6th edn. (1980)
18. J.C. Neu, S.W. Teitsworth, Unpublished (2017)

Part III
Mathematics for 2D Materials and Properties of Confined Nanostructures

Classical Density-Functional Theory Studies of Fluid Adsorption on Nanopatterned Planar Surfaces

Peter Yatsyshin and Serafim Kalliadasis

Abstract This contribution is based on our talk at the BIRS Workshop on "Coupled Mathematical Models for Physical and Biological Nanoscale Systems and Their Applications". Our aim here is to summarize and bring together recent advances in wetting of nanostructured surfaces, using classical density-functional theory (DFT). Classical DFT is an ab initio theoretical-computational framework with a firm foundation in statistical physics allowing us to systematically account for the fluid spatial inhomogeneity, as well as for the non-localities of intermolecular fluid-fluid and fluid-substrate interactions. The cornerstone of classical DFT, is to express the grand free energy of the system as a functional of its one-body density, thus generating a hierarchy of N-body correlation functions. Unconstrained minimization of a properly approximated free-energy functional with respect to the one-body density then yields the basic DFT equation. And since most macroscopic quantities of interest can often be cast as averages over a one-body distribution, this equation provides a very useful and accessible computational tool. Indeed, there has been a rapid growth of classical DFT applications across a broad variety of fields, including phase transitions in solutions of macromolecules, interfacial phenomena, and even nucleation. Here we attempt to give a taste of what simple equilibrium DFT models look like, and what they can and cannot capture, as far as wetting on chemically heterogeneous substrates is concerned. We review recent progress in the understanding of planar prewetting and interface unbending on such substrates and compute substrate-fluid interfaces and wetting isotherms.

P. Yatsyshin (✉) · S. Kalliadasis (✉)
Department of Chemical Engineering, Imperial College London,
London SW7 2AZ, UK
e-mail: p.yatsyshin@imperial.ac.uk

S. Kalliadasis
e-mail: s.kalliadasis@imperial.ac.uk

L. L. Bonilla et al. (eds.), *Coupled Mathematical Models for Physical and Biological Nanoscale Systems and Their Applications*, Springer Proceedings in Mathematics & Statistics 232, https://doi.org/10.1007/978-3-319-76599-0_10

1 Introduction

In the past few decades, high demand for nanoscale technological processes, as well as understanding of the intricate physics associated with the behaviour of matter at the nanoscale, has led to rapid growth of the field of soft condensed matter. Originally regarded as part of chemical and statistical physics, today soft condensed matter is a truly cross disciplinary area which involves physicists and applied mathematicians, as well as engineers, chemists and biologists on the academic front, and industrial scientists and engineers on the applied front. Soft matter problems range from liquid adsorption and wetting in fluids [1–3] and self-assembly and structural transitions in colloids, polymers, liquid crystals and charged fluids [4–7] to equilibrium and kinetics of phase separation and formation of interfaces and thin films [8, 9]. In chemical and bioengineering, statistical mechanical models incorporating fluid inhomogeneities at small scales are used to understand and improve various technological processes, e.g., design and operation of miniature lab-on-a-chip chemical reactors and nanofluidic devices [10–12], DNA separation [13], vapor-liquid-solid growth of semiconductor nanowires [14, 15], capillary filling of microchannels and adsorption in nanoporous materials [16, 17], design of nano-patterned superhydrophobic surfaces and surfaces with controllable wetting properties [18–20].

Soft matter systems typically involve multiple length and time scales. Therefore their description naturally calls for the development of statistical mechanical models which offer a middle ground between purely atomistic and continuum models, in terms of both computational efficiency but also fundamental scrutity. In general, equilibrium statistical mechanical systems of N constituent interacting particles are fully described by providing the configurational N-body distribution function. However, many observables of interest, such as the number density, magnetisation, charge distribution, pressure, compressibility and so on, can be quite adequately approximated as averages involving only one- and two-body distributions, defined as integrals of the N-body distribution over N-1 and N-2 sets of particle coordinates, respectively. Since in many cases, the two-body distribution can also be reliably approximated as a one-body distribution of a system subject to a specially defined external potential, a general framework for approximating the one-body distribution would be highly valuable for describing soft-matter systems. This framework is provided by classical density-functional theory (DFT).

The purpose of the present contribution—based on the presentation we gave at the BIRS workshop at Banff on "Coupled Models for Physical and Biological Nanoscale Systems and their Applications"—is to provide an example of constructing a computationally tractable, yet fully microscopic DFT approximation, and to illustrate its use for one of the most ubiquitous problems in soft matter—interfacial phase transitions in inhomogeneous fluids. We showcase the typical ingredients of a DFT approximation by constructing a simple, yet qualitatively reliable free-energy functional capable of capturing the properties of a liquid-gas interface of a classical fluid in contact with a solid, above the fluid bulk triple point. We show how the model constructed may be used to systematically investigate wetting and adsorption at the interface of a chemically heterogeneous wall in contact with undersaturated gas. Although the scope of this Chapter only allows us to discuss a particular regime, namely near the boundary of planar pre-wetting [21], it suffices to illustrate the potential and versatility of classical DFT methodology in soft matter and nano-science.

2 Classical DFT Philosophy

Historically, the original developments of the classical DFT formalism and its first applications dealt with interfacial problems in the physics of fluids [22]. The variational principle underlying classical and quantum DFTs was derived by Kohn and Hohenberg [23] for an inhomogeneous electron gas at zero temperatures, and generalised by Mermin [24] to include the effects of a finite temperature. According to a theorem due to Mermin [24], given the values of the bulk thermodynamic fields, temperature T and chemical potential μ, there exists a grand canonical free-energy functional $\Omega\left[\rho\left(\mathbf{r}\right), T, \mu\right]$, which is minimised by the equilibrium one-body distribution $\rho\left(\mathbf{r}\right)$ [25]. Moreover, the minimum of $\Omega\left[\rho\left(\mathbf{r}\right), T, \mu\right]$ over all the one-body distributions is the actual thermodynamic grand potential of the system. The many-body distributions are generated within DFT by taking the functional derivatives of $\Omega\left[\rho\left(\mathbf{r}\right), T, \mu\right]$ with respect to $\rho\left(\mathbf{r}\right)$ [26]. Therefore, the central problem of statistical mechanics—approximating the N-body distribution function—can be reformulated in terms of approximating $\Omega\left[\rho\left(\mathbf{r}\right), T, \mu\right]$.

Significant progress has been made in the past three decades in constructing reliable and computationally tractable approximations for $\Omega\left[\rho\left(\mathbf{r}\right), T, \mu\right]$ for different classes of soft-matter systems: from fluids and fluid mixtures with coarse-grained interactions and solutions of macromolecules, to porous media, biological systems and even active matter [26–29]. The accuracy with which DFT allows one to approximate various observables is often on par with fully atomistic molecular dynamics simulations. At the same time, the computational cost required for solving the DFT governing equations may be significantly less than that of simulations. In what follows, we consider a pedagogical example of a classical single component fluid.

Given μ, T and the potential $V\left(\mathbf{r}\right)$ of an external body force acting on the fluid molecules, the grand canonical free-energy functional is given by the Legendre transform of the "intrinsic" free energy functional [22, 26, 30]:

$$\Omega\left[\rho\left(\mathbf{r}\right)\right] = F_{\text{in}}\left[\rho\left(\mathbf{r}\right)\right] + \int \mathrm{d}\mathbf{r}\rho\left(\mathbf{r}\right)\left(V\left(\mathbf{r}\right) - \mu\right), \tag{1}$$

where the integration is carried out over the volume occupied by the fluid. The intrinsic free energy functional $F_{\text{in}}\left[\rho\right]$ does not depend on the external potential $V\left(\mathbf{r}\right)$.

For many realistic fluids, the coarse-grained interparticle interactions are pairwise, and given by the Lennard-Jones (LJ) potential:

$$\varphi^{\mathrm{LJ}}_{\varepsilon,\sigma}(r) = 4\varepsilon\left[-\left(\frac{\sigma}{r}\right)^{6} + \left(\frac{\sigma}{r}\right)^{12}\right], \tag{2}$$

where ε and σ control the well depth and the range of the potential, respectively. The Lennard-Jones (LJ) potential describes competing attractive-repulsive interparticle interactions, where the repulsions are sharp and short-ranged, and the long-ranged attractions decay algebraically with distance, as do the ubiquitous dispersive dipole–dipole interactions. The LJ potential (2) and its modifications can be used to describe, among other systems, noble gases, solutions of globular proteins and charged colloids [26, 28, 31, 32].

For attractive–repulsive interparticle potentials, $F_{\mathrm{in}}[\rho]$ is often approximated following a perturbative approach, where the free energy of a purely repulsive fluid of hard spheres (HS) of diameter σ is used as a reference, and the mean-field free energy of the pairwise attractions with the potential $\varphi_{\mathrm{attr}}(r)$ is added as a perturbation [22]. Thus, to first order in $\varphi_{\mathrm{attr}}(r)$, we can write

$$\begin{aligned} F_{\mathrm{in}}[\rho(\mathbf{r})] &= \int \mathrm{d}\mathbf{r}\, f_{\mathrm{id}}(\rho(\mathbf{r})) + F_{\mathrm{hs}}[\rho(\mathbf{r})] \\ &+\frac{1}{2}\int \mathrm{d}\mathbf{r}\int \mathrm{d}\mathbf{r}'\rho(\mathbf{r})\,\rho(\mathbf{r}')\,\varphi_{\mathrm{attr}}(|\mathbf{r}-\mathbf{r}'|), \end{aligned} \tag{3}$$

where $f_{\mathrm{id}}(\rho) = k_{\mathrm{B}}T\rho\left(\ln\left(\lambda^{3}\rho\right) - 1\right)$ is the configurainal free-energy density of the ideal gas, k_{B} and λ are the Boltzmann constant and the thermal wavelength, respectively, and $F_{\mathrm{hs}}[\rho(\mathbf{r})]$ is the HS free energy functional. The expression for $\varphi_{\mathrm{attr}}(r)$ can be obtained from the full LJ potential (2) following the perturbative treatment. We will use a simplification of the Barker–Henderson expansion [33]:

$$\varphi_{\mathrm{attr}}(r) = \begin{cases} 0, & r \le \sigma \\ \varphi^{\mathrm{LJ}}_{\varepsilon,\sigma}, & r > \sigma \end{cases}, \tag{4}$$

where σ is kept constant and equal to the HS diameter in $F_{\mathrm{hs}}[\rho(\mathbf{r})]$.

The exact HS free-energy functional $F_{\mathrm{hs}}[\rho(\mathbf{r})]$ is not known, but numerous approximations have been developed over the years, which allow one to balance computational complexity with accuracy [26]. Some of the most accurate approximations for $F_{\mathrm{hs}}[\rho(\mathbf{r})]$ are built upon the semi-phenomenological Carnahan-Starling equation of state for uniform HS fluids [34], which is equivalent to the following expression for the configurational part of the free energy density:

$$\psi(\rho) = k_{\mathrm{B}}T\frac{\eta(4-3\eta)}{(1-\eta)^{2}}, \quad \eta = \pi\sigma^{3}\rho/6. \tag{5}$$

Approximations for $F_{\text{hs}}[\rho(\mathbf{r})]$ based on Eq. (5) can be written in the following general form [26, 35]:

$$F_{\text{hs}}[\rho] = \int d\mathbf{r}\rho(\mathbf{r})\,\psi(\bar{\rho}(\mathbf{r}))\,, \quad \bar{\rho}(\mathbf{r}) = \int d\mathbf{r}' W\left(\mathbf{r}-\mathbf{r}'\right)\rho(\mathbf{r})\,, \tag{6}$$

where the integration is carried out over the volume occupied by the fluid. The weight $W(\mathbf{r})$ can be chosen to reproduce the essential features of the HS pair correlation function in the uniform limit. The latter can formally be obtained from Eq. (6) by functional differentiation, and is also known from integral equation theories, such as, e.g., the hypernetted chain theory, which gives rise to the famous Percus-Yevick approximation for the HS pair-correlation function [36].

Since $F_{\text{hs}}[\rho(\mathbf{r})]$ is only a part of the full free-energy functional in Eq. (3), we can choose to ignore the HS correlations altogether by setting

$$W(\mathbf{r}) = \delta(\mathbf{r})\,, \tag{7}$$

where $\delta(\mathbf{r})$ is the delta-function. This is commonly known as the local-density approximation (LDA) [26, 30]. Although using LDA in the full functional (3) neglects the excluded volume intermolecular correlations, its computational simplicity makes it an attractive approximation for problems where the physics is determined by the long-range asymptote of the intermolecular potential. In particular, a whole plethora of physical phenomena involving wetting by liquid and capillarity are governed by the attractive tails of the LJ potential (2) [21, 37], and using LDA in the HS part of $F_{\text{in}}[\rho(\mathbf{r})]$, allows one to investigate these phenomena with qualitative accuracy, and at a low computation cost [38, 39].

On the other hand, the simplest DFT, which captures the essential feature of the pair correlation function—the jump at the distance σ—involves a simple weight function $W(\mathbf{r}) = \frac{3}{4\pi\sigma^3}\,\Theta(\sigma - r)$ [40]. Including the excluded volume correlations into the full functional (3) is necessary for describing layering and freezing in fluids. The general method of systematically refining the weight $W(\mathbf{r})$ to achieve a better approximation for the uniform HS pair correlation function is known as the family of weighted density approximations (WDAs) [26, 30, 36]. The weight functions of highly accurate WDAs decay very slowly, and the procedure for obtaining them is not straightforward [36]. The most accurate and theoretically tractable description of HS fluids is achieved by the fundamental measure theory [41, 42], which uses three weighting functions and can be applied to single- and multi-component HS fluids. Moreover, the method can be used to construct accurate free-energy functionals for fluids with rigid non-spherical particles.

For simplicity, the computations discussed in the following section are performed with LDA in the HS part of the full functional (3). The DFT approximation based on Eqs. (1), (3)–(7) belongs to the family of random-phase approximations. Such approximations become exact in the limit of soft interparticle potentials and high fluid densities [26]. In the context of problems involving transitions at liquid-gas interfaces, this approximation is known to capture qualitatively the physics of wetting at temperatures above the bulk triple temperature.

It is noteworthy that in recent years there have been several studies applying DFT to the modelling of non-equilibrium statistical mechanical systems [25, 26, 43, 44]. Involving the "local equilibrium" assumption that the two-body time-dependent correlation function in a non-equilibrium system is related to the free energy functional in the same way as its equilibrium counterpart, Marconi and Tarazona have shown that the diffusion-driven evolution of a statistical-mechanical system follows a generalised "model-B-like" [45] conservation law [46–48]. In a later work, Lutsko and Durán-Olivencia have demonstrated that the equations of dynamic DFT can be derived from the equations of fluctuating hydrodynamics of an overdamped colloidal particles in a thermal bath [31, 32, 49]. Currently, there is significant interest in generalising the dynamic DFT to include the effects of hydrodynamic interactions within the context of overdamped colloidal particle dynamics. In particular, the works of Goddard et al. have provided the unification of dynamic DFTs for colloidal fluids, to include inertia and hydrodynamic interactions [50–52]. A further recent extension has focused on dynamic DFTs for anisotropic colloidal fluids [53].

3 Wetting on Nano-patterned Planar Surfaces

Here we apply the DFT approximation discussed above to the problem of fluid adsorption on a plane decorated with a chemical stripe of a nanoscopic width. Given the fluid–fluid and fluid–substrate interparticle potentials, the only relevant thermodynamic fields in the problem are T and μ, which will be our control parameters. In general, when a solid substrate with a planar surface is brought in contact with a gas, the LJ fluid-solid interactions give rise to an attractive body force acting on the fluid molecules. The potential of this force can be obtained by integrating the LJ potential (2) over the volume of the substrate. For a homogeneous planar wall occupying a half-space $y \leq 0$, it is given by the following expression [35]:

$$V_0\left(y\right) = 4\pi \rho_{\mathrm{w}} \varepsilon_{\mathrm{w}} \sigma_{\mathrm{w}}^3 \left[-\frac{1}{6}\left(\frac{\sigma_{\mathrm{w}}}{H_0 + y}\right)^3 + \frac{1}{45}\left(\frac{\sigma_{\mathrm{w}}}{H_0 + y}\right)^9 \right], \tag{8}$$

where ρ_{w} is the average density of the substrate material, and ε_{w} and σ_{w} are its LJ parameters. Since the physics of wetting is controlled by the long-range asymptotes of the interparticle potentials, we can introduce the cut-off H_0 as a mathematical device to eliminate the singularity of $V_0\left(y\right)$ at zero. Physically, this corresponds to also "cutting-off" the unimportant wall–fluid short-range repulsion. More details on the role of H_0 can be found in, e.g., Appendix 1 of Ref. [54].

The potential (8) may be sufficient to condense the gas near the wall to a liquid film of small width, covering the wall. The properties of such liquid films are closely related to the Young contact angle θ of a sessile drop sitting on the substrate surface at bulk coexistence [55, 56]. According to Young's equation, the macroscopic contact angle at coexistence is related to the interfacial tensions of solid-gas, $\sigma_{\rm sg}$, solid-liquid, $\sigma_{\rm sl}$, and liquid-gas interfaces, $\sigma_{\rm lg}$, via a force balance:

$$\sigma_{\rm sg} = \sigma_{\rm sl} + \sigma_{\rm lg} \cos\theta. \tag{9}$$

In particular, at temperatures above the substrate wetting temperature $T_{\rm w}$ (defined via the condition $\theta\,(T_{\rm w}) = 0$), the transition to wetting is continuous with μ. It is manifested by the growing height h of the adsorbed film as μ approaches the bulk saturation value $\mu_{\rm sat}$, as $h = O\left((\mu_{\rm sat} - \mu)^{-1/3}\right)$. The non-universal exponent $-1/3$ describing complete wetting in systems with dispersive interactions is directly related to the long-range behaviour $V_0^{(1)}(y) = O(y^{-3})$ of the wall potential (8) [57]. Another remarkable property of adsorption on homogeneous planar substrates is the so-called pre-wetting transition, which refers to the jump from a thin film of liquid adsorbed at the wall to a thick layer of liquid, as the chemical potential is increased towards saturation. This is a first-order phase transition associated with the (first-order) wetting transition at $T_{\rm w}$ [21, 37, 58]. More technically, pre-wetting refers to a line of such thin-thick transitions which occur near $\mu_{\rm sat}$ and above $T_{\rm w}$, see, e.g., Fig. 1 [60].

In DFT studies of wetting, the one-body density distribution at given μ and T is obtained by minimising the DFT functional $\Omega\,[\rho\,(\mathbf{r})] \equiv \Omega\,[\rho\,(y)]$ given in Eqs. (3)–(7) over the one-body distributions $\rho(y)$. In order to compute phase transition curves, such as the pre-wetting line of a planar wall, we must find such distributions $\rho_1(y)$ and $\rho_2(y)$, which, along with minimising the grand free energy functional, satisfy the phase coexistence equation

$$\Omega\,[\rho_1\,(y)] = \Omega\,[\rho_2\,(y)]\,, \tag{10}$$

which is an expression of the fact that the coexisting fluid configurations must have the same energies.

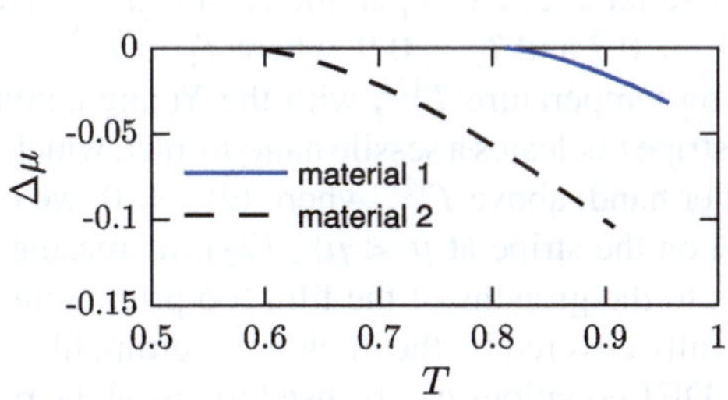

Fig. 1 Pre-wetting lines $\mu_{\rm pw}^{(i)}(T)$ of homogeneous planar walls of stronger (1, dashed black) and weaker (2, solid blue) materials (color online) [60]. The LJ parameters are, respectively, $\varepsilon_{\rm w}^{(1)} = 1$, $\varepsilon_{\rm w}^{(2)} = 0.6$, $\sigma_{\rm w}^{(1)} = \sigma_{\rm w}^{(2)} = 2$, and $H_0 = 5$ in (8). Note that the pre-wetting lines approach saturation ($\Delta\mu = 0$) tangentially at the respective wetting temperatures $T_{\rm w}^{(1)} \approx 0.60$ and $T_{\rm w}^{(2)} \approx 0.81$

We work in a system of units, where the hard core diameter σ and the well depth ε of the fluid–fluid LJ interactions are set as units of length and energy, respectively. The units of temperature are then given by $\varepsilon/k_{\mathrm{B}}$, and the bulk critical temperature is $T_{\mathrm{c}} = 1.006$. We further suppose that $\rho_{\mathrm{w}} = 1$ in (8). Consider the prewetting lines $\mu_{\mathrm{pw}}(T)$ of two materials differing in the depth of the LJ well, represented in Fig. 1, where $\Delta\mu = \mu - \mu_{\mathrm{sat}}$ is the deviation from saturation. These curves were computed by solving Eq. (10), together with the extremum condition for $\Omega\,[\rho\,(\mathbf{r})]$ for the coexisting density profiles and chemical potentials. Note that the wetting temperature of the stronger substrate (with larger ε_{w}) is lower than that of the weaker one. The pre-wetting lines approach coexistence tangentially at the respective wetting temperatures $T_{\mathrm{w}}^{(\mathrm{i})}$ and follow the asymptote [21, 38, 59]:

$$\mu_{\mathrm{sat}} - \mu_{\mathrm{pw}}^{(\mathrm{i})}\,(T) \propto \left(T - T_{\mathrm{w}}^{(\mathrm{i})}\right)^{3/2}. \tag{11}$$

Consider decorating a homogeneous wall made up of material i with a stripe of a different material j, so that the external potential acting on the fluid in contact with the wall is given by the following equation:

$$V_L\,(x, y) = V_0^{(i)}\,(y) - \rho_{\mathrm{w}}^{(i)} \int\limits_{\nu_L} \mathrm{d}\mathbf{r}'\varphi^{\mathrm{LJ}}_{\sigma_{\mathrm{w}}^{(i)},\varepsilon_{\mathrm{w}}^{(i)}}\left(\left|\mathbf{r} - \mathbf{r}'\right|\right) + \rho_{\mathrm{w}}^{(j)} \int\limits_{\nu_L} \mathrm{d}\mathbf{r}'\varphi^{\mathrm{LJ}}_{\sigma_{\mathrm{w}}^{(j)},\varepsilon_{\mathrm{w}}^{(j)}}\left(\left|\mathbf{r} - \mathbf{r}'\right|\right), \tag{12}$$

where $V_0^{(i)}\,(y)$ is Eq. (8) with the LJ parameters of material i. The integration in (12) is carried out over the volume ν_L of the stripe, excluding the coating: $\nu_L = \{(x, y, z) : -L/2 \le x \le L/2, -\infty < y \le -H_0, -\infty < z < \infty\}$. One effect that can be anticipated is that the stripe would serve to nucleate macroscopic configurations, such as sessile droplets or liquid films, below and above $T_{\mathrm{w}}^{(2)}$, respectively. Numerical experiments with DFT can yield valuable insight into the microscopic workings of these larger-scale phenomena. As an example, consider the computed fluid density profiles on a wall, made up of the stronger adsorbing material 1, with a stripe of the weaker material 2 (exact LJ parameters are given in the caption of Fig. 1) of width $L = 32$, at $T_1 = 0.7$ and $T_2 = 0.9$, where $T_1 < T_{\mathrm{w}}^{(2)} < T_2$, shown in Fig. 2. Below the wall wetting temperature $T_{\mathrm{w}}^{(2)}$, with the Young contact angle defined in Eq. (9) $\Theta^{(2)} > 0$, the stripe nucleates a sessile nanodroplet, which is metastable above saturation. On the other hand, above $T_{\mathrm{w}}^{(2)}$, where $\Theta^{(2)} = 0$, we observe a nucleating film, which is formed on the stripe at $\mu < \mu_{\mathrm{pw}}^{(2)}\,(T_2)$. Increasing μ towards the wall pre-wetting line leads to the growths of the film footprint along the wall, so that at $\mu = \mu_{\mathrm{pw}}^{(2)}$, the wall is fully covered by the thick pre-wetting film. Detailed numerical parametric studies of DFT equations can be used to reveal the physical mechanisms accompanying the formation of such nanodroplets and thin films.

In order to investigate wetting of a striped wall, we introduce the excess adsorption relatively to a homogeneous planar wall (without the stripe):

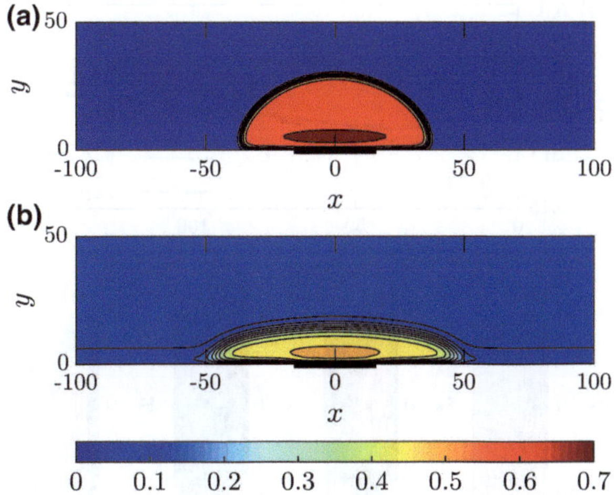

Fig. 2 Fluid density profiles near bulk coexistence on a wall of material 2 ($T_w^{(2)} \approx 0.81$) with a stripe of the stronger material 1 ($T_w^{(1)} \approx 0.60$), of width $L = 32$, located at $-16 \leq x \leq 16$, and designated by a black rectangle. The data in the contour plots is scaled according to the color bar provided, i.e., between light blue (gas-like densities) and dark red (liquid-like densities) (color online). **a** Metastable sessile nanodroplet at $T_1 = 0.7$ (the contact angle in Eq. (9) $\Theta^{(2)} = 0.63$) and $\Delta\mu_1 = 0.015$. **b** Nucleating flat pre-wetting film at $T_2 = 0.9$ ($\Theta^{(2)} = 0$) and $\Delta\mu_1 = -0.017 < \Delta\mu_{\mathrm{pw}}^{(2)}(T_2) = -0.016$

$$\Gamma = \int_{-\infty}^{\infty} dx \int_{0}^{\infty} dy \left[\rho_L(x, y) - \rho_0^{(\mathrm{w})}(y) \right], \tag{13}$$

where $\rho_L(x, y)$ and $\rho_0^{(\mathrm{w})}(x)$ are, respectively, the fluid density profiles on a wall with a stripe of width L, and on a homogeneous wall ($L = 0$), computed at the same values of T and μ. As it follows from Eq. (13), the adsorption Γ is a measure of matter adsorbed on the stripe, discounting the matter adsorbed on the rest of the wall. At a fixed temperature, the thermodynamic quantity conjugate to adsorption is the excess grand potential Ω_{ex}, which can be computed from the density profiles using (1) [38, 61]:

$$\Omega_{\mathrm{ex}} = \Omega\left[\rho_L(x, y)\right] - \Omega\left[\rho_0^{(\mathrm{w})}(x)\right]. \tag{14}$$

The relation between adsorption and excess grand potential is expressed by the Gibbs surface phase rule [30, 35, 38]:

$$\Gamma(\mu) = -\frac{1}{H}\left(\frac{\partial \Omega_{\mathrm{ex}}}{\partial \mu}\right)_T, \tag{15}$$

where H is the transverse dimension of the system (orthogonal to, e.g., Fig. 2). From (14) to (15) it follows, that if for the given T an adsorption isotherm $\Gamma(\mu)$ is computed

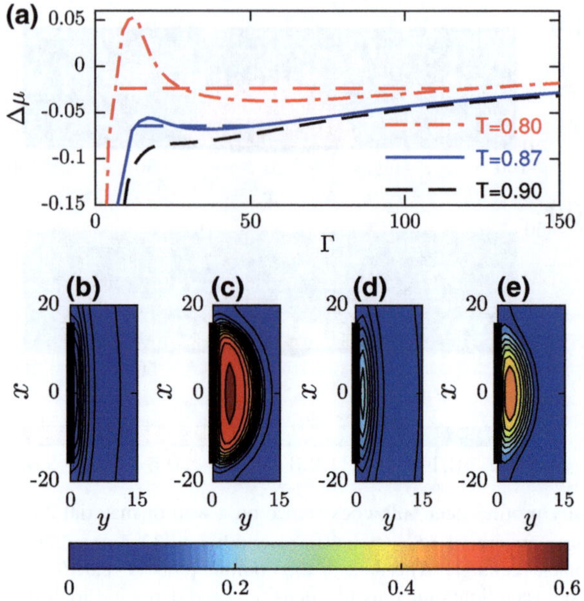

Fig. 3 Adsorption isotherms on a weaker wall of material 2 with a stripe of width $L = 32$ from the stronger material 1, at three different temperatures (**a**). At the two lower temperatures, unbending transitions occur at chemical potentials $\mu_u(T)$: $\Delta\mu_u(0.80) \approx -0.0236$ and $\Delta\mu_u(0.87) \approx -0.0645$, as determined by the equal area constructions (demarcated by the horizontal dashed lines). The isotherm at $T = 0.90$ is above the unbending critical temperature. Density profiles for the coexisting phases at $T = 0.8$ [(**b**) and (**c**)] and $T = 0.87$ [(**d**) and (**e**)] [60]

by finding the extrema of $\Omega[\rho(x, y)]$, the coexisting fluid configurations (those with equal grand free energies) can be found graphically, by an equal area construction performed on $\Gamma(\mu)$ [35, 38].

Figure 3 depicts adsorption isotherms for the case of the weak outer wall made up of material 2, and a strong stripe of material 1 (see caption to Fig. 1), at three different temperatures. At the two lower temperatures, an equal areas construction determines the coexisting adsorptions, which correspond to the so-called *unbending* transition [60, 62]. Note that in Fig. 3, for the highest temperature $T = 0.9$, such construction is impossible signalling that this isotherm is above the unbending critical temperature. Density profiles of the two coexisting phases at the lowest temperature $T = 0.8$ (where the loop is most pronounced) are also shown.

Unbending transition is a local condensation of liquid driven by the competition between the attractive substrate–fluid forces and the effect of the liquid–gas surface tension [62–65]. This phenomenon, closely related to the change from Wenzel to Cassie–Baxter states on rough surfaces, was first described for a corrugated wall, in which the troughs discontinuously fill with liquid below the wetting temperature,

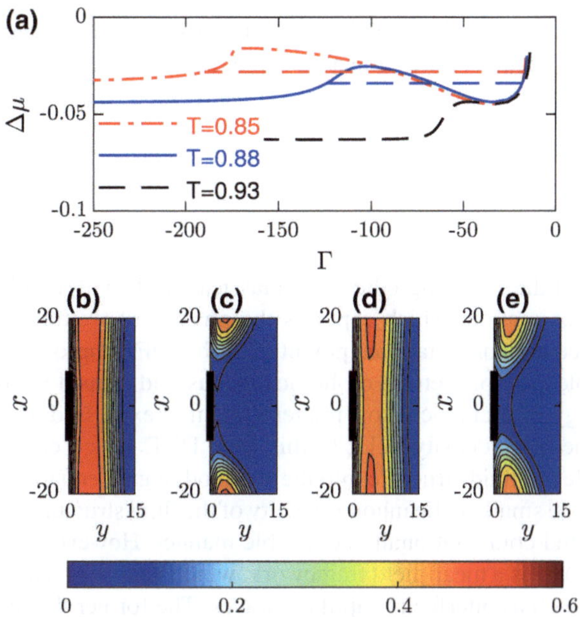

Fig. 4 Adsorption isotherms on a stronger wall ($\varepsilon_{\mathrm{w}} = 0.8$, $\sigma_{\mathrm{w}} = 2$, $H_0 = 5$) with a stripe of width $L = 16$ from the weaker material 2 (see Fig. 1), at three different temperatures (**a**). At the two lower temperatures, unbending transitions occur at chemical potentials $\mu_{\mathrm{u}}(T)$: $\Delta\mu_{\mathrm{u}}(0.85) \approx -0.0281$ and $\Delta\mu_{\mathrm{u}}(0.88) \approx -0.0340$, as determined by the equal area constructions (dashed lines). The isotherm at $T = 0.93$ is above the unbending critical temperature. Density profiles for the coexisting phases at $T = 0.85$ [(**b**) and (**c**)] and $T = 0.88$ [(**d**) and (**e**)]

thereby flattening the liquid–gas interface (hence, the term *unbending*) [66]. Similar phenomena were studied on chemically patterned substrates, including stripes and arrays of stripes [62, 67–70]. What was not appreciated in these earlier studies was how the pre-wetting transition is modified on patterned surfaces. Very interestingly, for walls that exhibit first-order wetting, unbending may act as a nucleation site for the thick pre-wetting film, and thus triggers the so-called *complete* pre-wetting, whereby the formation of the pre-wetting films occurs continuously along the substrate [60, 61]. In Fig. 3, the continuous formation of the pre-wetting film corresponds to the divergence of adsorption, which follows the scaling [60]

$$\Gamma \propto \left(\mu_{\mathrm{pw}}^{(1)} - \mu\right)^{-1/4}. \tag{16}$$

In Fig. 4 we present the unbending transition on a wall with a weaker stripe of width $L = 16$. As can be seen from the figure, the wetting scenario discussed above is essentially inverted, with the stripe nucleating the thin pre-wetting film, which also continuously unbinds as the boundary of pre-wetting is approached. Note that using the same computational mesh for all the three isotherms presented, we were able to follow the isotherm at the highest $T = 0.93$ only up to about $\Gamma \approx 160$. This is due to the fact that at high temperatures, the proportionality coefficient, implicit in Eq. (16) is large.

4 Conclusion

We have outlined the basic ingredients of classical DFT, with emphasis on a non-local DFT approximation, which captures the physics of wetting and adsorption, and demonstrated the computational potential of this DFT approximation by applying it to a simple case of wetting on homogeneous and striped walls. In DFT, the grand free-energy of a classical soft-matter system is expressed as a functional of the system's one-body density field. In this way, DFT, which can be viewed as a means to include the fluid structure into the thermodynamic equation of state, beautifully captures the small-scalle inhomogeneity of the fluid structure in a theoretically self-consistent and computationally accessible manner. However, it is important to remember that DFT is a mean-field framework, which neglects local thermal density fluctuations, as well as interfacial capillary-waves. The former fluctuations can lead to the rounding of the interfacial transitions, while the latter broaden the mean-field interfaces. Nevertheless, DFT investigations of wetting are valuable and instructive due to the high rigidity of liquid–gas interfaces, which essentially makes mean-field predictions valid for most practical purposes.

Clearly classical fluids on structured substrates can exhibit non-trivial wetting phenomenology and a multitude of new and exciting effects caused by the fluid inhomogeneity at the nanoscale and the non-locality of the intermolecular fluid–fluid and fluid–substrate interactions. In the examples we looked at, the surface exhibits two different types of wetting transitions: the first-order transition to wetting occurring at the wetting temperature of the outer wall (and the associated line of the pre-wetting transitions), and the unbending transition associated with the local jump of the interface height near the substrate inhomogeneity. Moreover, the interplay of unbending and pre-wetting leads to a change in the order of the pre-wetting transition making it a continuous phenomenon. As well as being of fundamental interest to the modern statistical mechanical theory of inhomogeneous fluids, the transitions discussed here should be also relevant to a wide spectrum of technological applications such as nanofluidics, chemical- and bioengineering, and the design of lab-on-a-chip devices.

Acknowledgements PY is grateful to Dr. Miguel A. Durán-Olivencia from the Chemical Engineering Department of Imperial College (IC) for numerous stimulating discussions. We acknowledge financial support from the Engineering and Physical Sciences Research Council (EPSRC) of the UK through Grants No. EP/L027186 and EP/L020564 as well as an EPSRC-IC Pathways to Impact-Impact Acceleration Award, Grant No. EP/K503733, European Research Council via Advanced Grant No. 247031 and the European Framework 7 via Grant No. 214919 (Multiflow).

References

1. X. Xu, G. Vereecke, C. Chen, G. Pourtois, S. Armini, N. Verellen, W.-K. Tsai, D.-W. Kim, E. Lee, C.-Y. Lin, P.V. Dorpe, H. Struyf, F. Holsteyns, V. Moshchalkov, J. Indekeu, S.D. Gendt, Capturing wetting states in nanopatterned silicon. ACS Nano **8**, 885 (2014)
2. S. Herminghaus, M. Brinkman, R. Seeman, Wetting and dewetting of complex surface geometries. Annu. Rev. Mater. Res. **38**, 101 (2008)
3. D. Lohse, X. Zhang, Surface nanobubbles and nanodroplets. Rev. Mod. Phys. **87**, 981 (2015)
4. A. Calvo, B. Yameen, F.J. Williams, G.J.A.A. Soler-Illia, O. Azzaroni, Mesoporous films and polymer brushes helping each other to modulate ionic transport in nanoconfined environments. An interesting example of synergism in functional hybrid assemblies. J. Am. Chem. Soc. **131**, 10866 (2009)
5. M.G. Knepley, D.A. Karpeev, S. Davidovits, R.S. Eisenberg, D. Gillespie, An efficient algorithm for classical density functional theory in three dimensions: ionic solutions. J. Chem. Phys. **132**, 124101 (2010)
6. J. Bleibel, A. Dominguez, M. Oettel, S. Dietrich, Capillary attraction induced collapse of colloidal monolayers at fluid interfaces. Soft Matter **10**, 4091 (2014)
7. P.E. Theodorakis, A. Chremos, Morphologies of bottle-brush block copolymers. ACS Nano Lett. **3**, 1096 (2014)
8. A. Checco, B.M. Ocko, M. Tasinkevych, S. Dietrich, Stability of thin wetting films on chemically nanostructured surfaces. Phys. Rev. Lett. **109**, 166101 (2012)
9. Z. Gou, W. Liu, Biomimic from the superhydrophobic plant leaves in nature: binary structure and unitary structure. Plant Sci. **172**, 1103 (2007)
10. T.M. Squires, S. Quake, Microfluidics: fluid physics at the nanoliter scale. Rev. Mod. Phys. **77**, 977 (2005)
11. M. Rauscher, S. Dietrich, Wetting phenomena in nanofluidics. Annu. Rev. Mater. Res. **38**, 143 (2008)
12. K. Binder, Modelling of wetting in restricted geometries. Annu. Rev. Mater. Res. **38**, 123 (2008)
13. H. Craighead, Future lab-on-a-chip technologies for interrogating individual molecules. Nature **442**, 387 (2006)
14. K.W. Schwarz, J. Tersoff, From droplets to nanowires: dynamics of vapor-liquid-solid growth. Phys. Rev. Lett. **102** (2009)
15. R.E. Algra, M.A. Verheijen, L.-F. Feiner, G.G.W. Immink, W.J.P. van Enckevort, E. Vlieg, E.P.A.M. Bakkers, The role of surface energies and chemical potential during nanowire growth. Nano Lett. **11**, 1259 (2011)
16. C. Rascón, A.O. Parry, Geometry-dominated fluid adsorption on sculpted solid substrates. Nature **407**, 986 (2000)
17. O. Gang, K.J. Alvine, M. Fukuto, P.S. Pershan, C.T. Black, B.M. Ocko, Liquids on topologically nanopatterned surfaces. Phys. Rev. Lett. **95**, 217801 (2005)
18. R. Seeman, M. Brinkman, E.J. Kramer, F.F. Lange, R. Lipowsky, Wetting morphologies at microstructured surfaces. Proc. Natl. Acad. Sci. USA **102**, 1848 (2005)
19. M. Nosonvsky, B. Bhushan, Superhydrophobic surfaces and emerging applications: non-adhesion, energy, green engineering. Curr. Opin. Colloid Interface Sci. **14**, 270 (2009)
20. Z. Wang, J. Wei, P. Morse, J.G. Dash, O.E. Vilches, D.H. Cobden, Phase transitions of adsrobed atoms on the surface of a carbon nanotube. Science **327**, 552 (2010)
21. W.F. Saam, Wetting, capillary condensation and more. J. Low Temp. Phys. **157**, 77 (2009)
22. R. Evans, The nature of the liquid-vapour interface and other topics in the statistical mechanics of non-uniform, classical fluids. Adv. Phys. **28**, 143 (1979)
23. P. Hohenberg, W. Kohn, Inhomogeneous electron gas. Phys. Rev. **136**, B864–B871 (1964)
24. N.D. Mermin, Thermal properties of the inhomogeneous electron gas. Phys. Rev. **137**, A1441–A1443 (1965)
25. H. Lowen, Density functional theory: from statics to dynamics. J. Phys. Condens. Matter **15**, V1 (2003)

26. J.F. Lutsko, Recent developments in classical density functional theory. Adv. Chem. Phys. Wiley (2010), p. 1
27. J. Landers, J.Y. Gor, A.V. Neimark, Density functional theory methods for characterization of porous materials. Colloid Surf. A **437**, 3 (2013)
28. J. Wu, Density functional theory for chemical engineering: from capillarity to soft materials. AIChE J. **52**, 1169 (2006)
29. L.J.D. Frink, A.G. Salinger, M.P. Sears, J.D. Weinhold, A.L. Frischknecht, Numerical challenges in the application of density functional theory to biology and nanotechnology. J. Phys. Condens. Matter **14**, 12167 (2002)
30. R. Evans, Fundamentals of inhomogeneous fluids, in *Chapter Density Functionals in the Theory of Nonuniform Fluids* (Dekker, New York, 1992), p. 85
31. J.F. Lutsko, M.A. Durán-Olivencia, A two-parameter extension of classical nucleation theory. J. Phys. Condens. Matter **27**, 235101 (2015)
32. J.F. Lutsko, M.A. Durán-Olivencia, Classical nucleation theory from a dynamical approach to nucleation. J. Chem. Phys. **138**, 244908 (2013)
33. J.A. Barker, D. Henderson, Perturbation theory and equation of state for fluids. II. A successful theory of liquids. J. Chem. Phys. **47**, 4714 (1967)
34. N.F. Carnahan, K.E. Starling, Equation of state for nonattracting rigid spheres. J. Chem. Phys. **51**, 635 (1969)
35. P. Yatsyshin, N. Savva, S. Kalliadasis, Density functional study of condensation in capped capillaries. J. Phys. Condens. Matter **27**, 275104 (2015)
36. A. Mulero, *Theory and Simulation of Hard-Sphere Fluids and Related Systems*. Lecture Notes in Physics, vol. 753 (Springer, Berlin, Heidelberg, 2008)
37. D.E. Sullivan, M.M. Telo da Gama, Wetting transitions and multilayer adsorption at fluid interfaces, in *Fluid Interfacial Phenomena*, ed. by C.A. Croxton (Wiley, New York, 1986), p. 45
38. P. Yatsyshin, N. Savva, S. Kalliadasis, Wetting of prototypical one- and two-dimensional systems: thermodynamics and density functional theory. J. Chem. Phys. **142**, 034708 (2015)
39. P. Yatsyshin, N. Savva, S. Kalliadasis, Geometry-induced phase transition in fluids: capillary prewetting. Phys. Rev. E **87**, 020402(R) (2013)
40. P. Tarazona, R. Evans, A simple density functional theory for inhomogeneous liquids. Wetting by gas at a solid-liquid interface. Mol. Phys. **52**, 847 (1984)
41. P. Tarazona, J.A. Cuesta, Y. Martinez-Raton, Density functional theories of hard particle systems, in *Theory and Simulations of Hard-Sphere Fluids and Related Systems*, ed. by A. Mulero. Lecture Notes in Physics, vol. 753 (Springer, Berlin, Heidelberg, 2008), p. 251
42. R. Roth, Fundamental measure theory for hard-sphere mixtures: a review. J. Phys. Condens. Matter **22**, 063102 (2010)
43. P. Yatsyshin, N. Savva, S. Kalliadasis, Spectral methods for the equations of classical density-functional theory: relaxation dynamics of microscopic films. J. Chem. Phys. **136**, 124113 (2012)
44. A. Archer, R. Evans, Dynamical density functional theory and its application to spinodal decomposition. J. Chem. Phys. **121**(9), 4246–4254 (2004)
45. P. Hohenberg, B.I. Halperin, Theory of dynamic critical phenomena. Rev. Mod. Phys. **49**, 435 (1977)
46. U.M.B. Marconi, P. Tarazona, Nonequilibrium inertial dynamics of colloidal systems. J. Chem. Phys. **124**(16), 164901–164911 (2006)
47. U.M.B. Marconi, P. Tarazona, Dynamic density functional theory of fluids. J. Phys. Condens. Matter **12**, A413–A418 (2000)
48. U.M.B. Marconi, P. Tarazona, Dynamical density functional theory of fluids. J. Chem. Phys. **110**, 8032–8044 (1999)
49. M.A. Durán-Olivencia, J.F. Lutsko, Mesoscopic nucleation theory for confined systems: a one-parameter mode. Phys. Rev. E **91**, 022402 (2015)
50. B.D. Goddard, A. Nold, S. Kalliadasis, Multi-species dynamical density functional theory. J. Chem. Phys. **138**, 144904 (2013)

51. B.D. Goddard, A. Nold, N. Savva, G.A. Pavliotis, S. Kalliadasis, General dynamical density functional theory for classical fluids. Phys. Rev. Lett. **109**, 120603 (2012)
52. B.D. Goddard, A. Nold, N. Savva, P. Yatsyshin, S. Kalliadasis, Unification of dynamic density functional theory for colloidal fluids to include inertia and hydrodynamic interactions: derivation and numerical experiments. J. Phys. Condens. Matter **25**, 035101 (2013)
53. M.A. Durán-Olivencia, B.D. Goddard, S. Kalliadasis, Dynamical density functional theory for orientable colloids including inertia and hydrodynamic interactions. J. Stat. Phys. **164**, 785 (2016)
54. P. Yatsyshin, S. Kalliadasis, Mean-field phenomenology of wetting in nanogrooves. Mol. Phys. **114**, 2688 (2016)
55. D. Bonn, D. Ross, Wetting transitions. Rep. Prog. Phys. **64**, 1085 (2001)
56. S. Dietrich. Wetting phenomena, in *Phase Transitions and Critical Phenomena*, ed. by C. Domb, J.L. Lebowitz, vol. 12 (Academic Press, 1988), p. 2
57. R. Evans, A.O. Parry, Liquids at interfaces: what can a theorist contribute? J. Phys. Condens. Matter **2**, SA15 (1990)
58. G. Forgacs, R. Lipowsky, T.M. Nieuwenhuizen, The behaviour of interfaces in ordered and disordered systems, in *Phase Transitions and Critical Phenomena*, ed. by C. Domb, J.L. Lebowitz, vol. 14 (Academic Press, 1991), p. 135
59. E.H. Hauge, M. Schick, Continuous and first-order wetting transition from the van der Waals theory of fluids. Phys. Rev. B **27**, 4288 (1983)
60. P. Yatsyshin, A.O. Parry, C. Rascón, S. Kalliadasis, Classical density functional study of wetting transitions on nanopatterned surfaces. J. Phys. Condens. Matter **29**, 094001 (2017)
61. P. Yatsyshin, A.O. Parry, S. Kalliadasis, Complete prewetting. J. Phys. Condens. Matter **28**, 275001 (2016)
62. C. Rascón, A.O. Parry, Surface phase diagrams for wetting on heterogenous substrates. J. Chem. Phys. **115**, 5258 (2001)
63. C. Bauer, S. Dietrich, A.O. Parry, Morphological phase transitions of thin fluid films on chemically structured substrates. Europhys. Lett. **47**, 474 (1999)
64. C. Rascón, A.O. Parry, Wetting on non-planar and heterogeneous substrates. J. Phys. Condens. Matter **12**, A369 (2000)
65. C. Bauer, E. Dietrich, Phase diagram for morphological transitions of wetting films on chemically structured substrates. Phys. Rev. E **61**, 1664 (2000)
66. C. Rascón, A.O. Parry, A. Sartori, Wetting at nonplanar substrates: unbending and unbinding. Phys. Rev. E **59**, 5697 (1999)
67. C. Bauer, S. Dietrich, Phase diagram for morphological transitions of wetting films on chemically structured substrates. Phys. Rev. E **61**, 1664 (2000)
68. W. Koch, S. Dietrich, M. Napiorkowski, Morphology and line tension of liquid films adsorbed on chemically structured substrates. Phys. Rev. E **51**, 3300 (1995)
69. C. Bauer, S. Dietrich, Quantitative study of laterally inhomogeneous wetting films. Eur. Phys. J. B **10**, 767 (1999)
70. C. Bauer, S. Dietrich, Wetting films on chemically heterogeneous substrates. Phys. Rev. E **60**, 6919 (1999)

Modeling Metastability in CdTe Solar Cells Due to Cu Migration

Da Guo, Daniel Brinkman, Abdul R. Shaik, I. Sankin, D. Krasikov, Christian Ringhofer and Dragica Vasileska

Abstract Thin-film modules of all technologies often suffer from performance degradation over time. Some of the performance changes are reversible and some are not, which makes deployment, testing, and energy-yield prediction more challenging. Manufacturers devote significant empirical efforts to study these phenomena and to improve semiconductor device stability. Still, understanding the underlying reasons of these instabilities remains clouded due to the lack of ability to characterize materials at atomistic levels and the lack of interpretation from the most fundamental material science. The most commonly alleged causes of metastability in CdTe device, such as "migration of Cu," have been investigated rigorously over the past fifteen years. Still, the discussion often ended prematurely with stating observed correlations between stress conditions and changes in atomic profiles of impurities or CV doping concentration. *Multiple hypotheses suggesting degradation of CdTe solar cell devices due to interaction and evolution of point defects and complexes were proposed, and none of them received strong theoretical or experimental confirmation.* It should be noted that atomic impurity profiles in CdTe provide very little intelligence on active

D. Guo · A. R. Shaik · C. Ringhofer · D. Vasileska (✉)
Arizona State University, Tempe, AZ, USA
e-mail: vasileska@asu.edu

D. Guo
e-mail: da.guo@asu.edu

A. R. Shaik
e-mail: arshaik@asu.edu

C. Ringhofer
e-mail: ringhofer@asu.edu

D. Brinkman
San Jose State University, San Jose, CA, USA
e-mail: daniel.brinkman@sjsu.edu

I. Sankin · D. Krasikov
First Solar, Tempe, AZ, USA
e-mail: isanking@firstsolar.com

D. Krasikov
e-mail: dmitry.krasikov@firstsolar.com

L. L. Bonilla et al. (eds.), *Coupled Mathematical Models for Physical and Biological Nanoscale Systems and Their Applications*, Springer Proceedings in Mathematics & Statistics 232, https://doi.org/10.1007/978-3-319-76599-0_11

doping concentrations. The same elements could form different energy states, which could be either donors or acceptors, depending on their position in crystalline lattice. Defects interact with other extrinsic and intrinsic defects; for example, changing the state of an impurity from an interstitial donor to a substitutional acceptor often is accompanied by generation of a compensating intrinsic interstitial donor defect. Moreover, all defects, intrinsic and extrinsic, interact with the electrical potential and free carriers so that charged defects may drift in the electric field and the local electrical potential affects the formation energy of the point defects. *Such complexity of interactions in CdTe makes understanding of temporal changes in device performance even more challenging and a closed solution that can treat the entire system and its interactions is required.* In this book chapter we first present validation of the tool that is used to analyze Cu migration in single crystal (sx) CdTe bulk. Since the usual diffusion analysis has limited validity, our simulation approach presented here provides more accurate concentration profiles of different Cu defects that lead to better understanding of the limited incorporation and self-compensation mechanisms of Cu in CdTe. Finally, simulations are presented that study Cu ion's role in light soaking experiments of CdTe solar cells under zero-bias and forward-bias stress conditions.

Keywords Thin-film solar cells · CdTe · Defect migration · Cu metastability · reliability

1 Introduction

The push towards thin-film technology has been driven largely by predictions of future economic viability [1–4]. Traditional single-crystal solar cells, such as Si and GaAs, demonstrate very high efficiencies (20–30%), but the production of crystalline material is expensive. The original reason thin-film materials were pursued was because they use much less material, which is directly related to the cost of production. Two of the leading thin-film materials are CdTe and $CuInGaSe_2$, chosen because their direct bandgaps require a smaller absorption length than Si (requires less thickness for optimum performance). CdTe is a nearly ideal material for terrestrial solar cell production, as its band gap of 1.45 eV (room temperature) yields the maximum theoretical efficiency for a solar cell, of about 32% (see Fig. 1 for the Schockley-Queisser limit for single cell under AM1.5 illumination.) The current record one-of-a-kind laboratory research cell was fabricated in 2017 by First Solar (FSLR) and has an efficiency of 22.1% [5].

Despite overwhelming advances in thin-film CdTe technology in recent years, performance of thin-film CdTe devices is still a subject to various metastable phenomena that could be characterized by temperature-dependent time constants (activation energies). Most of these metastable changes in CdTe are known to be reversible and require different recovery procedures; however, based on experimental and theoretical investigations at First Solar, metastabilities in CdTe device cannot be explained

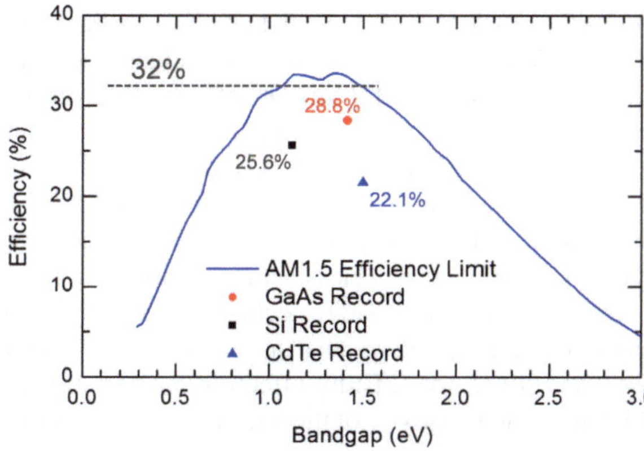

Fig. 1 Schockley-Queisser limit

solely by electronic capture-emission phenomena assuming fixed distributions of point defects.

Many of the physical properties of crystalline solids are determined by the presence of native or foreign point defects. In pure compound crystals the native defects are atoms missing from lattice sites where, according to the crystal structure, atoms should be (vacancies); atoms present at sites where atoms should not be (interstitials); and atoms occupying sites normally occupied by other atoms (substitutional). In addition, there may be defects in the electronic structure: quasi-free electrons in the conduction band or electrons missing from the valence band (holes). In impure or doped crystals there are also defects involving the foreign atoms. These may occupy normal lattice sites (substitutional foreign atoms) or interstitial sites (interstitial foreign atoms). In elemental crystals similar point defects occur; only misplaced atoms are missing.

In addition to point defects, the performance of CdTe solar cells is affected by extended defects which include dislocations, stacking faults, grain boundaries (GBs) and inclusions of second phases. Dislocations and GBs are well known to attract impurities, and to promote diffusion [6]. Such effects might be expected to lead to instability in devices, or to have an influence on the thermal processing conditions chosen to fabricate certain devices. An example is the inter-diffusion of CdTe and CdS in polycrystalline solar cells for which the grain boundary diffusion coefficient has been measured [7]. Grain boundary segregation is well known in metals (e.g. Cu in Pb [8]), the driving force being strain reduction at the boundary plane. Decoration of grain and twin boundaries in CdTe with Te inclusions is widely reported. Minor component impurities in CdTe have also been shown to segregate out to grain boundary regions [9] and to dislocation arrays [10].

The electrical states associated with grain boundaries and dislocations can have a number of adverse effects on the performance of CdTe solar cells. Firstly, the deep states associated with extended defects can promote undesirable recombination. Secondly, grain boundaries act as charge transport barriers. This is attributed to the grain boundary manifold being a charged interface, causing it to present an electrical barrier to current transport. Such barriers have been observed directly for CdTe using the so-called 'remote' electron beam induced current (EBIC) method [11] and are considered responsible for limiting effects in polycrystalline solar cells [12]. Thirdly, grain boundaries and dislocations may act as conduits for current transport rather than barriers. The impact of the grain boundaries is likely to depend on their position in the layer; near surface grain boundaries are likely to be Te-rich (i.e. conducting) on account of the etching used to prepare contacts, whereas those remote from the free surface may nevertheless act as recombination centers.

In today's thin-film CdTe technology, Cu is the key dopant that defines major performance parameters such as open-circuit voltage (Voc), short-circuit current (Jsc), and fill-factor (FF) by affecting built-in potential of the junction, collection efficiency, and resistivity of the back contact [13]. However, fast diffusion of Cu from the back contact toward the main junction is believed to contribute to degradation observed in long-term stability studies [14]. It was determined that while modest amounts of Cu enhance cell performance, excessive amounts degrade device quality and reduce performance [15]. Evolution of Cu-related point defects and complexes in CdTe grains and at grain boundaries (GBs) is expected to cause pronounced effect on device performance leading to observed metastable phenomena.

Interactions of Cu in CdTe involve multiple intrinsic and extrinsic point defects and complexes, and as a result, cannot be analyzed in isolation from the rest of the system. Although the other defects in CdTe system could be assumed relatively slow diffusors at typical operating and storage conditions of CdTe device, direct measurement of their distributions is very challenging. Moreover, rapid development of technology involves frequent changes of film growth/activation conditions and, therefore, resulting setup for Cu-related point defects and complexes. All of the above makes quantitative understanding of metastable phenomena in CdTe device virtually impossible with the off-the-shelf tools that researchers have currently.

2 Theoretical Model

As already noted in the Introduction part of this book chapter, the evolution of Cu-related point defects and complexes in CdTe grains and at GBs is expected to cause a pronounced effect on device performance leading to observed metastable phenomena. Interactions of Cu in CdTe involve multiple intrinsic and extrinsic point defects and complexes, and, as a result, cannot be analyzed in isolation from the rest of the system. Understanding the fundamentals behind performance and metastability of CdTe devices requires a model that captures and describes most relevant processes at the lowest level specific to CdTe system and could be confirmed with experiment.

We consider such a model in a form of a self-consistent system of time-dependent reaction-diffusion equations [16–18] describing the interactions and the evolution of point defects and complexes coupled with the Poisson equation (see Fig. 2). Because device performance is uniquely defined by device geometry, band-structure of semiconductors, and distributions of charge and recombination centers, the Unified Solver

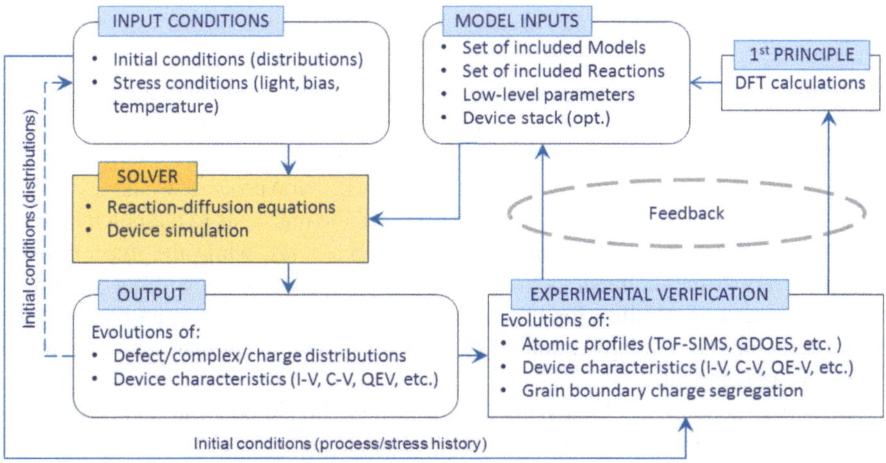

Fig. 2 Schematic block-diagram that illustrates the use of the Unified Solver to tune the model and study CdTe device metastability

has to be able to simulate macroscopic device characteristics such as current, capacitance and quantum efficiency as a function of the applied bias, DC light intensity, and ambient temperature. Such capabilities establish a tighter link between the microscopic core of the model and macroscopic experimental verification. The core of the solution is a multi-level solver that combines macroscopic diffusion-reaction equations describing sub-systems of the point defects with the global Poisson equation to form a closed system that is solved in time domain and quasi-3D space utilizing grains of specific shape. The characteristic length scale of the features is large enough that the semi-classical approximation (implicitly assumed by using reaction-diffusion equations) is valid. The developed Unified Solver offers flexibility in choosing (turning "on" and "off") models and reactions that involve selected point defects and complexes in individual materials or domains.

The species under investigation are described by sets of low-level parameters that include their formation energy, ionization energies and diffusion coefficients for different charged states, solubility limits, grain boundary segregation parameters. The evolution of the system for a given set of stressors (temperature, light, and bias) is calculated based on provided initial conditions (distributions). The solver outputs the distributions of charged and neutral dopants and recombination states. Given the device geometry and band structure of semiconductors, the solver uses these distributions to simulate I-V, C-V and QEV trends that could be confirmed experimentally on real device structures.

2.1 Physical Model

A specific example of importance is the penetration of Cu into CdTe absorber using a high temperature diffusion anneal. The Cu penetration involves several processes, including the Cu-Cd exchange reaction and the drift/diffusion of mobile defects, namely Cu_i^+ and Cd_i^{+2}. To simulate such reaction-diffusion process we use a standard multi-compartment approach [19], which assumes the reactions to happen inside the finite homogeneous compartments (well-stirred reactors), while the mass transfer between compartments happens via pure drift/diffusion process without any effect of defect reactions. Within this multi-compartment approach, we can describe the reactions using the standard chemical kinetics formalism in application to the defect chemistry in solid state.

We write the main Cu-Cd exchange reaction of the Cu penetration process as a bimolecular exchange reaction facilitated by knock-off:

$$Cu_{Cd}^- + Cd_i^{+2} \underset{K^b}{\overset{K^f}{\rightleftharpoons}} Cu_i^+ + Cd_{Cd} \tag{1}$$

where K^f and K^b are reaction rate constants. In the forward exchange reaction (1), the double donor interstitial Cd defect (Cd_i^{+2}) reacts with the acceptor substitutional Cu defect (Cu_{Cd}^-), kicks-out Cu into an interstitial position and occupies its place forming regular Cd-on-cation-site lattice atom (Cd_{Cd}). Such exchange reaction is one of the three types of elementary bimolecular defect reactions, while two others being the formation of a complex pair defect of two isolated point defects and the recombination of interstitial point defect with vacancy [20]. The rate of such bimolecular exchange reaction is proportional to the concentrations of both reactants (Cu_{Cd}^- and Cd_i^{+2}).

The backward reaction, knock-off of Cd by Cu_i^+, is treated as monomolecular with only one reactant (Cu_i^+) because in the uniform binary CdTe matrix in the diluted concentrations approximation, this backward reaction does not require the reactants (Cu_i^+ and Cd_{Cd}) to find each other (the regular Cd lattice atoms are available nearby any Cu_i). Therefore, the rate of the backward reaction (1) is proportional only to the concentration of one reactant Cu_i^+.

The above considerations allow us to define the reaction rates (R^f, R^b) in the following form:

$$\begin{cases} R^f = K^f[Cu_{Cd}^-][Cd_i^{+2}] \\ R^b = K^b[Cu_i^+] \end{cases} \tag{2}$$

where the superscripts f and b indicate the forward and backward reactions, $[X]$ is the concentration of defect X and $K^{f,b}$ is the corresponding reaction rate constant.

As no reliable experimental methods exist for determination of the rates of the reactions between individual point defects, we use first principles-based analysis to estimate these reaction rates. Following the standard approach for defect chemistry in solids, we consider two major contributions to the Gibbs free energy of point defects, namely the formation enthalpy and the configuration entropy. The calculation of the formation enthalpy of different atomic configurations in the supercell approach allows analyzing the potential energy landscapes and finding the most favorable states of single neutral and charged defects as well as the minimum energy pathways for single reactions. The inclusion of the configuration entropy changes in defect reactions allows accounting for the effect of temperature and defect concentrations on the reaction rates. More details on first-principles calculations can be found in recent works [20, 21].

Analysis of the potential energy landscape for reacting Cu_{Cd}^{-} and Cd_i^{+2} defects shows that the highest energy barrier in the forward reaction (1) is the Cd_i^{+2} diffusion barrier and the formation enthalpy of the products is lower than that of reactants [20, 22]. This allows us to write the forward rate constant as a steady-state rate constant of diffusion-controlled reaction between non-interacting uniformly distributed species (see e.g. [22]). To derive the backward rate constant we use the principle of the detailed balance:

$$\begin{cases} K^f = 4\pi D R_{capt} \\ K^b = K^f K^{eq} C_s \end{cases} \tag{3}$$

In Eq. (3), R_{capt} represents the capture radius of reactants, here for reaction between the attractive oppositely charged defects with charges q_1, q_2 we use the Onsager capture radius $R_{capt}^{q_1q_2} = {q_1q_2}/{4\pi\varepsilon\varepsilon_0 kT}$ [23]; D is the sum of diffusivities of reactants dominated by the diffusivity of mobile Cd_i^{+2} defect in our case; $K^{eq} = \exp\left(-{\Delta H}/{kT}\right)$ is equilibrium constant of reaction (1), ΔH is the change of formation enthalpy in the forward reaction; $C_s = 1.48 \times 10^{22}$ cm^{-3} is the concentration of regular lattice sites in CdTe. This approach of calculating the reaction constants delivers equilibrium distribution of defects if the simulation time is long enough.

The overall time-space evolution of point defects involved in Cu penetration is described by the following set of reaction-diffusion equations:

$$\begin{cases} \frac{d[Cu_i^+]}{dt} = -\nabla \cdot J_{Cu_i^+} + R^f - R^b \\ \frac{d[Cd_i^{+2}]}{dt} = -\nabla \cdot J_{Cd_i^{+2}} - R^f + R^b \\ \frac{d[Cu_{Cd}^-]}{dt} = -R^f + R^b \end{cases} \tag{4}$$

Note that in Eq. (4), diffusion of Cu_{Cd} has been ignored due to very small diffusion coefficient [22]. The fluxes J_X in Eq. (4) result from both the diffusion due to concentration gradients and from the drift due to external electrostatic field:

$$J_X = -D_X \nabla[X] + \upsilon_X[X] \tag{5}$$

Assuming Boltzmann statistics (valid for diluted concentrations) and the charge θ carried by the defect, its drift velocity υ_x in electric field F equals to $\mu_X F$, where mobility μ_X is expressed via diffusion coefficient using Einstein relation $\mu_X = \theta\frac{D_X}{kT}$.

In other words:

$$\upsilon_X = \theta \frac{D_X}{kT} F \tag{6}$$

The ionization states of the defects can be calculated from first principle calculations. The diffusion coefficients of point defects have temperature dependence provided by Arrhenius expression:

$$D = D_0 \exp\left(-\frac{\Delta E_D}{kT}\right) \tag{7}$$

where D_0 and ΔE_D are diffusivity prefactor and the energy barrier of elementary diffusion jumps, respectively. These parameters for many important point defects in CdTe have been recently calculated from first principles in works [22, 24, 25].

The electric field in the film is determined not only by the electron/hole concentrations, but also by the ionized defect distributions inside the grain and its boundaries (surfaces), and can be found by solving the Poisson equation:

$$\nabla \cdot (\varepsilon_S \nabla V) = -q\left(p - n + [Cu_i^+] + 2[Cd_i^{+2}] - [Cu_{Cd}^-]\right) \tag{8}$$

In Eq. (8), p stands for the hole concentration, n is the electron concentration and ε_S is the spatially varying dielectric constant of the materical that comprise the CdTe solar cell.

Boundary conditions for diffusion-reaction Eq. (8) are provided by the sink of defects near the boundary given by:

$$J_{Gb} = -\sigma_X\left(C_X - C_{eq}\right). \tag{9}$$

In Eq. (9), σ_X is the recombination/generation rate and C_{eq} is the equilibrium concentration of the defect at the boundary. Usually, the velocity σ_X is proportional to the diffusion constant of the defect. In most cases, when the strong sink of defects at grain boundary is assumed, C_{eq} is considered to be constant. Beyond this approximation, the effect of grain boundary segregation has to be taken into account. The accumulation of impurity atoms at boundaries changes the chemical potential and hence makes C_{eq} floating. By introducing the floating chemical potential of impurities, the properties of the grain boundary could be taken into account. According to theories of grain boundary segregation, the bulk concentration at grain surface is given by:

$$C_{surface}(t) = \exp\left(-\frac{\Delta G}{kT}\right)\frac{C_s^b(t)}{1 - C_s^b(t)}. \tag{10}$$

Eq. (10) provides connection between the impurity concentrations at grain boundary and bulk by introducing ΔG—the atomic energy difference between the grain boundary and in the bulk. C_s^b in Eq. (10) represents the fraction of the grain boundary sites occupied by impurity atoms.

One can generalize the above notation and compactly write the Poisson and the diffusion-reaction equations as:

$$
\begin{aligned}
-\nabla \cdot (\varepsilon_s \nabla V) &= q(p - n + \sum_i \theta_i X_i) \\
\partial_t p - \nabla \cdot (D_p \nabla p + \mu_p p \nabla \phi) &= R(n, p) \\
\partial_t n - \nabla \cdot (D_n \nabla n + \mu_n n \nabla \phi) &= R(n, p) \\
\partial_t X_i - \nabla \cdot (D_i \nabla X_i + \mu_i \theta_i X_i \nabla \phi) &= R_i(\bar{X})
\end{aligned}
\tag{11}
$$

where X_i is the concentration of the ith type of defect and θ_i is the charge of the ith-type of defect. We also define $\phi = V + V_{bi}$ where V_{bi} corresponds not only to the usual built-in-voltage, but also includes terms that account for heterojunctions. In future analysis this term will include effects due to the change in chemical potentials for charge carriers in grain boundaries, to account for different materials, or both. (Note that for neutral particles ($\theta_i = 0$) this will require including a term in the fourth equation which is not proportional to θ). We assume that V_{bi} is constant with respect to time and therefore does not require an additional partial differential equation.

2.2 *Numerical Methods*

The major challenge in numerically solving the system of diffusion-reaction equations that are coupled to a global Poisson equation solver—given by the Eq. (11)—is the presence of vastly different time and spatial scales. For example, Cu is a fast diffusor, whereas the diffusion of Cl occurs on a longer time scale and both of these are slow compared to energetically favorable ionization reactions.

For the 2D case, we have previously developed a Unified Solver based on the first order implicit Euler method for the time integration and a Slotboom Finite Element method in space [19].

2.2.1 Time Splitting

Leaving aside for a moment the issue of space discretization, we developed a time discretization scheme which allows us to decouple the equations (in the sense that each equation can be solved independently for a given time step). In particular, there are two types of coupling. The electrical coupling is the most involved, with every (charged) defect appearing in Poisson's equation and the derivative of the potential appearing in each of the (charged) defect equations. This drift term involves only a single species of defect, but is nonlocal in space. The other coupling is through the $R_i(\hat{X})$ term, which simultaneously couples all of the reaction-diffusion equations, but acts locally in space.

The usual first step in the numerical solution of Eq. (11) is to use a Gummel-type iteration alternating between calculation of Poisson's equation and the reaction-diffusion equation [26]. This replaces the solution of a large system of coupled equations by the repeated solution of many smaller equations. When the number of degrees of freedom is large (as is usually the case in 2D or 3D), the iterative scheme will solve the system more efficiently for any reasonable iteration tolerance.

As first step we reformulate the transport equations, using so-called Slotboom variables of the form $u_i = e^{\frac{\theta_i \phi_i}{U_T}} X_i$. This transforms the transport equations for the defects into

$$\partial_t X_i - \nabla \cdot \left(\mu_i U_T e^{-\frac{\theta_i \phi_i}{U_T}} \nabla u_i \right) = 0 \tag{12}$$

The advantage of the formulation is that the spatial differential operator is self-adjoint in the variables u_i, yielding a stable disretization of the transport equation for large electric fields and, consequently, large spatial variations in the potential ϕ. We note that the Slotboom variables, u_i, will exhibit a large dynamic range due to the exponential in the variable transform. Thus, the primary variables will always be the concentrations X_i and the transport equations will always be discretized in the form

$$\partial_t X_i - \nabla \cdot \left(\mu_i U_T e^{-\frac{\theta_i \phi_i}{U_T}} \nabla \left(e^{\frac{\theta_i \phi_i}{U_T}} X_i \right) \right) = 0 \tag{13}$$

where the spatial derivatives are approximated numerically by differentiating $u_i = e^{\frac{\theta_i \phi_i}{U_T}} X_i$.

We utilize an operator splitting (a fractional step-method) for the time discretization of the defect equations, separating the diffusion from the reaction terms. The operator splitting method is of the following form: Given X_i at time t_K and a time step Δt

- Step 1: Solve

$$\partial_t X_i^{(1)} - \nabla \cdot \left(\mu_i U_T e^{-\frac{\theta_i \phi_i}{U_T}} \nabla \left(e^{\frac{\theta_i \phi_i}{U_T}} X_i^{(1)} \right) \right) = 0, \tag{14}$$

  for $t_k \le t \le t_k + \Delta t, \quad X_i^{(1)}(t_k) = X_i(t_k)$
- Step 2: Solve

$$\partial_t X_i^{(2)} = R_i \left(\vec{X}^{(2)} \right), \quad for\ t_k \le t \le t_k + \Delta t, \quad X_i^{(2)}(t_k) = X_i^{(1)}(t_k + \Delta t) \tag{15}$$

- Step 3: Set

$$X_i(t_k + \Delta t) = X_i^{(2)}(t_k + \Delta t) \tag{16}$$

A simple Taylor expansion argument yields that the above method is first order in the time step Δ_t. A second order method can be obtained immediately by using a standard modification such as Strang splitting [19], but the advantages of the approach are limited by the nature of the Gummel iteration.

The advantage of the operator splitting approach is the following: An optimal simultaneous implementation of the transport and the reaction terms would result in the solution of a block sparse matrix vector multiplication problem. The matrix will be $NN_{grid} \times N\ N_{grid}$ where N is the number of defects and N_{grid} is the number of grid points or finite elements. Each block has size $N \times N$ and refers to interactions of defects at a single point in space. Depending on the reaction network structure, each block may be fully dense. For a regular triangular finite element mesh (or a rectilinear finite difference or finite volume grid), the matrices will be block-pentadiagonal with a bandwidth of $N \cdot \sqrt{N_{grid}}$ with N_{grid} the number of gridpoints or finite elements. Many algorithms exist for solving such banded sparse systems, but a reasonable lower bound for the computational complexity for an $M \times M$ matrix with bandwidth K is $O(KM)$. For our matrix, this gives a computation time of $O\left(N^2 N_{grid}^{\frac{3}{2}}\right)$. Whereas a simultaneous implementation may be computationally feasible in one spatial dimension, the operator splitting approach is essential in higher dimensions when N_{grid} is the product of the number of grid points in each direction.

In contrast to the above mentioned approach, using the splitting method we need to solve N different matrix-vector multiplication problems for $N_{grid} \times N_{grid}$ matrices which have the usual banded structure of a 2D finite element problem. (Many defects exist at lattices cites and do not diffuse, reducing the number of required solves.) Using the same estimates as above, we predict a speed up factor of N. However, modern linear algebra packages have optimized solvers for matrices with structures arriving from discretizing PDEs and our pentadiagional matrices will be solved with complexities approaching $O\left(N_{grid} \log\left(N_{grid}\right)\right)$. For large values of N_{grid}, the speed-up here is considerable. Indeed, the speed-up also occurs in 1D where fast tridiagonal solvers will run in $O(3\ N_{grid})$ time.

We must still deal with Poisson's equation, but since it too is linear in X_i, we can consider it in the same step as the diffusion equation for the X_i, retaining the decoupled structure. Note that we are left with significant freedom in the structure of the iteration. In particular, we note that the reaction equation does not a-priori respect the boundary conditions of the problem. For realistic final results, we should therefore choose to calculate the diffusion step last, or proceed by using the Strang-splitting method discussed above. The general time-splitting scheme is shown in Fig. 3.

2.2.2 Drift-Diffusion Implementation

Due to our choice of splitting methods above, our spatial discretization scheme can be optimized for solving only the following problem:

$$\partial_t X_i - \nabla \cdot \left(\mu_i U_T e^{-\frac{\theta_i \phi_i}{U_T}} \nabla u_i\right) = 0 \tag{17}$$

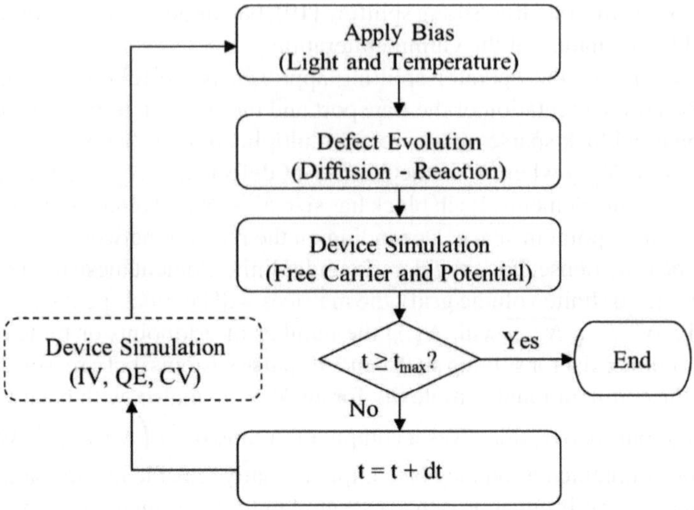

Fig. 3 Flow-chart of the Unified Solver. Device simulation is performed for each time step to obtain self-consistent and real-time electric field and carrier distributions during the diffusion-reaction simulation. Device simulation for IV, QE and CV is presented in dashed line box and is currently employed for the modeling of Cu-related metastabilities of CdTe solar cells

This spatial discretization of the problem is flexible, since none of its properties are dictated by the time-splitting scheme discussed above. In 1D, Scharfetter-Gummel [26] is appropriate (and in some cases optimal). The 1D version of the Unified Solver (obviously without grain boundaries) has already been demonstrated [27, 28].

In 2D, however, many of the schemes in the literature [29] do not work as well as desired. Any scheme which relies upon a-priori knowledge information about the structure of the devices runs into significant difficulties when the doping can change throughout time. The doping at many points in the device changes from p-type to n-type over time. In particular, schemes which require edges perpendicular to the direction of the electric field will not deal well with the nonlinear wave-fronts associated with grain boundary diffusion, let alone complicated grain boundary geometries. To handle these complicated geometries, we choose to use a finite element method for the 2D Unified Solver. Instead of relying upon an elaborate grid scheme, we directly discretize the Slotboom variables and use the properties of the scheme to allow exponential fitting of the charge carriers and defects between nodes.

Spatial Discretization

In order to deal with the complicated geometries which arise from the grain boundaries, we seek to solve this equation using finite elements. In the usual way [30], we will triangulate our domain $\Omega = \bigcup T_h$ with small triangles on the grain boundaries to handle large changes in the gradients. We then take the approximations $u_i \in V_h$ of

u_i, which are piecewise linear on the triangles, and continuous on the whole domain, but still satisfy the boundary conditions. Multiplying our equation by a test function v in the same space and integrating over the domain yields:

$$\int_{\Omega} \partial_t X_i v dx = \int_{\Omega} \nabla \cdot \left(\mu_i U_T e^{-\frac{\theta_i \phi_i}{U_T}} \nabla u_i \right) v dx \tag{18}$$

Integrating by parts and separating the terms allows us to calculate the stiffness and mass matrices:

$$\boldsymbol{A}_i := \sum_{K \in T_h} \mu_i U_T \int_K -e^{-\frac{\theta_i \phi_i}{U_T}} \nabla \psi_j \cdot \nabla \psi_l dx \tag{19}$$

$$\boldsymbol{M}_i := \sum_{K \in T_h} \int_K \psi_j \psi_l dx \tag{20}$$

Details can be found in any finite element method introduction [30]. Combining the previous results, we can reformulate our PDE by:

$$\begin{aligned} \int_{\Omega} \partial_t X_i v dx &= \int_{\Omega} \nabla \cdot \left(\mu_i U_T e^{-\frac{\theta_i \phi_i}{U_T}} \nabla u_i \right) v dx \\ \boldsymbol{v}\boldsymbol{M}(\partial_t X_i) &= \boldsymbol{v} A_i \boldsymbol{u} \ \forall v \in V_h \end{aligned} \tag{21}$$

One advantage of finite elements is that zero-flux boundary conditions are incorporated naturally. Because we normally want to conserve atomic species during a simulation, we choose to use zero-flux conditions for all defects. For the carriers we use Ohmic front and back contacts and the matrices $\boldsymbol{A}_i$ must be modified. The only complication over traditional methods [29, 30] is that if the doping changes over time, the boundary conditions must also be updated.

In the usual manner for finite elements, we can assume that this relationship holds for all admissible functions v and write the equation as a function of u and X only. Inserting the definition of u to close the equation yields:

$$\boldsymbol{M}(\partial_t X_i) = A_i \left(X_i e^{\frac{\theta_i \phi}{U_T}} \right) \tag{22}$$

Note that for piecewise linear finite elements, the exponential inside the stiffness matrix $\boldsymbol{A}$ will be evaluated at the barycenter of the finite elements. In contrast, the exponential on the left-hand side must be evaluated at the grid points. Thus, for each entry of the stiffness matrix, this will yield exponentials of the form:

$$e^{-\dfrac{\theta_i \phi_i(\hat{x}_K)}{U_T}} e^{\dfrac{\theta_i \phi_i(P_i)}{U_T}} = e^{\dfrac{\theta_i (\phi_i(P_i) - \phi_i(\hat{x}_K))}{U_T}}. \tag{23}$$

Note that this form bears great resemblance to the 1D Scharfetter-Gummel exponential [26]. It plays a similar role in allowing a linear problem to approximate exponential fitting between nodes. We can finish the solution by assuming implicit Euler for the time derivative:

$$M\frac{1}{\Delta t}(X_i^{k+1} - X_i{}^k) = A_i\left(X_i{}^{k+1}e^{\frac{\theta_i\phi}{U_T}}\right) \tag{24}$$

$$\left(\frac{M}{\Delta t} - A_i\boldsymbol{\phi}\right)X_i{}^{k+1} = \frac{M}{\Delta t}X_i{}^k \tag{25}$$

where the matrix $\boldsymbol{\phi} = \operatorname{diag}\left(e^{\frac{\theta_i\phi}{U_T}}\right)$. As discussed previously, this scales the stiffness matrix so that $A_i\boldsymbol{\phi}$ is no longer symmetric). Formally, our solution is given by:

$$X_i{}^{k+1} = \left(\frac{\boldsymbol{M}}{\Delta t} - A_i\boldsymbol{\phi}\right)^{-1}\frac{\boldsymbol{M}}{\Delta t}X_i{}^k \tag{26}$$

(In practice, we will always work with the matrix equation directly instead of actually inverting the matrix.)

Reaction Step

The system of ordinary differential equations, $\partial_t X_i = R_i(\hat{x})$, in Step 2 of the operator splitting algorithm is actually quite involved. In general we will have an arbitrary number of defects and an arbitrary number of reactions. However, these reactions are all of two very specific forms, either representing the bimolecular reaction of two defects or the formation/decay of a single defect. We therefore consider only binary reactions of the form $1+2 \rightleftharpoons 3+4$, $1+2 \rightleftharpoons 3$, $1 \rightleftharpoons 2+3$ the reaction terms can be written as a quadratic form:

$$R_i\left(\vec{X}\right) = \vec{X}^T A_i\vec{X} + b_i^T\vec{X} \tag{27}$$

with the matrices A_i and the column vectors b_i containing the reaction rates. Increasing the number of defects may increase the total number of possible reactions, but only 3 or 4 defects will be involved in any particular reaction.

There are two strategies for implementing individual reactions. Each reaction can be implemented independently or simultaneously.

- A sequential implementation would involve solving each reaction independently. The jth reaction generates a matrix $A_{i,j}$ and a vector $b_{i,j}$. Note that for bimolecular reactions $A_{i,j}$ has two nonzero entries and $b_{i,j}$ is the zero vector. For formation/decay reactions, both $A_{i,j}$ and $b_{i,j}$ have one nonzero entry each. For each j, we have the following ODE:

$$\partial_t X_i = R_i\left(\hat{x}\right) = \vec{X}^T(t_k)\, A_{i,j} \vec{X}(t) + b_{i,j}^T\, \vec{X}, \tag{28}$$

Each of these reactions can be solved using implicit Euler—see section "Iteration by Reaction".

- A simultaneous implementation of the reaction equations will require the numerical solution of the ODE system by, for example, the implicit Euler method together with a Newton method for the large quadratic system. The Jacobian $\frac{\partial \vec{R}}{\partial \vec{X}}$, can then be easily computed from Eq. (27).

In both cases, there is no need to use a higher order method since the operator splitting method described in Section A is only of first order.

Iteration by Reaction

Because every considered reaction is of either single molecule formation/dissociation or bimolecular type, we can solve every reaction network using just the explicit formulas given below. Simply iterating through all possible reactions in any order gives a first order unconditionally stable method for the reaction step. Experimental results demonstrate that changing this order has no significant impact on the results for the case of copper migration.

For more complex reaction networks complications can occur. These are discussed elsewhere [28].

1. **Single Molecule**
   Let us consider the reaction given by the rate $R_1^{2,3}$:

$$\begin{aligned}\partial_t X_1 &= K_{2,3}^1 X_2 X_3 - K_1^{2,3} X_1\\ \partial_t X_2 &= K_1^{2,3} X_1 - K_{2,3}^1 X_2 X_3\\ \partial_t X_3 &= K_1^{2,3} X_1 - K_{2,3}^1 X_2 X_3\end{aligned} \tag{29}$$

   This could, for instance, represent the knock-off equation in Eq. (1) where interstitial Copper ($X_1 = [Cu_i]$) replaces lattice Cadmium, resulting in Copper atom at a Cadmium site ($X_2 = [Cu_{Cd}]$) and an interstitial Cadmium ($X_3 = [Cd_i]$). Note that the right hand sides are identical (up to a sign) and that the quantities $(X_1 + X_2)$ and $(X_1 + X_3)$ are conserved (corresponding to conservation of copper and conservation of lattice sites in our example). In particular, if the superscript refers to the time-step, we have that $X_1^{k+1} + X_2^{k+1} = X_1^k + X_2^k$ and $X_1^{k+1} + X_3^{k+1} = X_1^k + X_3^k$. We will use these equations to decouple the ODEs. First, let us use the Implicit Euler method to write the first ODE for the time step k as an algebraic expression:

$$\frac{1}{\Delta t}\left(X_1^{k+1} - X_1^k\right) = K_{2,3}^1 X_2^{k+1} X_3^{k+1} - K_1^{2,3} X_1^{k+1}. \tag{30}$$

Since Δt should be small, we multiply it out to avoid numerically unstable division by a small number. Using conservation laws, we can also write X_2^{k+1} and X_3^{k+1} in terms of X_1^{k+1} and the known values of X as time step k:

$$X_2^{k+1} = X_1^k + X_2^k - X_1^{k+1} \tag{31}$$

$$X_3^{k+1} = X_1^k + X_3^k - X_1^{k+1} \tag{32}$$

Substitution yields:

$$X_1^{k+1} - X_1^k = K_{2,3}^1 \Delta t \left(X_1^k + X_2^k - X_1^{k+1}\right)\left(X_1^k + X_3^k - X_1^{k+1}\right) - K_1^{2,3} \Delta t X_1^{k+1} \tag{33}$$

Careful examination reveals that this is a quadratic equation for X_1^{k+1}. Rearrangement and expansion yields the three coefficients as:

$$\begin{aligned} A &= \quad K_{2,3}^1 \Delta t \\ B &= -\left(1 + K_1^{2,3}\Delta t + K_{2,3}^1 \Delta t (2X_1^k + X_2^k + X_3^k)\right) \\ C &= \quad X_1^k + K_{2,3}^1 \Delta t \left(X_1^k + X_2^k\right)\left(X_1^k + X_3^k\right) \end{aligned} \tag{34}$$

We can then use the quadratic formula to yield:

$$X_1^{k+1} = \frac{-B \pm \sqrt{B^2 - 4AC}}{2A} = \frac{2C}{\sqrt{B^2 - 4AC} - B}. \tag{35}$$

We note that the middle expression has A in the denominator, but that since A is proportional to Δt it may be small. We therefore rearrange the expression to avoid numerical division by a small number. Because B is always negative for positive concentrations, we can also determine the appropriate sign of the square root to obtain our final solution. Finally, since X_1^{k+1} is now known, we can immediately substitute back into our conservation laws to obtain X_2^{k+1} and X_3^{k+1}.

2. **Bimolecular**
Let us consider the reaction $R_{1,2}^{3,4}$:

$$\begin{aligned} \partial_t X_1 &= K_{3,4}^{1,2} X_3 X_4 - K_{1,2}^{3,4} X_1 X_2 \\ \partial_t X_2 &= K_{3,4}^{1,2} X_3 X_4 - K_{1,2}^{3,4} X_1 X_2 \\ \partial_t X_3 &= K_{1,2}^{3,4} X_1 X_2 - K_{3,4}^{1,2} X_3 X_4 \\ \partial_t X_4 &= K_{1,2}^{3,4} X_1 X_2 - K_{3,4}^{1,2} X_3 X_4. \end{aligned} \tag{36}$$

This could correspond to interstitial Copper ($X_1 = [\mathrm{Cu_i}]$) interacting with Zinc on a Cadmium lattice cite ($X_2 = [\mathrm{Zn_{Cd}}]$), resulting in interstitial Zinc ($X_3 = [\mathrm{Zn_i}]$) and Copper on a Cadmium site ($X_4 = [\mathrm{Cu_{Cd}}]$). Following the method of the previous section, we can develop three conservation laws and rearrange:

$$
\begin{aligned}
X_2^{k+1} &= X_2^k - X_1^k + X_1^{k+1} \\
X_3^{k+1} &= X_3^k + X_1^k - X_1^{k+1} \\
X_4^{k+1} &= X_4^k + X_1^k - X_1^{k+1}
\end{aligned} \tag{37}
$$

Discretizing the equation for X_1 using the Implicit Euler method and substituting the conservation laws will yield a quadratic equation for X_1^{k+1} in the same manner as in the previous section. The coefficients are:

$$
\begin{aligned}
A &= \left(K_{3,4}^{1,2} - K_{1,2}^{3,4}\right)\Delta t \\
B &= -1 - K_{3,4}^{1,2}\Delta t(2X_1^k + X_3^k + X_4^k) - K_{1,2}^{3,4}\Delta t(X_2^k - X_1^k) \\
C &= X_1^k + K_{3,4}^{1,2}(X_1^k + X_3^k)(X_1^k + X_4^k)
\end{aligned} \tag{38}
$$

Comparison with Newton's Method

Particular care needs to be taken in order to have a computationally stable steady-states. These states are vital for standard device simulation results such as IV and CV measurements. By the principle of detailed balance, at equilibrium every individual reaction will be in equilibrium. This indicates that we will obtain a stable equilibrium for any iteration order. As noted above, the iteration scheme works well for the copper diffusion case.

However, some problems can occur in far from equilibrium conditions. This behavior becomes even more noticeable when one defect is unstable and has multiple dissociation pathways. For this case, whichever dissociation reaction occurs first in the iteration will occur preferentially to the later reaction, regardless of the relative size of the reaction rates.

This behavior does not occur for the Newton iteration scheme. Because the reactions are considered simultaneously, each reaction pathway will occur with relative frequency exactly governed by the ratio of the reaction rates. In practice, we find that the Newton method is vastly superior for complicated reaction networks. Furthermore, the method does not present a significant computational issue because the reactions are local in space and the Jacobian can be generated independently for each grid point.

3 Verification of 1D Unified Solver

3.1 *Verification of 1D Device Simulation Routine Versus SCAPS Device Simulator*

As discussed in Sect. 2, in 1D version of the Unified Solver, the device simulation is performed on each time step to obtain self-consistent electric field and carrier distributions during the diffusion-reaction simulation. Besides, the device simulation routine serves to obtain the steady-state electric current during the simulation of current-voltage characteristics of the device. The device simulation routine can use the defect profiles obtained from the consistent solution of 1D kinetic reaction-diffusion problem as well as the artificially set uniform or nonuniform defect profiles similar to other solar cell simulators. In order to verify the accuracy of the device simulation routine, we performed its rigorous comparison with the widely used in the community SCAPS solar cell simulation code [31].

In order to perform such comparison, we used the same device structures (highly p-doped back layer, lightly p-doped CdTe layer, highly n-doped front layer), properties of materials, light absorption spectra and illumination spectra both in 1D Unified Solver and SCAPS solar cell simulator. By comparing the simulated energy bands, distributions of free carrier in the dark and under the stress conditions, as well as JV curves for different doping levels and recombination rates, we conclude that the device simulation routine of 1D Unified Solver reliably produces accurate solution of the drift-diffusion problem for free carriers under different bias conditions (Fig. 4).

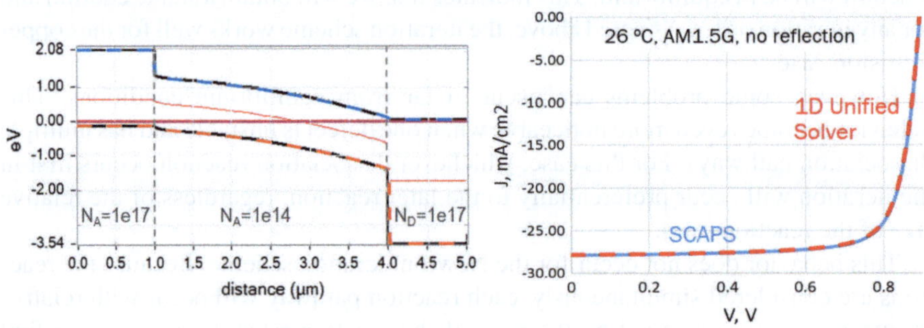

Fig. 4 Comparison of band diagram (left) and the IV curve calculated with 1D Unified Solver (dashed lines) and SCAPS solar cell simulator (solid lines) for one of the test device structures considered in this study

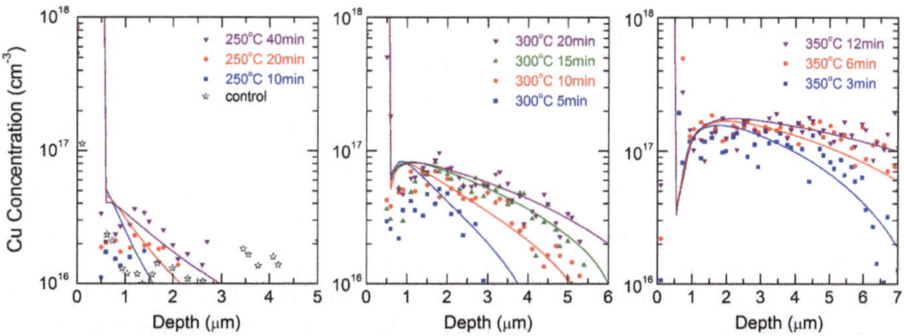

Fig. 5 Atomic Cu profiles achieved with different annealing recipes. Black pentagrams represent the control sample without any annealing. Solid lines represent the simulated Cu profiles

3.2 Verification of the Defect Reaction-Diffusion Model on Experimental Data

3.2.1 Experiment on Cu Penetration

In order to verify the reaction-diffusion model of Cu penetration into CdTe, experimental study of Cu migration in single crystal CdTe was done at First Solar Inc. (Perrysburg, OH). In this experiment, a thin Cu-containing ZnTe layer was deposited on sx-CdTe substrates provided by JX Nippon. After deposition, to drive Cu into CdTe, anneals with different durations were performed at 250, 300 and 350 °C.

Measured Cu profiles show strong dependence on the annealing temperature and duration: Cu penetrates deeper into CdTe as we increase the annealing time/duration (Fig. 5). The high concentration of Cu appearing in the first 0.5 μm is the residual Cu concentration from the ZnTe layer. Abnormally high concentration of Cu ($>10^{18}$ cm^{-3}) in the region ($0.5 < x < 1$ m) obtained after very high annealing temperatures (350 °C) or long annealing durations (20 min at 300 °C), was ascribed to the broadening of ZnTe or formation of ZnCdTe, caused by Zn diffusion [32] and will not be further addressed here.

Another important finding in this experiment is a dip of Cu concentration at the edge of CdTe adjacent to ZnTe, while the peak of the Cu concentration in the CdTe layer is situated 1–2 m beneath the interface. This is interpreted as a back diffusion of Cu during the cool down process (SIMS measurements are performed at room temperature after cooling down the sample).

3.2.2 Simulation

To model this experiment, we created a simulation domain with 15 μm of initially undoped CdTe and 0.5 μm of p-doped ZnTe:Cu layer. 10^{21} cm^{-3} initial concentration of Cu_i was defined in ZnTe:Cu source layer and a constant 10^{17} cm^{-3} p-type doping was maintained in this layer during the entire simulation. A 0.45 eV difference in standard formation energy of Cu_i in CdTe and ZnTe layers (more favorable in ZnTe) was used to obtain the best correspondence of the Cu concentration difference across the interface as achieved by different annealing recipes. This energy difference is in a qualitative agreement with the results of first principles calculations. We used Neumann boundary conditions for both ends of the simulation domain to maintain the conservation of all defects. For carrier transport properties and material properties we used widely accepted values of polycrystalline CdTe [33]. Although better carrier lifetime and material quality can be expected in sx-CdTe [34], there should be no considerable impact on the resulting Cu profiles since no (or negligible) electric current flows through the samples during the annealing process. To simulate this experiment on Cu penetration, we included the primary exchange reaction (1) as well as diffusion of Cu_i and Cd_i into the reaction-diffusion model.

The original built-in electric field between highly p-type doped ZnTe and intrinsic CdTe layers prevents Cu_i^+ from moving into the CdTe region in large amount. When a small amount of Cu_i is able to diffuse into CdTe, it quickly knocks Cd atoms off generating an immobile Cu_{Cd}^- acceptor and mobile Cd_i^{2+} donor (backward reaction (1). Part of Cd_i^{2+} is drifted into ZnTe layer under the same built-in electric field across the interface and, as a result, p-type region starts to form in CdTe. We note that since the charge is conserved in all reactions, achieving p-type doping without Cd_i^{2+} moving out of CdTe is very difficult within this model. We found that during the annealing process, the concentration of Cu_{Cd} does not show a monotonic increase, a saturation effect is observed instead.

The saturation behavior can be explained by analyzing the distribution of defect concentrations and the band diagram during annealing (Fig. 5 left bottom). As Cu forms acceptors in CdTe, an electric field is generated between the Cu occupied p-type region and the intrinsic CdTe region without Cu, which again prevents further movement of Cu_i into the intrinsic region of CdTe. Once the distribution of defects gets close to the balance of all involved reactions, such as $0.5 < x < 6$m region in Fig. 6, less Cu_{Cd} will be generated. Hence more Cu_i can travel through this saturated region to occupy Cd sites in the newly formed p-i junction area ($x > 6$m in Fig. 6).

As we have mentioned before, the measured Cu profiles after cooldown show a dip at CdTe/ZnTe interface. Here we analyze how such dip is formed according to our model. As we cool down a sample to room temperature, Cu_i starts to diffuse back into ZnTe due to the 0.45 eV difference in the standard formation energies of Cu_i between CdTe and ZnTe, and reaction (1), which is out of equilibrium now begins to produce more Cu_i by kick-out Cu_{Cd} with available Cd_i. After reaction (1) consumes most of the Cd_i, additional Cd_i from ZnTe penetrates into CdTe and continues to kick-out Cu_{Cd} to reach the new equilibrium among involved reactions. Since only the interface region can get supplemental Cd_i, Cu reduction is more obvious near the interface (Fig. 6 right). As temperature further decreases, both reaction and diffusion rates become quite slow. However, this process is not be completely stopped right after the cooling process. Experimental evidence of Cu movements in CdTe devices stressed at 65 °C has been reported recently [35].

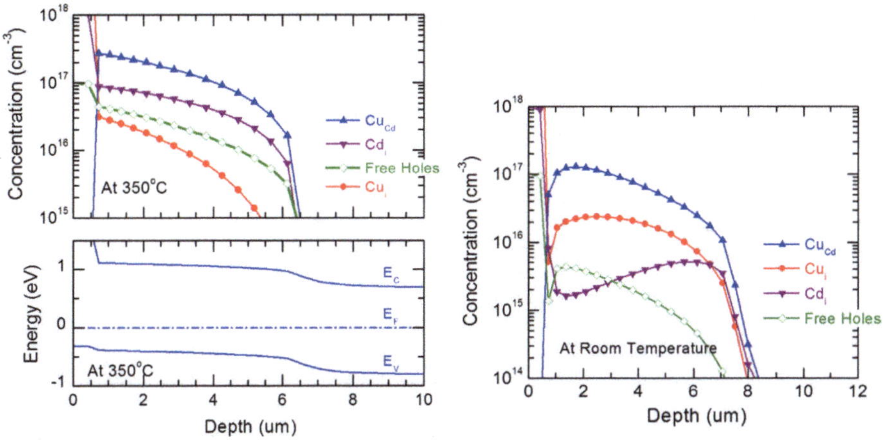

Fig. 6 Left: Simulated profiles of major Cu-related defects and free carriers (top panel) and band diagram in the ZnTe:Cu/sx-CdTe structure after 3 min of annealing at 350 °C (bottom panel). Right: Simulated profiles of major Cu-related defects in the sample after 3 min annealing at 350 °C and an extra 12 min cooling process

Another important outcome of this simulation is the insight into the formation and evolution of doping during the annealing and cooldown. At high temperature, a partial compensation between Cu_{Cd}^{-} and Cd_i^{2+} is achieved as the atomic Cu concentration is around 2×10^{17} cm^{-3} while the hole concentration is one order of magnitude less (see the line with diamonds (green in the online version) in Fig. 6 left, labeled as "Free Holes"). In addition, partial ionization of the Cu_{Cd} acceptors plays a role in the compensation mechanism since the acceptor level of Cu_{Cd} is not that shallow. About 90% of Cu_{Cd} acceptors in the saturated area are ionized at 350 °C according to detailed results from our simulation. After cooldown, the free carrier concentration drops below 3×10^{15} cm^{-3} level (Fig. 6 right) with only a smaller reduction in atomic concentration of Cu at room temperature. More importantly, the new compensation is mostly achieved between Cu_{Cd}^{-} and Cu_i^{+}. Therefore during cooling, the compensation mechanism is changed in this case. The observed change is a complex process determined by diffusion, drift, reactions, and temperature-dependent Fermi-Dirac statistics both for free carriers and Cu_{Cd} acceptors. Again, the resulting room temperature hole density depends crucially not only on donor-acceptor compensation but also on the possibility of the ionization of Cu_{Cd} acceptors.

Simulated free carrier concentration in the saturation region after cooldown are about an order of magnitude higher than the carrier concentration in poly-crystalline Cu, Cl-treated CdTe solar cells from Colorado State University [35] (Fig. 7). This difference stems from a number of factors: the difference in annealing temperatures and Cu concentration in ZnTe source, not accounting for defect interactions in our simulations, additional doping compensation by Cl and even the effect of grain boundaries on the measured CV doping profiles in px-CdTe films. Therefore, even though, our simulation provides a qualitative understanding of the formation and

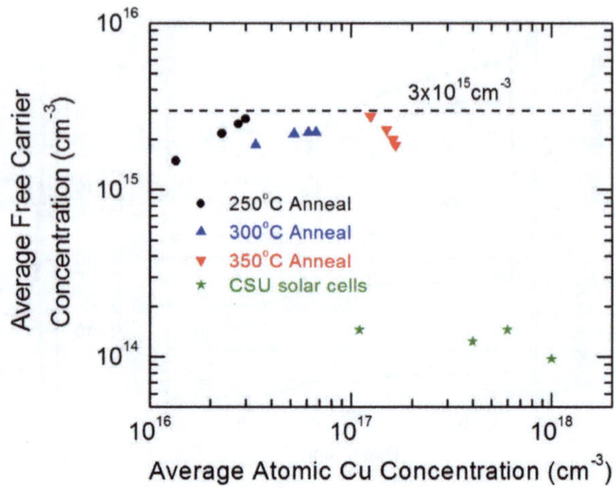

Fig. 7 Simulated and experimental free carrier concentrations versus atomic Cu concentration at room temperature. Green pentagrams are measured free carrier concentrations in px-CdTe solar cells fabricated by Colorado State University

Table 1 Fitted diffusion parameters with calculated theoretically in work [25] (in parenthesis)

Defect	D_0 (cm^2/s)	ΔE_D (eV)
Cu_i (+)	4.9×10^{-3} (6.3×10^{-3})	0.72 (0.46)
Cu_i (2+)	3.0×10^{-4} (3.2×10^{-3})	0.72 (0.47)

compensation of doping in CdTe, it is expected that more targeted experiments and simulations will be performed in the future to explain the peculiarities of doping formation in px-CdTe.

Table 1 shows the comparison of diffusion parameters used to obtain the best correspondence with experimental curves (Fig. 5) and parameters calculated from first principles. It is seen, that in our 1D simulation, the diffusion barrier height of the major specie responsible for Cu penetration (Cu_i^+) is about 0.26 eV higher that theoretically predicted one. We attribute this discrepancy to the fact that we did not account for the interactions between the defects and the formation of complexes, which may influence the overall diffusion kinetics. This effect is worth investigating in the future studies.

We conclude that the Cu penetration mechanism implemented in the developed reaction-diffusion model is capable to reproduce the experimental dependence of the Cu profile on the annealing conditions, to reproduce some specific features of experimental atomic Cu profiles (dip at the CdTe/ZnTe interface), as well as to provide a theoretical insight into the process of formation and compensation of Cu doping in CdTe layer.

4 Modeling of Light Soaking Effect in CdTe Solar Cells

Nearly all PV technologies exhibit changes in device performance under extended illumination, or "light soaking", although the magnitude and the trend of these changes are not always the same among different technologies. Experiments on both commercial modules and research cells based on CdTe technology have shown improvement of cell performance under light stress conditions for up to 20 h [36]. Many accredited such phenomena to the passivation of traps and migration of Cu ions, however no simulations were previously done to confirm any of these mechanisms. We use the developed 1D Solver to simulate self-consistently the effects of illumination, bias and temperature on the evolution of defect profiles during stress and the resulting performance changes of the device.

In this simulation, we employ a simplified dopant compensation model. Namely, 10^{16} cm^{-3} Cu_{Cd}^-, 0.4×10^{16} cm^{-3} Cu_i^+ and 0.5×10^{16} cm^{-3} background donor is assumed as the initial defect distribution in this simulation, resulting in $\sim 10^{15}$ cm^{-3} equilibrium hole concentration in the quasi-neutral region of CdTe absorber layer (typical doping in px-CdTe absorber). We use a standard ZnTe/CdTe/CdS device structure with common electronic properties.

The light stress is typically performed after the dark storage of solar devices. Therefore, in order to prepare the initial system for light soaking simulations, first, we simulate the equilibrium of the defects in CdTe cells under dark without any bias at 65 °C (Fig. 8 (left)). In equilibrium, most of the Cu_i^+ is pushed away from the depletion region, due to the built-in potential of the p-n diode. As a result, more uncompensated acceptors appear in the junction area. Cu_{Cd} acceptor is partially ionized in the quasi-neutral region, while it is completely ionized in the depletion region due to low density of holes in that area.

Under light stress condition under applied forwards bias of 0.8 V (maximum power point), the defect equilibrium changes in the following way: (1) Cu_i^+ moves closer to the main junction due to the forward bias, reducing the uncompensated acceptor doping density in the depletion region, (2) part of ionized defects (Cu_i^+ and Cu_{Cd}^-) are converted into the neutral state after capture of light-generated free carriers, (3) the zero-bias depletion region width increases because of the reduced p-doping in the junction area (Fig. 8 (right)), (4) carrier collection efficiency improves, thus increasing the performance.

In order to isolate the effect of ionization degree change due to excess carriers capture on Cu_i^+ and Cu_{Cd}^- defects and the effect defect drift under the applied voltage bias, we simulate different stress conditions with different combinations of illumination and forward bias (Table 2). Figure 9 shows the efficiency changes of the solar cells as a function of soaking time with different modes. The results of our simulations suggest that both Cu_i^+ migration towards the junction area and the injection of excess holes in the depletion region can cause the enhancement of device performance. Simulations with defect migration resulted in the strongest performance enhancement effect. Moreover, the effect of defect migration in the dark and with

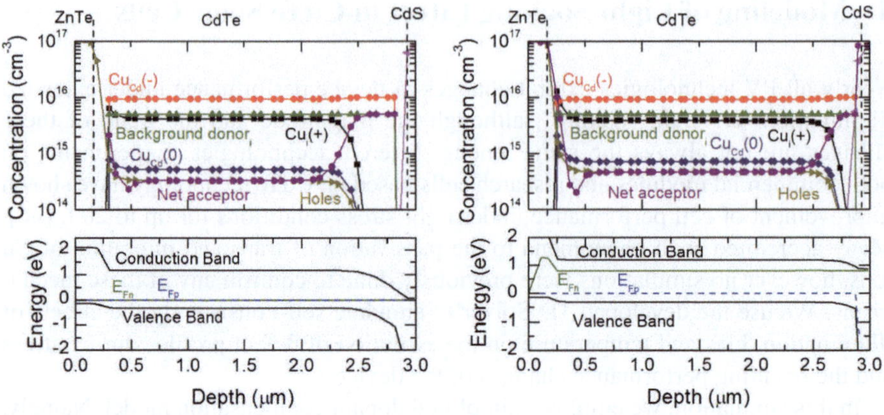

Fig. 8 Simulated concentrations of defects and free carriers and device energy bands in the dark (left) and under light soak conditions (right) at 65 °C

Table 2 List of stress conditions

Soak	Illumination	Voltage bias (V)	Mechanism
A	1 Sun	0.8	Both
B	1 Sun	0.8	Passivation
C	Dark	0.8	Migration
D	Dark	0.8	None
E	1 Sun	0.8	No migration
G	1 Sun	0.8	10^{-12} cm^2/s
G	1 Sun	0.8	10^{-13} cm^2/s

applied forward bias is almost the same as the overall effect of applied light and forward bias.

The kinetics of the performance increase upon such dark to light switching depends on the diffusivity of mobile species involved in the transformation. Using high Cu_i^+ diffusion coefficient (9×10^{-10} cm^2/s at 65 °C) calculated from the first-principles [25], only 0.03 h of light stress is enough for device to reach a new steady-state (see Soak A of Fig. 9). Experimentally observed performance transients often happen on multi-hour timescale [37, 38]. By decreasing the diffusion coefficient of Cu_i^+ down to 10^{-13} cm^2/s, we were able to obtain better agreement with the experimentally observed 10 h long device performance stabilization. This indicates that while our model correctly captures the response of the system to the external stress, the actual kinetic mechanisms behind the redistribution of species may be more complex than simple diffusion of Cu_i^+. This result provides the evidence that more complicated diffusion mechanisms beyond simple diffusion of Cu_i^+ may be involved in the defects redistribution in CdTe. For example, the formation of pair complex defects can influence the diffusivities of species by slowing down the fast

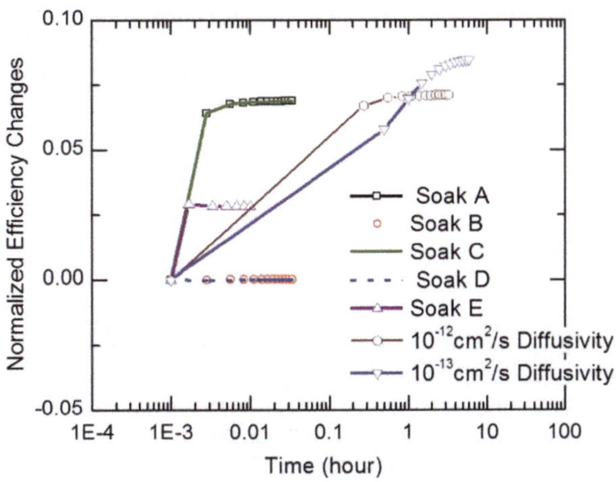

Fig. 9 Device performance changes as a function of soaking time with various conditions

diffusion of interstitials as well as introducing additional mechanisms of diffusion of weakly mobile substitutional defects [20]. We expect that inclusion of such advanced mechanisms into reaction-diffusion simulations will allow to reproduce, understand and even predict the kinetics of transient performance changes of the real devices.

5 Conclusions

In summary, we have implemented (in MATLAB) self-consistent 2D numerical solvers for simulating defects reactions and migration in CdTe material. We have verified the 2D solver using the widely used code for solar cell simulation (SCAPS) as well as the experimental data on Cu penetration into CdTe. By doing so, we have shown that the Cu penetration mechanism based on the Cu-Cd exchange reaction and implemented in 2D reaction-diffusion solver is able to reproduce the experimental dependence of Cu profiles in CdTe on the annealing temperature and time. Using the developed 2D reaction-diffusion solver, we have revealed the new mechanism of a transient response of CdTe-based solar cell to the applied light-bias conditions. According to our simulations, the experimentally observed performance enhancement under illumination and/or applied forward bias may be, at least partially, caused by the migration of defects changing the distribution of doping profile in the absorber. Based on simulation results, we discuss the possible explanations for limited incorporation and compensation mechanisms of Cu dopants inside CdTe bulk as well as the possible defect migration mechanisms beyond the simple diffusion of interstitials. We expect that the discrepancies between simulations and experiments will be reduced by inclusion more detailed defect chemistry models with all important defects and interactions between them and making use of more complicated 2D simulation framework that includes the effects of grain boundaries.

Acknowledgements This work was supported by the Department of Energy SunShot Program, PREDICTS Award DE-EE0006344 and PVRD Award DE-EE0007536. The authors would also like to thank Dr. Su-Huai Wei and Dr. Ji-Hui Yang from National Renewable Energy Laboratory (NREL) for providing some of the initial first principle calculated defect parameters and for the discussions related to this work.

References

1. D.A. Jenny, R.H. Bube, Phys. Rev. **96**(5), 1190 (1954)
2. R.H. Bube, Proc. IRE **43**(12), 1836–1850 (1955)
3. D.A. Cusano, Solid State Electron. **6**(3), 217 (1963)
4. B. Goldstein, Phys. Rev. **109**(2), 601 (1958)
5. http://en.wikipedia.org/wiki/First_Solar
6. A.F.W. Willoughby, in *Narrow-Gap II-VI Compounds for Optoelectronic and Electromagnetic Applications*, vol. 3, ed. by P. Capper (Chapman and Hall, London, 1997), pp. 268–290
7. D.W. Lane, J.D. Painter, M.A. Cousins, G.J. Conibeer, K.D. Rogers, Thin Solid Films **431**, 73–77 (2003)
8. H. Gleiter, Acta Metall. **18**, 117–121 (1970)
9. S.A. Galloway, P.R. Edwards, K. Durose, Inst. Phys. Conf. Ser. No **157**, 579–582 (1997)
10. L.O. Bubulac, J. Bajaj, W.E. Tennant, P.R. Newman, D.S. Lo, J. Cryst. Growth **86**(1–4), 536–543 (1988)
11. K. Durose, J.R.E. Sadler, A.J.W. Yates, A. Szczerbakow, in *28th IEEE Photovoltaic Specialists Conference* (2000), pp. 487–490
12. J.D. Major, CdTe solar cells: Growth phenomena and device performance. Ph.D. thesis, Durham University, UK, 2008
13. B. Tetali, V. Viswanathan, D. L. Morel and C. S. Ferekides, in *Proceedings of 30th IEEE Photovoltaic Specialist Conference* (2002), pp. 600–603
14. S.E. Asher, F.S. Hasoon, T.A. Gessert, M.R. Young, P. Sheldon, J. Hiltne and J. Sites, in *Proceedings of 28th IEEE Photovoltaic Specialists Conference* (2000), pp. 479–482
15. G.B. Abdullaev, T.D. Dzhafarov, *Atomic Diffusion in Semiconductor Structures* (Harwood Academic, New York, 1987)
16. E. Hairer, G. Wanner, *Solving Ordinary Differential Equations II: Stiff and Differential-Algebraic Problems (2nd*, revised edn. (Springer, Berlin, 1996)
17. U.M. Ascher, L.R. Petzold, *Computer Methods for Ordinary Differential equations and Differential-Algebraic equations*, ISBN 0-89871-412-5 (1998)
18. G. Fabian, D.A. van Beek, J.E. Rooda, Math. Comput. Model. Dyn. Syst. **7**(2), 173187 (2001)
19. D. Brinkman, D. Guo, R. Akis, C. Ringhofer, I. Sankin, T. Fang, D. Vasileska, Self-consistent simulation of CdTe solar cells with active defects. J. Appl. Phys. **118**, 035704 (2015)
20. D. Krasikov, I. Sankin, J. Mater. Chem. A **5**, 3503–3513 (2017)
21. J.-H. Yang, W.-J. Yin, J.-S. Park, J. Ma, S.-H. Wei, Semicond. Sci. Technol. **31**, 083002 (2016)
22. D. Krasikov, A. Knizhnik, B. Potapkin, S. Selezneva, T. Sommerer, Thin Solid Films **535**, 322–325 (2013)
23. L. Onsager, Phys. Rev. **15**, 554–557 (1938)
24. V. Lordi, J. Cryst. Growth **379**, 84–92 (2013)
25. J.-H. Yang, J.-S. Park, J. Kang, S.-H. Wei, Phys. Rev. B **91**, 075202 (2015)
26. D. Scharfetter, H. Gummel, IEEE Trans. Electron Devices **16**, 64 (1969)
27. D. Guo, R. Akis, D. Brinkman, I. Sankin, T. Fang, D. Vasileska, C. Ringhofer, in *2014 IEEE 40th Photovoltaic Specialist Conference (PVSC)*, (IEEE, 2014), pp. 2011–2015
28. D. Guo, T. Fang, A. Moore, D. Brinkman, R. Akis, I. Sankin, C. Ringhofer, D. Vasileska, IEEE J. Photovolt. **6**(5), 1286–1291 (2016)

29. F. Brezzi, L. Marini, S. Micheletti, P. Pietra, R. Sacco, and S. Wang, in *Numerical Methods in Electromagnetics, Handbook of Numerical Analysis*, ed. by W. Schilders, E. ter Maten, Vol. 13 (Elsevier, 2005)
30. C. Johnson, Numerical Solution of Partial Differential Equations by the Finite Element Method (Dover Publications, 2009)
31. http://scaps.elis.ugent.be/
32. J. Li, D.R. Diercks, T.R. Ohno, C.W. Warren, M.C. Lonergan, J.D. Beach, C.A. Wolden, Solar Energy Mater. Solar Cells **133**, 208 (2015)
33. N. Amin, K. Sopian, M. Konagai, Solar Energy Mater. Solar Cells **91**(13), 1202 (2007)
34. A.P. Kirk, M.J. DiNezza, S. Liu, X.-H. Zhao, Y.-H. Zhang, in *Proceedings of the IEEE 39th Photovoltaic Specialists Conference* (IEEE, Tampa, FL, 2013) 2015
35. A. Moore, T. Fang, J. Sites, in *Proceedings of 42th IEEE Photovoltaic Specialist Conference* (IEEE, New Orleans, LA, 2015)
36. M. Gostein, L. Dunn, in *2011 37th IEEE Photovoltaic Specialists Conference (PVSC)* (2011), pp. 3126–3131
37. C. Deline, J. del Cueto, D.S. Albin, C. Petersen, L. Tyler, G. TamizhMani, in *2011, IEEE 37th Photovoltaic Specialists Conference (PVSC)* (Seattle, WA, June 2011), pp. 003113–003118
38. T.J. Silverman, M.G. Deceglie, B. Marion, S.R. Kurtz, in *2014, IEEE 40th Photovoltaic Specialist Conference (PVSC)* (Denver, CO, June 2014), pp. 3676–3681

A Multiscale Molecular Dynamics and Coupling with Nonlinear Finite Element Method

Shingo Urata and Shaofan Li

Abstract In this work, we have developed a multiscale coupling method between the multiscale micromorphic molecular dynamics (MMMD) and the nonlinear finite element method. The multiscale micromorphic molecular dynamics (MMMD) is a three-scale non-equilibrium molecular dynamics that span from microscale to mesoscale and to macroscale. A multiscale computational algorithm is formulated to couple the continuum scale equations of motion in terms of finite element method (FEM) with the coarse scale molecular dynamics of MMMD. To validate the computational formulation, we apply the multiscale coupling method to simulate nano-indentation of silicon crystals.

Keywords Continuum mechanics · Molecular dynamics · Multiscale simulation Non-equilibrium molecular dynamics · Phase transition

1 Introduction

One of the focuses of current multiscale simulation research is how to set up the multiscale boundary condition to couple the fine scale atomistic motion with the coarse scale continuum displacement field. In recent years, various multiscale boundary conditions have been proposed e.g. "handshaking condition", or "non-reflection boundary condition", e.g. Rudd and Broughton [1], Wagner and Liu [2], Xiao and Belytschko [3], To and Li [4], Li et al. [5–7], Park et al. [8], Karpov et al. [9], Namilae et al. [10], Miller and Tadmor [11] and among others. On the other hand, there have been many efforts to formulate multiscale coarse-grained molecular dynamics. Among them, Rudd and Broughton [12], Chen and Lee [13], Rudd and Broughton [14], Liu and Li [15], Li [16] are probably most notable contributions. In particular,

S. Urata
Innovative Technology Research Center, Asahi Glass Co., Ltd., 1150 Hazawa-cho, Kanagawa-ku, Yokohama, Kanagawa 221-8755, Japan

S. Li (✉)
Department of Civil and Environmental Engineering, University of California, Berkeley, CA 94720, USA
e-mail: shaofan@berkeley.edu

L. L. Bonilla et al. (eds.), *Coupled Mathematical Models for Physical and Biological Nanoscale Systems and Their Applications*, Springer Proceedings in Mathematics & Statistics 232, https://doi.org/10.1007/978-3-319-76599-0_12

in a series work Chen and Lee [17], Chen [18], Xiong et al. [19] have proposed a multiscale micromorphic dynamics. In fact, the classical micromorphic continuum theory is also a multiscale theory e.g. Eringen [20].

Instead of designing a compatible multiscale interface boundary, in this work, we would like to re-formulate classical molecular dynamics theory, which, we believed, has already contained an intrinsic multiscale structure and information that are needed in multiscale simulation. In early 1980s, Andersen [21] first repartitioned the conventional molecular dynamics Lagrange, and proposed a multiscale isoenthalpic-isobaric ensemble of molecular dynamics (MD) allowing the volume of a cubic lattice cell to vary. Subsequently, Parrinello and Rahman [22, 23] extended Andersen's formalism to the anisotropic case allowing both the volume and the shape of a molecular dynamics (MD) cell to vary, which is in fact an early form of multiscale molecular dynamics. In recent years, there have been renewed interests in revising APR molecular dynamics, e.g. Laio et al. [24], Martonik et al. [25, 26], Podio-Guidugli [27], Ulz [28, 29], which attempted to extend APR MD to non-equilibrium condition or to macroscale simulation.

In a series of recent work [30–32], by following the Andersen-Parrinello-Rahman approach, the present authors have repartitioned the Lagrange of the first-principle atomistic molecular dynamics into different scales under the local equilibrium assumption; by doing so, we have discovered that there exists a universal micromorphic multiscale structure in atomistic molecular dynamics, which is an intrinsic mathematical structure of conventional molecular dynamics. The repartitioned molecular dynamics Lagrange yields three coupled particle dynamics at three different scales. The MMMD method allows us to: (1) Apply *macroscale boundary conditions* to finite-size molecular dynamics systems, which usually are in the non-equilibrium condition; (2) Calculate non-uniform and inhomogeneous distribution of macroscale field variables of the MD system under applied macroscale boundary conditions, and (3) Bridge microscale molecular dynamics with macroscale continuum mechanics. As an example, in the present work, we hope to demonstrate how to use MMMD to couple molecular dynamics and continuum mechanics.

The paper is arranged into five Sections. In Sect. 2, a detailed multiscale partition of first-principle MD Lagrange and the derivation of the multiscale micromorphic molecular dynamics (MMMD) are presented, in which we focus on the discussion of micromorphic multiplicative decomposition. In Sect. 3, we derive the dynamical equations of the multiscale micromorphic molecular dynamics (MMMD). In Sect. 4, we discuss the coupling of MMMD and finite element method (FEM) addressing the multiscale boundary condition problem. We then use the coupled MMMD-FEM method to simulate nano-indentation of crystalline silicon materials, which validates the proposed multiscale molecular dynamics.

2 Multiscale Partition of First-Principle MD Lagrangian

We first partition the Lagrangian of first-principle molecular dynamics based on domain decomposition and multiscale kinematics of the atom motion.

2.1 Scale Decomposition

The first part of multiscale partition is the domain decomposition, i.e. we divide the finite size MD simulation domain into a finite number of supercells or local ensembles, and we choose the α-cell as the representative supercell, to illustrate formulation, and in the whole simulation domain, we have $\alpha = 1, 2, \ldots, N$ number of cells.

Assume that in the α-th cell, it has N_α number of atoms. The position of the center of mass (COM) in this cell is defined as,

$$\mathbf{r}_\alpha(t) = \frac{1}{\sum_i m_i} \sum_i m_i \mathbf{r}_i(t) \,. \tag{1}$$

where $\mathbf{r}_i, i = 1, 2, \ldots, N_\alpha$ is the spatial position of each atom in the current cell configuration at current time t. Inversely, we can express the spatial atomic position of i-th atom in the α-th MD cell as,

$$\mathbf{r}_i = \mathbf{r}_\alpha + \mathbf{r}_{\alpha i}, \quad \alpha = 1, 2, \ldots, N; \quad i = 1, 2, \ldots, N_\alpha \tag{2}$$

where $\mathbf{r}_{\alpha i}$ is the relative position for the i-th atom in α-th cell, and by its definition,

$$\sum_i m_i \mathbf{r}_{\alpha i} = \mathbf{0} \,, \tag{3}$$

which leads to the condition,

$$\sum_i m_i \mathbf{S}_{i_\alpha} = \mathbf{0} \,. \tag{4}$$

Here we may use i_α to distinguish the numbering index in different supercells. However, if without confusion, we denote i_α as i or $i \in \alpha$ in the rest of paper.

Now we introduce a so-called micromorphic multiplicative decomposition to describe the relative position of each atom,

$$\mathbf{r}_{\alpha i} = \boldsymbol{\phi}_\alpha \cdot \mathbf{S}_i, \quad \text{and} \quad \boldsymbol{\phi}_\alpha := \mathbf{F}_\alpha \cdot \boldsymbol{\chi}_\alpha \tag{5}$$

where $\boldsymbol{\phi}_\alpha$ is the total deformation gradient of the α-th supercell; $\mathbf{S}_i$ are the scaled atom position vectors for every atoms inside the α-th cell (we use $\mathbf{S}_i$ instead of $\mathbf{S}_{\alpha i}$ for simplicity); the second order tensor $\boldsymbol{\chi}_\alpha$ is the micro deformation tensor of the α-th cell, and its physical meaning may be interpreted as the shape tensor of α-th cell,

$$\boldsymbol{\chi}_\alpha(t)\mathbf{S}_i(t) = \xi_i(t)\mathbf{a}(t) + \eta_i(t)\mathbf{b}(t) + \zeta_i(t)\mathbf{c}(t), \quad i = 1, 2, \ldots, N_\alpha$$

where $\mathbf{a}$, $\mathbf{b}$, and $\mathbf{c}$ are the MD cell edge vectors, and ξ_i, η_i, and ζ_i are the components of the local position vector $\mathbf{S}_i$ projecting onto the MD cell edge. They can be expressed into the following matrix form,

$$\boldsymbol{\chi}_\alpha(t) = \begin{pmatrix} a_x(t) & b_x(t) & c_x(t) \\ a_y(t) & b_y(t) & c_y(t) \\ a_z(t) & b_z(t) & c_z(t) \end{pmatrix}, \quad \mathbf{S}_i(t) = \begin{pmatrix} \xi_i(t) \\ \eta_i(t) \\ \zeta_i(t) \end{pmatrix}$$

If we place the center of the supercell at the origin of the local coordinate, the components of the local position vector of $\mathbf{S}_i$ oscillate in the range,

$$-0.5 \leq \xi_i(t), \eta_i(t), \zeta_i(t) \leq 0.5,$$

which characterizes the internal degrees of freedom of atom motions. When the total number of atoms increases, the motion of $\mathbf{S}_i$ may become chaotic. Therefore, we refer the local position vectors $\mathbf{S}_i$ as the statistical variables. Borrowing the terminology in macroscale continuum mechanics, we call the assemble of $\{\mathbf{S}_i\}$ as the statistical parametric configuration, i.e. $\mathbf{S}_i \in \boldsymbol{\mathcal{B}}_S$.

The key part of the multiplicative micromorphic decomposition is the introduction of a coarse scale deformation gradient $\mathbf{F}_\alpha = \mathbf{F}_\alpha(\{\mathbf{r}_\beta\})$ that are determined by the overall motion of all centers of mass of every cells, which are completely determined by the spatial distribution of centers of mass (implied by the dependence of $\{\mathbf{r}_\beta\}$) of all cells.

Fig. 1 shows a schematic spatial cell division and the corresponding distribution of centers of mass. $\mathbf{F}_\alpha$ together with $\boldsymbol{\chi}_\alpha$ constitute the total deformation gradient $\boldsymbol{\phi}_\alpha$, whereas $\mathbf{F}_\alpha$ are determined by the distribution of $\{\mathbf{r}_\alpha\}$. For example, we can determine $\mathbf{F}_\alpha$ by using an approach adopted in the reproducing kernel particle method Li and Liu [33] or the state-based peridynamics Silling et al. [34].
Denote

$$\mathbf{r}_{\alpha\beta}(t) = \mathbf{r}_\beta(t) - \mathbf{r}_\alpha(t), \quad \mathbf{R}_\alpha := \mathbf{r}_\alpha(0), \quad \text{and } \mathbf{R}_{\alpha\beta} = \mathbf{R}_\beta - \mathbf{R}_\alpha\,, \tag{6}$$

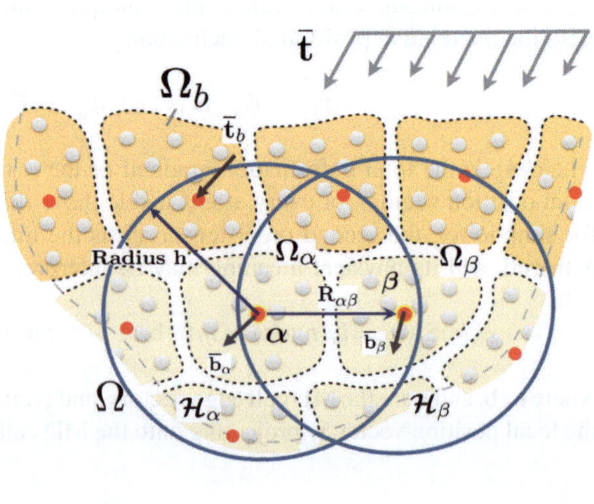

Fig. 1 Schematic illustration of supercell domain decomposition; the dark points are centers of mass of supercells that may be subjected external body force; the dark supercells are boundary cells that may be subjected external traction. The center of mass for each supercell has a support $\mathcal{H}_\alpha$ that contains some other centers of mass of adjacent supercells

where the Greek subscripts are indices of the centers of mass of MD cells. As shown in Fig. 1, each supercell center has a compact support, say $\mathcal{H}_\alpha$ for the cell α. We can then construct the so-called moment matrix or shape tensor for the cell center α by calculating following discrete sum over the other cell centers inside the support $\mathcal{H}_\alpha$ (See [33]),

$$\mathbf{M}_\alpha := \sum_{\beta=1}^{N_h} \omega(|\mathbf{R}_{\alpha\beta}|)\mathbf{R}_{\alpha\beta} \otimes \mathbf{R}_{\alpha\beta}\Omega_{\beta 0}, \tag{7}$$

where $\otimes$ is the tensor product operator; N_h is the number of supercell centers inside $\mathcal{H}_\alpha$; $\Omega_{\beta 0}$ is the volume of the β-th cell in the referential configuration $\mathcal{B}_R$, and $\omega(|\mathbf{R}_{\alpha\beta}|)$ is a localized positive window function.

To construct the macroscale deformation gradient tensor, we first define a two-point second order tensor by the following discrete sum,

$$\mathbf{N}_\alpha := \Big(\sum_{\beta=1}^{N_h} \omega(|\mathbf{R}_{\alpha\beta}|)\mathbf{r}_{\alpha\beta} \otimes \mathbf{R}_{\alpha\beta}\Omega_{\beta 0}\Big) . \tag{8}$$

Considering the Cauchy-Born rule Tadmor et al. [35],

$$\mathbf{r}_{\alpha\beta} = \mathbf{F}_\alpha \cdot \mathbf{R}_{\alpha\beta} \quad \rightarrow \quad \mathbf{N}_\alpha = \mathbf{F}_\alpha \cdot \mathbf{M}_\alpha \ ,$$

we can then derive the coarse scale deformation gradient as

$$\mathbf{F}_\alpha = \Big(\sum_{\beta=1}^{N_h} \omega(|\mathbf{R}_{\alpha\beta}|)\mathbf{r}_{\alpha\beta} \otimes \mathbf{R}_{\alpha\beta}\Omega_{\beta 0}\Big) \cdot \mathbf{M}_\alpha^{-1} . \tag{9}$$

where $\omega(|\mathbf{R}_{\alpha\beta}|)$ is a localized positive window function, and a common choice is the Gaussian function,

$$\omega_h(\mathbf{x}) = \frac{1}{(\pi h^2)^{d/2}} \exp\Big(-\frac{\mathbf{x}\cdot\mathbf{x}}{h^2}\Big) , \tag{10}$$

here d is the number of space dimension, and h is the radius of the support. Obviously, $\mathbf{F}_\alpha$ depends on the relative positions of all other atoms in the support of the α-atom, i.e. Ω_α. Note that in the configurations $\mathcal{B}_R$ and $\mathcal{B}_I$ the center of the supercell occupies the same position, i.e. $\mathbf{R}_\alpha \equiv \mathcal{R}_\alpha$, but not for each atom, i.e. $\mathbf{R}_{\alpha i} \neq \mathcal{R}_{\alpha i}$.

At the beginning of the coarse scale deformation, the initial position of the supercell center $\mathbf{r}_\alpha(0) = \mathbf{R}_\alpha$, therefore,

$$\mathbf{F}_\alpha = \Big(\sum_{\beta=1}^{N_h} \omega(|\mathbf{R}_{\alpha\beta}|)\mathbf{r}_{\alpha\beta} \otimes \mathbf{R}_{\alpha\beta}\Omega_{\beta 0}\Big) \cdot \mathbf{M}_\alpha^{-1} = \Big(\sum_{\beta=1}^{N_h} \omega(|\mathbf{R}_{\alpha\beta}|)\mathbf{R}_{\alpha\beta} \otimes \mathbf{R}_{\alpha\beta}\Omega_{\beta 0}\Big) \cdot \mathbf{M}_\alpha^{-1} = \mathbf{I},$$

where $\mathbf{I}$ is the unit of second order tensor.

Initially, the supercell shape tensor, χ_0, which is spanned by the three edges of the MD cell, $\mathbf{a}(0)$, $\mathbf{b}(0)$, and $\mathbf{c}(0)$, may be expressed as,

$$\chi_0 := \chi(0) = [\mathbf{a}(0), \mathbf{b}(0), \mathbf{c}(0)] \, .$$

For all the atoms in a representative MD cell, say the α-th cell, its referential position are defined as

$$\mathbf{R}_i := \mathbf{R}_\alpha + \mathbf{R}_{\alpha i}, \quad \text{with } \mathbf{R}_{\alpha i} = \chi_0 \mathbf{S}_i \, . \tag{11}$$

In the rest of this paper, we refer this molecular configuration as the referential configuration $\mathcal{B}_R$. One can find that this definition is consistent with kinematic assumption,

$$\mathbf{r}_i = \mathbf{r}_\alpha + \mathbf{r}_{\alpha i} = \mathbf{r}_\alpha + \mathbf{F}_\alpha \cdot \chi_\alpha \cdot \mathbf{S}_i \, . \tag{12}$$

Based on Eq. (12), we may define an intermediate referential configuration,

$$\mathcal{R}_i = \mathbf{R}_\alpha + \mathcal{R}_{\alpha i} = \mathbf{R}_\alpha + \chi_\alpha(t) \cdot \mathbf{S}_i, \tag{13}$$

so that the coarse scale deformation gradient is a continuum scale deformation gradient.

In the proposed multiscale method, we have introduced four different configurations in a representative cell α: (1) The spatial configuration, i.e. $\mathcal{B}_c(\mathbf{r})$-configuration, in which $\mathbf{r}_i \in \Omega_\alpha$; (2) the intermediate equilibrium configuration, i.e. $\mathcal{B}_I(\mathcal{R})$-configuration; (3) the referential equilibrium configuration, i.e. $\mathcal{B}_I(\mathbf{R})$-configuration, in which $\mathbf{R}_i \in \Omega_{\alpha 0}$, and (4) the statistical configuration $\mathcal{B}_S$, in which the scaled atom coordinate $\mathbf{S}_i$ are primary variables. Figure 2 shows a deformation map that connects these four configuration spaces.

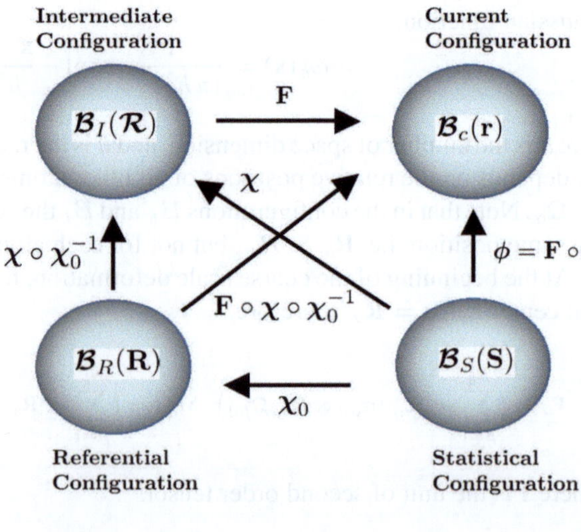

Fig. 2 Deformation map that connects four different kinematic configuration spaces: (1) Statistical configuration; (2) Referential configuration; (3) Intermediate configuration, and (4) Current configuration.

Since for the monoatom systems $\sum_i \mathbf{S}_i = 0$ and $\mathbf{S}_i$ are oscillating in the range $-0.5 \leq \mathbf{S}_i \leq 0.5$, i.e. $\|\mathbf{S}_i\| < 1$, as atoms oscillate around their equilibrium positions, we may interpret $\mathbf{S}_i$ as a statistical variable. For a given macroscale kinematic variable or point, $\mathbf{r}_\alpha(t)$ (the center of cell), there are many sets of $\{\mathbf{S}_i\}$ corresponding to it, and this is the case because as long as $\sum_i m_i \mathbf{S}_i = 0$ satisfies. Therefore, a given macroscale material point $\mathbf{r}_\alpha(t)$ does not correspond to a single set of $\{\mathbf{S}_i\}$ in a unique one-to-one manner, i.e. such correspondence is not unique. A macroscale state $\mathbf{r}_\alpha$ corresponds to many microstates $\{\mathbf{S}_i\}$, and a microstate is determined by the distribution of statistical variable $\mathbf{S}_i$.

It may be also noted that on the configuration spaces $\mathcal{B}_I$, $\mathcal{B}_R$, the center of mass coordinate are the same, i.e. $\mathbf{R}_\alpha$, whereas the configuration space $\mathcal{B}_S$ is defined cell by cell, and the coordinate of the center of mass of each cell is zero.

A pictorial illustration of three physical configurations is displayed in Fig. 3, where physical interpretations of different kinematic mappings at different scales are graphically illustrated. In passing we note that the continuum compatibility condition is a macroscale condition, and at both microscale and mesoscale this condition is not necessarily satisfied. For crystalline solids, this may be linked to the defect states or quasi-crystal states. Even though the referential configuration is not essential in the fine scale calculations, but its information is needed in coarse scale computations. To clearly specify displacement decomposition at each scale, we consider the following atomic position decomposition,

$$\mathbf{r}_i = \mathbf{r}_\alpha + \mathbf{r}_{\alpha i} \;\; \Rightarrow \;\; \mathbf{r}_i = \mathbf{r}_\alpha + \mathbf{F}_\alpha \cdot \mathcal{R}_{\alpha i} = \mathbf{r}_\alpha + \mathbf{F}_\alpha \cdot \boldsymbol{\chi}_\alpha \cdot \mathbf{S}_i \tag{14}$$

where

$$\mathcal{R}_{\alpha i} = \boldsymbol{\chi}_\alpha \cdot \mathbf{S}_i \; .$$

These relations are schematically illustrated in Fig. 4. Subsequently, we can decompose the atomistic displacement into three different scales,

$$\bar{\mathbf{u}} = \mathbf{r}_\alpha - \mathbf{R}_\alpha; \;\; \tilde{\mathbf{u}}_i = \Big(\mathbf{F}_\alpha \cdot \boldsymbol{\chi}_\alpha - \boldsymbol{\chi}_\alpha\Big)\mathbf{S}_i, \;\; \text{and} \;\; \mathbf{u}'_i = \Big(\boldsymbol{\chi}_\alpha - \boldsymbol{\chi}_{\alpha 0}\Big)\mathbf{S}_i, \tag{15}$$

where $\bar{\mathbf{u}}$, $\tilde{\mathbf{u}}_i$, $\mathbf{u}'_i$ denote macroscale, mesoscale, and microscale displacements. Apparently, three of them together constitute the total displacement $\mathbf{u}_i$, which can be expressed as,

$$\mathbf{u}_i = \bar{\mathbf{u}} + \tilde{\mathbf{u}}_i + \mathbf{u}'_i = \mathbf{r}_i - \mathbf{R}_i \; . \tag{16}$$

In the scale decomposition, we select three independent kinematic variables, $\{\mathbf{S}_i, \boldsymbol{\chi}_\alpha$, and $\mathbf{r}_\alpha\}$, to represent three different scales. The novelty of the proposed multiscale decomposition is the **Multiplicative Multiscale Decomposition**,

$$\boldsymbol{\phi}_\alpha = \mathbf{F}_\alpha \cdot \boldsymbol{\chi}_\alpha. \tag{17}$$

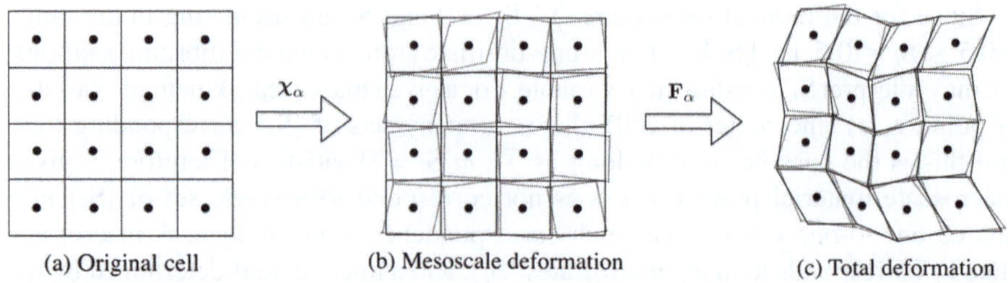

Fig. 3 Cell Deformation in different configurations: **a** the original undeformed system which is divided into several unit cells. The atomic positions are given by $\mathbf{R}_i$. The solid circles represent the centers of mass; **b** the cells (dashed parallelograms) further undergoes microscale deformation $\boldsymbol{\chi}_\alpha$ around their own center of mass separately without connection (with the cell center fixed). This is because that continuity may not be valid at fine scale. **c** The configuration undergoes macroscale deformation $\mathbf{F}_\alpha$, i.e. now the cell centers start to move. Notice that the continuity is a macroscale concept. Based on this concept, we can use the positions of all cell centers to construct, e.g. interpolate, a coarse scale displacement field

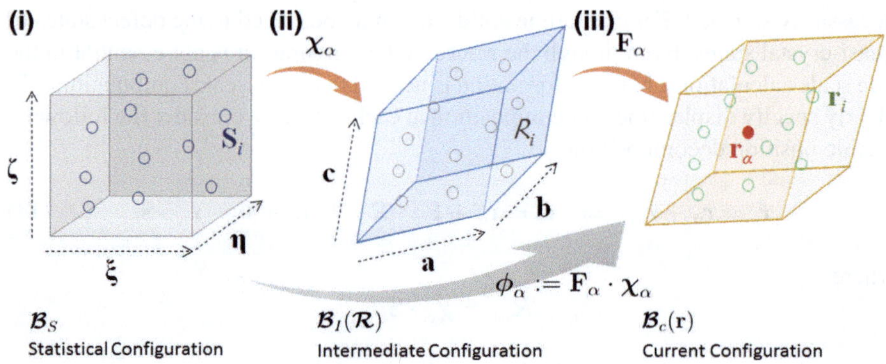

Fig. 4 Schematic illustrations of relations among **i** statistical configuration, **ii** intermediate configuration and **iii** current configuration

This decomposition is in sharp contrast with the additive multiscale decomposition proposed by Wagner and Liu [2]. The total deformation gradient $\boldsymbol{\phi}_\alpha$ consists of a macroscale deformation gradient $\mathbf{F}_\alpha$ and a micromorphic (meso) deformation gradient $\boldsymbol{\chi}_\alpha$. The macroscale deformation gradient at a local position is completely determined by the local distribution of the supercell centers, which can be quantified, for example, by Eq. (9). On the other hand, the microscale deformation gradient, which is often called as the micro-deformation in the literature of micromorphic theory, is an independent mesoscale field variable that quantifies the independent supercell rotation or microscale deformation. The macro deformation $\mathbf{F}_\alpha$ may be considered to be **Long-range order**, whereas the micro-deformation $\boldsymbol{\chi}_\alpha$ may be considered to be **Short-range order**.

2.2 Statistical Conditions

Before deriving the dynamic equations of MMMD, we would like to exam the kinetic energy of the α-th cell:

$$
\begin{aligned}
K_\alpha &= \frac{1}{2}\sum_i m_i \dot{\mathbf{r}}_i \cdot \dot{\mathbf{r}}_i = \frac{1}{2}\sum_i m_i (\dot{\mathbf{r}}_\alpha + \dot{\boldsymbol{\phi}}_\alpha \cdot \mathbf{S}_i + \boldsymbol{\phi}_\alpha \cdot \dot{\mathbf{S}}_i) \cdot (\dot{\mathbf{r}}_\alpha + \dot{\boldsymbol{\phi}}_\alpha \cdot \mathbf{S}_i + \boldsymbol{\phi}_\alpha \cdot \dot{\mathbf{S}}_i) \\
&= \underbrace{\frac{M_\alpha}{2}\dot{\mathbf{r}}_\alpha \cdot \dot{\mathbf{r}}_\alpha}_{K_1} + \underbrace{\frac{1}{2}\dot{\boldsymbol{\phi}}_\alpha^T \dot{\boldsymbol{\phi}}_\alpha \sum_i m_i \mathbf{S}_i \otimes \mathbf{S}_i}_{K_2} + \underbrace{\frac{1}{2}\sum_i m_i \dot{\mathbf{S}}_i \cdot \mathbf{C} \cdot \dot{\mathbf{S}}_i}_{K_3} \\
&\quad + \underbrace{\frac{1}{2}\dot{\boldsymbol{\phi}}_\alpha^T \boldsymbol{\phi}_\alpha \sum_i m_i \mathbf{S}_i \otimes \dot{\mathbf{S}}_i + \frac{1}{2}\boldsymbol{\phi}_\alpha^T \dot{\boldsymbol{\phi}}_\alpha \sum_i m_i \dot{\mathbf{S}}_i \otimes \mathbf{S}_i}_{K_4}
\end{aligned}
\tag{18}
$$

where $M_\alpha = \sum_i m_i$ is the mass of the cell, and $\mathbf{C} = \boldsymbol{\phi}_\alpha^T \boldsymbol{\phi}_\alpha$ is the multiscale right Cauchy-Green tensor.

We postulate the following statistical condition,

$$
\mathbf{J}_\alpha^S = \sum_i m_i \mathbf{S}_i \otimes \mathbf{S}_i = \boldsymbol{\chi}_{\alpha 0}^{-1} \cdot \mathbf{J}_\alpha \cdot \boldsymbol{\chi}_{\alpha 0}^{-T} = constant\ tensor. \tag{19}
$$

where

$$
\mathbf{J}_\alpha = \sum_i m_i \mathbf{R}_{\alpha i} \otimes \mathbf{R}_{\alpha i}, \tag{20}
$$

is a second order tensor that is similar to the moment of inertia tensor of the cell in referential space. In mechanics, the moment of inertia tensor of a rigid body is defined as,

$$
\mathbf{I}_{Euler} := \int_V \rho\Big(\|\mathbf{R}\|^2 \mathbf{I}^{(2)} - \mathbf{R} \otimes \mathbf{R}\Big) dV
$$

where ρ is mass density, and $\mathbf{I}^{(2)}$ is the second order unit tensor. It is obvious that $\mathbf{I}_{Euler}$ is dependent on both the shape and size of the rigid body. Similarly the tensor $\mathbf{J}_\alpha$ is also dependent on both the shape and size of the micromorphic MD cell. On the other hand, the second order tensor $\mathbf{J}_\alpha^S$ is not dependent on the MD cell size and shape. In fact $\mathbf{J}_\alpha^S$ is the push-back of $\mathbf{J}_\alpha$ Marsden and Hughes [36], and $\boldsymbol{\chi}_{\alpha 0} = \boldsymbol{\chi}(0)$ is a constant second order tensor, i.e. the geometric shape tensor of the original cell.

We can replace the discrete summation (19) with a continuous integration, i.e.

$$
\mathbf{J}_\alpha^S = \int_\omega \rho \mathbf{S} \otimes \mathbf{S} d\omega_S \tag{21}
$$

where ω is the volume of a unit cube. Let $\mathbf{E}_I$ be the principal axes of $\mathbf{J}^S_\alpha$. We can then have the spectral decomposition of $\mathbf{J}^S_\alpha$,

$$\mathbf{J}^S_\alpha = \sum_{I=1}^{3} \left(\int_\omega \rho S_I^2 d\omega \right) \mathbf{E}_I \otimes \mathbf{E}_I \ . \tag{22}$$

If we assume that $\mathbf{J}^S_\alpha$ is a spherical tensor, we have

$$\int_\omega \rho S_1^2 d\omega = \int_\omega \rho S_2^2 d\omega = \int_\omega \rho S_3^2 d\omega =: W_\alpha \ . \tag{23}$$

It should be noted that the supercell is not a necessarily unit cube, and it is related to the physical shape of the lattice representative cell. Therefore, the related Euler's moment of inertia tensor, $\mathbf{I}_{Euler}$ is in general anisotropic. On the other hand, the moment tensor $\mathbf{J}^S_\alpha$ in (21) is defined in the statistical configuration. The assumption that $\mathbf{J}^S_\alpha$ is a spherical tensor is based on the general ergodicity postulate that the space and time that a molecular dynamics simulation is involved are statistical isotropic and homogeneous, which is one of the fundamental hypothesis in statistical physics and molecular physics. Note that we are discussing a quantity in a statistical configuration $\mathcal{B}_S$, not the lattice configuration $\mathcal{B}_R$, $\mathcal{B}_I$ nor the current configuration $\mathcal{B}_c$. The anisotropy of a material in the lattice space is not in conflict with the ergodic assumption stated here.

Next we consider the fourth term of the kinetic energy K_4. Since we may label $\mathbf{S}_i$ statistical variables, we may find that $\mathbf{J}^S_\alpha$ is in fact a tensorial autocorrelation function of statistical positions,

$$\mathbf{J}^S(\tau) =< \mathbf{S}_i(t) \otimes \mathbf{S}_i(t+\tau) >:= \sum_i m_i \mathbf{S}_i(t) \otimes \mathbf{S}_i(t+\tau) \tag{24}$$

One can prove that the autocorrelation tensor is an even function of τ, if $\mathbf{J}^S$ is spherical. This implies the derivation at the origin $\tau = 0$ is zero. Hence,

$$\frac{d}{d\tau} \mathbf{J}^S_\alpha |_{\tau=0} = \sum_i m_i \mathbf{S}_i(t) \otimes \dot{\mathbf{S}}_i(t) = \mathbf{0}, \tag{25}$$

and similarly,

$$\sum_i m_i \dot{\mathbf{S}}_i(t) \otimes \mathbf{S}_i(t) = \mathbf{0} \ . \tag{26}$$

Equations (25) and (26) imply that $K_4 = \mathbf{0}$. In fact, the Second statistical condition states that $\mathbf{J}^S_\alpha$ is a spherical tensor, which implies that

$$\mathbf{J}^S_\alpha(\tau) =< \mathbf{S}_i(t) \otimes \mathbf{S}_i(t+\tau) >= \sum_i m_i \mathbf{S}_i(t) \otimes \mathbf{S}_i(t+\tau) = \Big(\sum_i m_i S_i(t) S_i(t+\tau)\Big) \mathbf{E}_I \otimes \mathbf{E}_I$$

where $\mathbf{E}_I$, $I = 1, 2, 3$ are the basis of a Cartesian coordinate. It then follows,

$$\begin{aligned}
\mathbf{J}^S_\alpha(-\tau) &= < \mathbf{S}_i(t) \otimes \mathbf{S}_i(t-\tau) >= \sum_i m_i \mathbf{S}_i(t) \otimes \mathbf{S}_i(t-\tau) \\
&= \Big(\sum_i m_i S_i(t) S_i(t-\tau)\Big) \mathbf{E}_I \otimes \mathbf{E}_I = \Big(\sum_i m_i S_i(t+\tau) S_i(t)\Big) \mathbf{E}_I \otimes \mathbf{E}_I \\
&= < \mathbf{S}_i(t) \otimes \mathbf{S}_i(t+\tau) >= \mathbf{J}^S_\alpha(\tau)
\end{aligned}$$

Utilizing the two statistical conditions, we can write the Lagrangian of the MMMD system as,

$$\begin{aligned}
\mathcal{L}_m &= \frac{1}{2}\sum_\beta M_\beta \dot{\mathbf{r}}_\beta \cdot \dot{\mathbf{r}}_\beta + \frac{1}{2}\sum_\beta \mathbf{J}^S_\beta : (\dot{\boldsymbol{\phi}}^T_\beta \dot{\boldsymbol{\phi}}_\beta) \\
&+ \frac{1}{2}\sum_\beta \sum_i m_i \dot{\mathbf{S}}_i \cdot \mathbf{C}_\beta \cdot \dot{\mathbf{S}}_i - \frac{1}{2}\sum_\beta \sum_\gamma \sum_{i\in\beta, j\in\gamma} V(r_{ij}) - \sum_\beta \sum_{i\in\beta} \mathbf{b}_i \cdot \mathbf{r}_i \qquad (27)
\end{aligned}$$

where β, γ are cell indices, and the abbreviation $i \in \beta$ means that the i-th atom belongs to the β-th cell. $\mathbf{C}_\beta := \boldsymbol{\phi}^T_\beta \boldsymbol{\phi}_\beta$ is the micro right Cauchy-Green tensor for total deformation.

In the rest of the paper, we choose three independent field variables, $\mathbf{r}_\alpha, \boldsymbol{\phi}_\alpha$, and $\mathbf{S}_i$ representing kinematic variables for three scales, instead of the set of original variables $\mathbf{r}_\alpha, \boldsymbol{\chi}_\alpha, \mathbf{S}_i$, which will be equivalent each other. Thus $\mathcal{L}_m = \mathcal{L}_m(\mathbf{r}_\alpha, \boldsymbol{\phi}_\alpha, \mathbf{S}_i)$. The equations of motion are stated as

$$\frac{d}{dt}\frac{\partial \mathcal{L}_m}{\partial \dot{\mathbf{r}}_\alpha} - \frac{\partial \mathcal{L}_m}{\partial \mathbf{r}_\alpha} = 0; \quad \frac{d}{dt}\frac{\partial \mathcal{L}_m}{\partial \dot{\boldsymbol{\phi}}_\alpha} - \frac{\partial \mathcal{L}_\alpha}{\partial \boldsymbol{\phi}_\alpha} = 0, \text{ and } \frac{d}{dt}\frac{\partial \mathcal{L}_m}{\partial \dot{\mathbf{S}}_i} - \frac{\partial \mathcal{L}_\alpha}{\partial \mathbf{S}_i} = 0\,. \qquad (28)$$

3 Multiscale Micromorphic Molecular Dynamics (MMMD)

In this Section, we shall derive the dynamic equations for multiscale micromorphic molecular dynamics.

3.1 *Coarse Scale Dynamic Equation*

We start by deriving some relations that are needed in subsequent derivations. Considering

$$\mathbf{F}_\beta = \mathbf{F}_\beta(\{\mathbf{r}_\alpha\}) \text{ and } \dot{\mathbf{F}}_\beta = \dot{\mathbf{F}}_\beta(\{\mathbf{r}_\alpha\}, \{\dot{\mathbf{r}}_\alpha\}), \qquad (29)$$

we then have

$$\dot{\mathbf{F}}_\beta = \sum_\alpha \frac{\partial \mathbf{F}_\beta}{\partial \mathbf{r}_\alpha} \dot{\mathbf{r}}_\alpha, \tag{30}$$

which leads to the relation,

$$\frac{\partial \dot{\mathbf{F}}_\beta}{\partial \dot{\mathbf{r}}_\alpha} = \frac{\partial \mathbf{F}_\beta}{\partial \mathbf{r}_\alpha} . \tag{31}$$

Moreover

$$\ddot{\mathbf{F}}_\beta = \sum_\alpha \left(\frac{d}{dt} \left(\frac{\partial \mathbf{F}_\beta}{\partial \mathbf{r}_\alpha} \right) \dot{\mathbf{r}}_\alpha + \frac{\partial \mathbf{F}_\beta}{\partial \mathbf{r}_\alpha} \ddot{\mathbf{r}}_\alpha \right) , \tag{32}$$

on the other hand,

$$\ddot{\mathbf{F}}_\beta = \sum_\alpha \left(\frac{\partial \dot{\mathbf{F}}_\beta}{\partial \mathbf{r}_\alpha} \dot{\mathbf{r}}_\alpha + \frac{\partial \dot{\mathbf{F}}_\beta}{\partial \dot{\mathbf{r}}_\alpha} \ddot{\mathbf{r}}_\alpha \right) \tag{33}$$

Comparing Eqs. (32) and (33) and utilizing (31), we obtain

$$\frac{d}{dt} \left(\frac{\partial \mathbf{F}_\beta}{\partial \mathbf{r}_\alpha} \right) = \frac{\partial \dot{\mathbf{F}}_\beta}{\partial \mathbf{r}_\alpha} . \tag{34}$$

Furthermore considering,

$$\dot{\boldsymbol{\phi}}_\beta = \dot{\mathbf{F}}_\beta \boldsymbol{\chi}_\beta + \mathbf{F}_\beta \cdot \dot{\boldsymbol{\chi}}_\beta, \tag{35}$$

we can obtain,

$$\frac{\partial \dot{\boldsymbol{\phi}}_\beta}{\partial \dot{\mathbf{r}}_\alpha} = \frac{\partial \dot{\mathbf{F}}_\beta}{\partial \dot{\mathbf{r}}_\alpha} \boldsymbol{\chi}_\beta + \frac{\partial \mathbf{F}_\beta}{\partial \dot{\mathbf{r}}_\alpha} \dot{\boldsymbol{\chi}}_\beta = \frac{\partial \mathbf{F}_\beta}{\partial \mathbf{r}_\alpha} \boldsymbol{\chi}_\beta = \frac{\partial \boldsymbol{\phi}_\beta}{\partial \mathbf{r}_\alpha} . \tag{36}$$

Here, we assume that $\frac{\partial \mathbf{F}_\beta}{\partial \dot{\mathbf{r}}_\alpha} = 0$, according to Eq. (29).

By virtue of Eqs. (34)–(36), we have

$$\begin{aligned} \frac{\partial \dot{\boldsymbol{\phi}}_\beta}{\partial \mathbf{r}_\alpha} &= \frac{\partial \dot{\mathbf{F}}_\beta}{\partial \mathbf{r}_\alpha} \boldsymbol{\chi}_\beta + \frac{\partial \mathbf{F}_\beta}{\partial \mathbf{r}_\alpha} \dot{\boldsymbol{\chi}}_\beta = \frac{\partial \mathbf{F}_\beta}{\partial \mathbf{r}_\alpha} \dot{\boldsymbol{\chi}}_\beta + \frac{d}{dt} \left(\frac{\partial \mathbf{F}_\beta}{\partial \mathbf{r}_\alpha} \right) \boldsymbol{\chi}_\beta \\ &= \frac{d}{dt} \left(\frac{\partial \boldsymbol{\phi}_\beta}{\partial \mathbf{r}_\alpha} \right) = \frac{d}{dt} \left(\frac{\partial \dot{\boldsymbol{\phi}}_\beta}{\partial \dot{\mathbf{r}}_\alpha} \right) . \end{aligned} \tag{37}$$

This relation is needed in the subsequent derivation. Considering the coarse scale Lagrangian equation and utilizing the above relation, we have

$$
\begin{aligned}
\frac{d}{dt}\Big(\frac{\partial \mathcal{L}_m}{\partial \dot{\mathbf{r}}_\alpha}\Big) &= \frac{d}{dt}\Big(M_\alpha \dot{\mathbf{r}}_\alpha + \sum_\beta \frac{\partial \mathcal{L}_m}{\partial \dot{\boldsymbol{\phi}}_\beta}\cdot\frac{\partial \dot{\boldsymbol{\phi}}_\beta}{\partial \dot{\mathbf{r}}_\alpha}\Big)\\
&= M_\alpha \ddot{\mathbf{r}}_\alpha + \sum_\beta \frac{d}{dt}\Big(\frac{\partial \mathcal{L}_m}{\partial \dot{\boldsymbol{\phi}}_\beta}\Big)\cdot\frac{\partial \dot{\boldsymbol{\phi}}_\beta}{\partial \dot{\mathbf{r}}_\alpha} + \sum_\beta \frac{\partial \mathcal{L}_m}{\partial \dot{\boldsymbol{\phi}}_\beta}\cdot\frac{d}{dt}\Big(\frac{\partial \dot{\boldsymbol{\phi}}_\beta}{\partial \dot{\mathbf{r}}_\alpha}\Big)\\
&= M_\alpha \ddot{\mathbf{r}}_\alpha + \sum_\beta \frac{\partial \mathcal{L}_m}{\partial \boldsymbol{\phi}_\beta}\cdot\frac{\partial \boldsymbol{\phi}_\beta}{\partial \mathbf{r}_\alpha} + \sum_\beta \frac{\partial \mathcal{L}_m}{\partial \dot{\boldsymbol{\phi}}_\beta}\cdot\frac{d}{dt}\Big(\frac{\partial \boldsymbol{\phi}_\beta}{\partial \mathbf{r}_\alpha}\Big)\\
&= M_\alpha \ddot{\mathbf{r}}_\alpha + \sum_\beta \frac{\partial \mathcal{L}_m}{\partial \boldsymbol{\phi}_\beta}\cdot\frac{\partial \boldsymbol{\phi}_\beta}{\partial \mathbf{r}_\alpha} + \sum_\beta \frac{\partial \mathcal{L}_m}{\partial \dot{\boldsymbol{\phi}}_\beta}\cdot\frac{\partial \dot{\boldsymbol{\phi}}_\beta}{\partial \mathbf{r}_\alpha}
\end{aligned}
\tag{38}
$$

Furthermore,

$$
\frac{\partial \mathcal{L}_m}{\partial \mathbf{r}_\alpha} = \sum_{\beta\neq\alpha}\sum_{i\in\alpha, j\in\beta} V'(r_{ij})\frac{\mathbf{r}_{ij}}{|\mathbf{r}_{ij}|} - \sum_{i\in\alpha}\mathbf{b}_i + \sum_\beta \frac{\partial \mathcal{L}_m}{\partial \boldsymbol{\phi}_\beta}\cdot\frac{\partial \boldsymbol{\phi}_\beta}{\partial \mathbf{r}_\alpha} + \sum_\beta \frac{\partial \mathcal{L}_m}{\partial \dot{\boldsymbol{\phi}}_\beta}\cdot\frac{\partial \dot{\boldsymbol{\phi}}_\beta}{\partial \mathbf{r}_\alpha}\,. \tag{39}
$$

Substituting Eqs. (38) and (39) into the coarse scale Lagrangian equation, we obtain

$$
\frac{d}{dt}\Big(\frac{\partial \mathcal{L}_m}{\partial \dot{\mathbf{r}}_\alpha}\Big) - \frac{\partial \mathcal{L}_m}{\partial \mathbf{r}_\alpha} = 0 \;\rightarrow\; M_\alpha \ddot{\mathbf{r}}_\alpha = \sum_{\beta\neq\alpha}\sum_{i\in\alpha, j\in\beta} V'(r_{ij})\frac{\mathbf{r}_{ij}}{|\mathbf{r}_{ij}|} + \mathcal{B}_\alpha \tag{40}
$$

where $\mathcal{B}_\alpha = \sum_i \mathbf{b}_i$. Note that $-V'(r_{ij})\dfrac{\mathbf{r}_{ij}}{|\mathbf{r}_{ij}|} = \mathbf{f}_{ij}$ is the pair force between atoms. If one would like to include the macroscale traction boundary condition as indicated in Fig. 1, one needs to add another term on the center of mass of the boundary cells. In this case, we can redefine the external force density as,

$$
\mathcal{B}_\alpha = S_\alpha \bar{\mathbf{t}}_\alpha + \Omega_\alpha \bar{\mathbf{b}}_\alpha, \tag{41}
$$

where S_α is the area of macroscale traction boundary of α-th cell, $\bar{\mathbf{t}}_\alpha$ is the traction vector on S_α; and $\bar{\mathbf{b}}_\alpha = \sum_i \mathbf{b}_i / \Omega_\alpha$.

Then we can rewrite the above equation as,

$$
M_\alpha \ddot{\mathbf{r}}_\alpha = -\sum_{\beta\neq\alpha}\sum_{i\in\alpha, j\in\beta} \mathbf{f}_{ij} + \mathcal{B}_\alpha \tag{42}
$$

The first term in the right hand side (RHS) of Eq. (42) represents the cell-to-cell interactions. We can also rewrite the first term of RHS of Eq. (42) as

$$
\sum_{\beta\neq\alpha}\sum_{i\in\alpha, j\in\beta} \mathbf{f}_{ij} = \sum_{\beta\neq\alpha}\mathbf{f}_{\alpha\beta}, \quad \text{where } \mathbf{f}_{\alpha\beta} := \sum_{i\in\alpha}\sum_{j\in\beta}\mathbf{f}_{ij}\,. \tag{43}
$$

Then Eq. (42) can be written in terms of coarse grain variables completely,

$$M_\alpha \ddot{\mathbf{r}}_\alpha = -\sum_{\beta \neq \alpha} \mathbf{f}_{\alpha\beta} + \mathcal{B}_\alpha \,. \tag{44}$$

where the Greek letters α, β denote the centers of mass of MD cells.

In Eq. (43), the cell-to-cell interaction term consists of all the atomic bond forces that passing through the surface of the cell, which is actually the divergence of the Cauchy stress ($\nabla \cdot \boldsymbol{\sigma}_\alpha$) generated by the atomic interaction from outside the cell multiply the volume Ω_α. This can be seen as follows,

$$-\sum_{\beta \neq \alpha} \sum_{i \in \alpha, j \in \beta} \mathbf{f}_{ij} \approx \int_{\partial \Omega_\alpha} \boldsymbol{\sigma} \cdot \mathbf{n} dS = \int_{\Omega_\alpha} \nabla \cdot \boldsymbol{\sigma} dV \approx \nabla \cdot \boldsymbol{\sigma}_\alpha \Omega_\alpha \,.$$

Thus Eq. (42) may be interpreted as,

$$M_\alpha \ddot{\mathbf{r}}_\alpha = (\nabla \cdot \boldsymbol{\sigma}_\alpha)\Omega_\alpha + \mathcal{B}_\alpha \tag{45}$$

which is a coarse-grain version of the Balance of Linear Momentum. Equation (45) not only establishes a link from molecular dynamics to continuum mechanics, but also provides an alternative computational formulation, which is explained as follows.

First we can rewrite the divergence of the Cauchy stress as

$$\begin{aligned}
\Omega_\alpha \nabla_r \cdot \boldsymbol{\sigma}_\alpha &= \tilde{\Omega}_\alpha J \nabla_{\mathcal{R}} \cdot \boldsymbol{\sigma}_\alpha \left(\frac{\partial \boldsymbol{\mathcal{R}}}{\partial \mathbf{r}}\right) = \tilde{\Omega}_\alpha \left(\nabla_{\mathcal{R}} \cdot \left(J \boldsymbol{\sigma}_\alpha \mathbf{F}_\alpha^{-T}\right) - \boldsymbol{\sigma}_\alpha \cdot \nabla_{\mathcal{R}} \cdot (J \mathbf{F}_\alpha^{-T})\right) \\
&= \tilde{\Omega}_\alpha \nabla_{\mathcal{R}} \cdot \left(J \boldsymbol{\sigma}_\alpha \mathbf{F}_\alpha^{-T}\right) = \tilde{\Omega}_\alpha \nabla_{\mathcal{R}} \cdot \boldsymbol{\mathcal{P}}_\alpha \,,
\end{aligned}$$

where $\tilde{\Omega}_\alpha$ is the volume of the MD cell in the intermediate configuration, i.e. $\boldsymbol{\mathcal{R}}$-configuration, and $J := det(\mathbf{F}_\alpha) = \Omega_\alpha / \tilde{\Omega}_\alpha$. In the above derivation, the Piola identity $\nabla_{\mathcal{R}} \cdot (J\mathbf{F}^{-T}) = 0$ is used, and moreover $\boldsymbol{\mathcal{P}}_\alpha = J \boldsymbol{\sigma}_\alpha \mathbf{F}_\alpha^{-T}$ is the first Piola-Kirchhoff stress with respect to the current configuration ($\mathbf{r}$-configuration) and the referential configuration ($\boldsymbol{\mathcal{R}}$-configuration), which can be shown as

$$\boldsymbol{\mathcal{P}}_\alpha = \boldsymbol{\mathcal{P}}_\alpha^{Virial} = \frac{1}{\tilde{\Omega}_\alpha} \sum_{i \in \alpha} \Big(-\boldsymbol{\phi}_\alpha m_i \dot{\mathbf{S}}_i \otimes \dot{\mathbf{S}}_i + \frac{1}{2} \sum_{j \in \alpha, j \neq i} \mathbf{f}_{ij} \otimes \mathbf{S}_{ij} \Big) \cdot \boldsymbol{\chi}_\alpha^T \,, \tag{46}$$

when the equilibrium state is reached. The detailed derivation of Eq. (46) will be discussed in the subsequent sections, and our purpose here is to derive the macroscale dynamic equation.

Now Eq. (45) can be re-written as

$$M_\alpha \ddot{\mathbf{r}}_\alpha = \tilde{\Omega}_\alpha \nabla_{\mathcal{R}} \cdot \boldsymbol{\mathcal{P}}_\alpha + \mathcal{B}_\alpha \,. \tag{47}$$

Recall that the coarse scale deformation gradient may be approximated as (9),

$$\mathbf{F}_\alpha = \nabla_{\mathcal{R}} \otimes \mathbf{r}_\alpha \approx \Big(\sum_{\beta=1}^{N_h} \omega(|\mathbf{R}_{\alpha\beta}|)\mathbf{r}_{\alpha\beta} \otimes \mathbf{R}_{\alpha\beta}\tilde{\Omega}_\beta\Big) \cdot \mathbf{M}_\alpha^{-1}\,, \tag{48}$$

which indicates that the structure of the discrete differential operator may be expressed as follows,

$$\nabla_{\mathcal{R}} \otimes \Big(\mathbf{f}_\alpha\Big) \approx \Big(\sum_{\beta=1}^{N_h} \omega(|\mathbf{R}_{\alpha\beta}|)(\mathbf{f}_\beta - \mathbf{f}_\alpha) \otimes \mathbf{R}_{\alpha\beta}\tilde{\Omega}_\beta\Big) \cdot \mathbf{M}_\alpha^{-1}\,.$$

Following the same procedure, we can define the discrete gradient operator as,

$$\nabla_{\mathcal{R}} \cdot \Big(\mathbf{f}_\alpha\Big) \approx \Big(\sum_{\beta=1}^{N_h} \omega(|\mathbf{R}_{\alpha\beta}|)(\mathbf{f}_\beta - \mathbf{f}_\alpha) \cdot \mathbf{R}_{\alpha\beta}\tilde{\Omega}_\beta\Big) \cdot \mathbf{M}_\alpha^{-1}\,. \tag{49}$$

Note that the center of mass positions, $\mathbf{R}_\alpha$ or $\mathbf{R}_\beta$, are the same in the intermediate configuration ($\mathcal{R}$) as well as in the referential configuration ($\mathbf{R}$).

Finally, we can approximate Eq. (47) as

$$M_\alpha \ddot{\mathbf{r}}_\alpha \approx \Big(\sum_{\beta=1}^{N_h} \omega(|\mathbf{R}_{\alpha\beta}|)(\mathcal{P}_\beta - \mathcal{P}_\alpha) \cdot \mathbf{R}_{\alpha\beta}\tilde{\Omega}_\beta\Big) \cdot \mathbf{M}_\alpha^{-1} + \mathcal{B}_\alpha\,. \tag{50}$$

This is an alternative computation formulation in comparison to Eq. (42), because one does not need to calculate interactions among different cells but only calculate the Virial stress in each cell. This formulation provides a natural passage bridging molecular dynamics and continuum mechanics coupling.

3.2 Mesoscale Dynamic Equations

We now exam the mesoscale Lagrangian equation. Considering the derivative terms with respect to the chosen mesoscale variable, i.e. $\dot{\boldsymbol{\phi}}_\alpha$ and $\boldsymbol{\phi}_\alpha$, we first have,

$$\frac{\partial \mathcal{L}_m}{\partial \dot{\boldsymbol{\phi}}_\alpha} = \dot{\boldsymbol{\phi}}_\alpha \cdot \mathbf{J}_\alpha^S \;\rightarrow\; \frac{d}{dt}\Big(\frac{\partial \mathcal{L}_m}{\partial \dot{\boldsymbol{\phi}}_\alpha}\Big) = \frac{d}{dt}(\dot{\boldsymbol{\phi}}_\alpha \cdot \mathbf{J}_\alpha^S) = \ddot{\boldsymbol{\phi}}_\alpha \cdot \mathbf{J}_\alpha^S, \tag{51}$$

and then we have,

$$
\begin{aligned}
\frac{\partial \mathcal{L}_m}{\partial \boldsymbol{\phi}_\alpha} &= \frac{1}{2}\sum_i m_i \dot{\mathbf{S}}_i \frac{\partial \mathbf{C}_\alpha}{\partial \boldsymbol{\phi}_\alpha} \dot{\mathbf{S}}_i - \frac{1}{2}\sum_{i,j\in\alpha} V'(r_{ij}) \frac{\mathbf{r}_{ij}}{|\mathbf{r}_{ij}|} \cdot \frac{\partial \mathbf{r}_{ij}}{\partial \boldsymbol{\phi}_\alpha} \\
&\quad - \sum_\beta \sum_{\substack{i\in\alpha \\ j\in\beta\neq\alpha}} V'(r_{ij}) \frac{\mathbf{r}_{ij}}{|\mathbf{r}_{ij}|} \cdot \frac{\partial \mathbf{r}_{ij}}{\partial \boldsymbol{\phi}_\alpha} - \sum_{i\in\alpha} \mathbf{b}_i \cdot \frac{\partial \mathbf{r}_i}{\partial \boldsymbol{\phi}_\alpha} \\
&= \boldsymbol{\phi}_\alpha \sum_i m_i \dot{\mathbf{S}}_i \otimes \dot{\mathbf{S}}_i - \frac{1}{2}\sum_{i,j\in\alpha} \mathbf{f}_{ij} \otimes \mathbf{S}_{ij} + \sum_\beta \sum_{\substack{i\in\alpha \\ j\in\beta\neq\alpha}} \mathbf{f}_{ij} \otimes \mathbf{S}_i + \sum_{i\in\alpha} \mathbf{b}_i \otimes \mathbf{S}_i \,.
\end{aligned}
\tag{52}
$$

Finally, the mesoscale Lagrangian equations have the form,

$$
\ddot{\boldsymbol{\phi}}_\alpha \cdot \mathbf{J}^S_\alpha = \boldsymbol{\phi}_\alpha \sum_i m_i \dot{\mathbf{S}}_i \otimes \dot{\mathbf{S}}_i - \frac{1}{2}\sum_{i,j\in\alpha} \mathbf{f}_{ij} \otimes \mathbf{S}_{ij} + \sum_\beta \sum_{\substack{i\in\alpha \\ j\in\beta\neq\alpha}} \mathbf{f}_{ij} \otimes \mathbf{S}_{ij} + \sum_{i\in\alpha} \mathbf{b}_i \otimes \mathbf{S}_i \,.
\tag{53}
$$

To understand physical meanings of the above equation, we can define the mesoscale 1st Piola-Kirchhoff stress tensors as,

$$
\mathbf{P}^{Virial}_\alpha := \sum_{i\in\alpha}\Big(-\boldsymbol{\phi}_\alpha m_i \dot{\mathbf{S}}_i \otimes \dot{\mathbf{S}}_i + \frac{1}{2}\sum_{j\in\alpha, j\neq i} \mathbf{f}_{ij} \otimes \mathbf{S}_{ij} \Big) \tag{54}
$$

$$
\mathbf{P}^{ext}_\alpha = \sum_{\beta\neq\alpha} \sum_{i\in\alpha, j\in\beta} \mathbf{f}_{ij} \otimes \mathbf{S}_{ij} = \boldsymbol{\phi}_\alpha \cdot \Big(\sum_{\beta\neq\alpha} \sum_{i\in\alpha, j\in\beta} V'(r_{ij}) \frac{\mathbf{S}_{ij} \otimes \mathbf{S}_{ij}}{r_{ij}} \Big), \tag{55}
$$

where $\mathbf{P}^{Virial}_\alpha$ and $\mathbf{P}^{ext}_\alpha$ are the first Piola-Kirchhoff stress of the α-th cell, which are the two-point tensors defined on the current configuration i.e **r**-configuration and the statistical configuration, i.e. **S**-configuration.

Figuratively speaking, these tensors have two legs, one is in **r**-configuration and the other is in **S**-configuration. If we push the leg in **S**-configuration forward to the intermediate configuration $\mathcal{R}$-configuration, i.e. $\mathcal{R}_i = \boldsymbol{\chi} \cdot \mathbf{S}_i$, the corresponding first Piola-Kirchhoff stress will have the following expression,

$$
\mathcal{P}^{Virial}_\alpha = \frac{1}{\tilde{\Omega}_\alpha} \sum_{i\in\alpha}\Big(-\boldsymbol{\phi}_\alpha m_i \dot{\mathbf{S}}_i \otimes \dot{\mathbf{S}}_i + \frac{1}{2}\sum_{j\in\alpha, j\neq i} \mathbf{f}_{ij} \otimes \mathbf{S}_{ij} \Big) \cdot \boldsymbol{\chi}^T_\alpha \,, \tag{56}
$$

which is used in Eq. (46).

With above definitions of stresses, we can recast the mesoscale dynamics equations as

$$
\ddot{\boldsymbol{\phi}}_\alpha \cdot \mathbf{J}^S_\alpha = -\Big(\mathbf{P}^{Virial}_\alpha - \mathbf{P}^{ext}_\alpha \Big) + \mathbf{M}_\alpha \tag{57}
$$

where $\mathbf{P}_\alpha^{Virial}$ is given by Eq. (54); and $\mathbf{P}_\alpha^{ext}$ is given by Eq. (55), and $\mathbf{M}_\alpha = \sum_{i\in\alpha} \mathbf{b}_i \otimes \mathbf{S}_i$ is the mesoscale external couple. Note that Eqs. (54) and (55) are insightful, because it resolves one of outstanding debates on the definition of the Virial Stress. Equation (54) is basically the mathematical definition of the Virial stress e.g. Irving and Kirkwood [37], Tsai [38]. However, Zhou [39] argued that the kinetic energy part should be dropped out in the stress calculation, even though many disagreed, e.g. Murdoch [40], Subramaniyan and Sun [41]. We now see from Eqs. (54) and (55) that if the stress is internally generated, the definition of the Virial stress is the original definition of the Virial stress; but if the stress is an external stress, then the kinetic energy part should drop out from its expression. This is because that the current formulation of the multiscale micromorphic molecular dynamics is a adiabatic formulation, which does not consider the heat exchange among the cells. If in Eq. (57), $\ddot{\boldsymbol{\phi}}_\alpha = 0$, we have $\mathbf{P}_\alpha^{ext} = \mathbf{P}_\alpha^{Virial} \rightarrow \mathcal{P}_\alpha^{ext} = \mathcal{P}_\alpha^{Virial}$, which is the proof of Eq. (46).

3.3 Microscale Dynamic Equations

After evaluating the fine scale Lagrange equation for $i \in \alpha$, we first have

$$\frac{d}{dt}\frac{\partial \mathcal{L}_m}{\partial \dot{\mathbf{S}}_i} = m_i\Big(\mathbf{C}_\alpha \cdot \ddot{\mathbf{S}}_i + \dot{\mathbf{C}}_\alpha \cdot \dot{\mathbf{S}}_i\Big). \tag{58}$$

The derivative of fine scale Lagrange equation with respect to $\mathbf{S}_i$ has two cases,

$$(a)\ \alpha = \beta : \quad \frac{\partial \mathcal{L}_m}{\partial \mathbf{S}_i} = -\frac{1}{2}\sum_{j\neq i}\Big(\frac{V'(r_{ij})}{r_{ij}}\mathbf{C}_\alpha \cdot \mathbf{S}_{ij}\Big)$$

$$(b)\ \alpha \neq \beta : \quad \frac{\partial \mathcal{L}_m}{\partial \mathbf{S}_i} = -\frac{1}{2}\sum_{\alpha\neq\beta}\sum_{j\neq i}\Big(\frac{V'(r_{ij})}{r_{ij}}\boldsymbol{\phi}_\alpha^T \cdot \mathbf{r}_{ij}\Big),$$

where $\mathbf{r}_{ij} = \mathbf{r}_{\alpha\beta} + \boldsymbol{\phi}_\beta \cdot \mathbf{S}_j - \boldsymbol{\phi}_\alpha \cdot \mathbf{S}_i$.

Combining the two equations and Eq. (58), we finally obtain

$$m_i\ddot{\mathbf{S}}_i + \frac{1}{2}\boldsymbol{\phi}_\alpha^{-1}\sum_{\beta}\sum_{i\neq j}\Big(\frac{V'(r_{ij})}{r_{ij}}(\mathbf{r}_{\alpha\beta} + \boldsymbol{\phi}_\beta \cdot \mathbf{S}_j - \boldsymbol{\phi}_\alpha \cdot \mathbf{S}_i)\Big) + m_i\mathbf{C}_\alpha^{-1}\dot{\mathbf{C}}_\alpha \cdot \dot{\mathbf{S}}_i + \boldsymbol{\phi}_\alpha^{-1} \cdot \mathbf{b}_i = \mathbf{0}, \tag{59}$$

where $i \in \alpha$. One may note that the second term in Eq. (59) contains both interaction of atoms within the α-th cell and between two different cells, i.e. the case $i \in \alpha,\ j \in \beta$ and $\beta \neq \alpha$.

In summary, the three scale governing equations of Multiscale Micromorphic Molecular Dynamics are as follows,

$$M_\alpha \ddot{\mathbf{r}}_\alpha = -\sum_{\beta \neq \alpha} \sum_{i \in \alpha, j \in \beta} \mathbf{f}_{ij} + S_\alpha \bar{\mathbf{t}}_\alpha + \Omega_\alpha \bar{\mathbf{b}}_\alpha \tag{60}$$

$$\ddot{\boldsymbol{\phi}}_\alpha \cdot \mathbf{J}^S_\alpha = -(\mathbf{P}^{Virial}_\alpha - \mathbf{P}^{ext}_\alpha) + \mathbf{M}_\alpha \tag{61}$$

$$m_i \ddot{\mathbf{S}}_i = -m_i \mathbf{C}^{-1}_\alpha \cdot \dot{\mathbf{C}}_\alpha \cdot \dot{\mathbf{S}}_i + \boldsymbol{\phi}^{-1}_\alpha \Big(\sum_{\beta} \sum_{j \in \beta \neq i \in \alpha} \mathbf{f}_{ji} + \mathbf{b}_i \Big) \tag{62}$$

where the micromorphic deformation tensor is $\boldsymbol{\phi}_\alpha = \mathbf{F}_\alpha \cdot \boldsymbol{\chi}_\alpha$. One can see from Eq. (60) that the macroscale traction boundary condition is embedded in the coarse scale dynamic equation.

4 Multiscale Coupling Algorithms

One of the main advantages of MMMD is that it provides a perfect means to design multiscale coupling strategy, because it automatically split conventional molecular dynamics into three scales. Therefore, if one would like to couple MD with continuum scale dynamics, S/he can directly connect the coarse scale MD with the continuum dynamics as a recent contribution Tong and Li [42] having shown that MMMD can perfectly couple with peridynamics.

We are now developing a multiscale coupling algorithm to couple nonlinear continuum mechanics based finite element method with MMMD.

In Fig. 5, we show a spatial coupling partition of finite element domain and molecular dynamics domain. In the molecular dynamics domain, we employ the multiscale micromorphic molecular dynamics (MMMD); while in the finite element domain we employ a Cauchy-Born based nonlinear finite element method to model and simulate the material. For self-completeness, we briefly discuss the Cauchy-Born rule based finite element in the following subsection.

4.1 Multiscale Finite Element Formulation

Based on continuum mechanics theory, The Hamiltonian principle for a continuum body can be expressed in terms of displacement variation,

$$\int_{t_0}^{t_1} \Big(\delta \mathcal{K} - (\delta \mathcal{W}_{int} + \delta \mathcal{W}_{ext}) \Big) dt = 0, \tag{63}$$

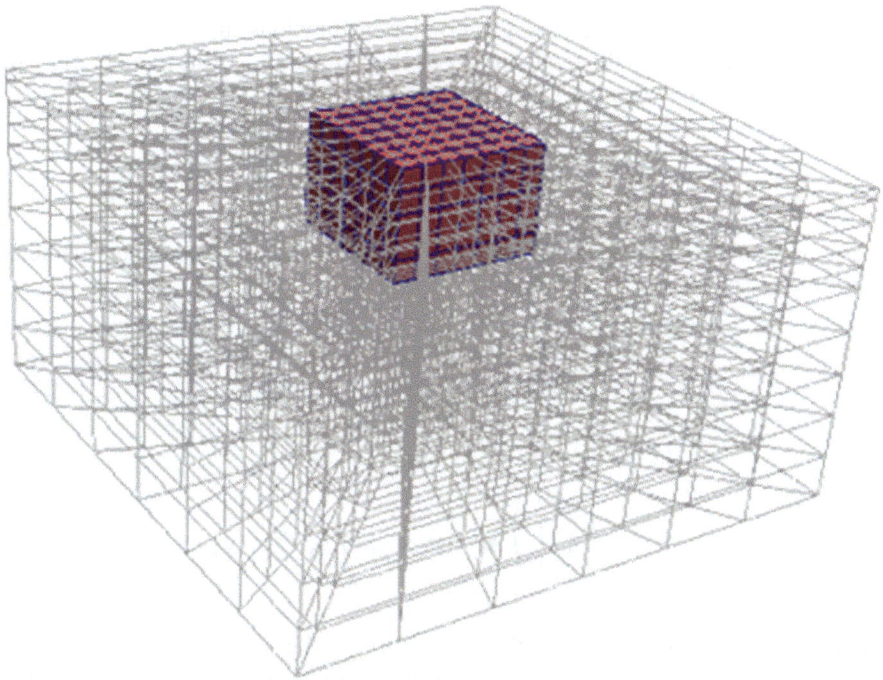

Fig. 5 Coupling between the finite element mesh and molecular dynamics domain

where

$$\int_{t_1}^{t_2} \delta\mathcal{K}dt = \int_{t_1}^{t_2}\int_{\Omega} \rho\dot{\mathbf{u}}\cdot\delta\dot{\mathbf{u}}dVdt = -\int_{t_1}^{t_2}\int_{\Omega} \rho\ddot{\mathbf{u}}\cdot\delta\mathbf{u}dVdt \tag{64}$$

$$\delta\mathcal{W}_{int} = \int_{\Omega} \frac{\partial W}{\partial \mathbf{F}} : \delta\mathbf{F}dV = \int_{\Omega} \mathbf{P} : \delta\mathbf{F}dV \tag{65}$$

$$\delta\mathcal{W}_{ext} = -\int_{\Omega} \mathbf{b}\cdot\delta\mathbf{u}dV - \int_{\delta\Omega_t} \bar{\mathbf{T}}\cdot\delta\mathbf{u}dS \tag{66}$$

where integration by parts is used in the line; W is the strain energy density; $\mathbf{P} = \dfrac{\partial W}{\partial \mathbf{F}}$ is the first Piola-Kirchhoff stress tensor; and $\mathbf{b}$ and $\bar{\mathbf{T}}$ are body force and prescribed traction vector.

Consider the finite element interpolation for the displacement field,

$$\mathbf{u}^h(\mathbf{X}) = \mathop{\mathbf{A}}_{e=1}^{nelem} \sum_{I=1}^{N_e} N_I(\mathbf{X})\mathbf{u}_I, \tag{67}$$

where $N_I(\mathbf{X})$ are the finite element interpolation functions, $\mathbf{u}_I$ are the nodal displacements, and $nelem$ is the total number of element, and N_e is the number of FEM shape functions in the element e.

The discrete equation of motion for FEM procedure can be expressed as,

$$\mathbf{M}\ddot{\mathbf{d}} = \mathbf{f}^{int}(\mathbf{d}) - \mathbf{f}^{ext} \tag{68}$$

where, $\mathbf{M}$ is the mass matrix. $\mathbf{f}^{int}$ and $\mathbf{f}^{ext}$ are force vectors from each element, respectively. They are defined as follows:

$$\mathbf{M} = \mathop{\mathbf{A}}_{e=1}^{nelem} \int_{\Omega_e} \rho_0 \mathbf{N}^{eT}\mathbf{N}^e dV \tag{69}$$

$$\mathbf{f}^{int} = \mathop{\mathbf{A}}_{e=1}^{nelem} \int_{\Omega_e} \mathbf{B}^{eT}\mathbf{P}^e(\mathbf{d}) dV \tag{70}$$

$$\mathbf{f}^{ext} = \mathop{\mathbf{A}}_{e=1}^{nelem} \left\{ \int_{\Omega_e} \mathbf{N}^{eT}\mathbf{B}^e dV + \int_{\Gamma_t} \mathbf{N}^{eT}\bar{\mathbf{T}}^e dS \right\} \tag{71}$$

where Ω_e is the domain for the e-th element; $\partial\Gamma_t$ is the traction boundary; $\mathbf{B}^e$ are the element strain-displacement matrix and the gradient of the strain-displacement matrix, respectively.

Crystal structure of silicon is diamond cubic, $Fd3m$, which is regarded as a set of two FCC lattices, α and β (Fig. 6). Hence, the inner displacements between two lattices can be defined as a vector $\mathbf{v}$ from the first lattice α to another β (see [43–45]).

When evaluating stress based the Tersoff potential, we only take the nearest four atoms into account to calculate interactions, the strain energy density and its derivative can be calculated by the pseudo unit cell composed of five atoms. The nearest four atoms, which composed of a tetrahedral structure with the central atom, all belong to a different FCC lattice unit cell β from the FCC lattice α to which the center silicon resides (Fig. 6). If we term the central atom as index 1, and others are from 2 to 5, distances of atoms are defined as,

$$r_{1j} = |\mathbf{r}_{1j}| = |\mathbf{r}_1 - \mathbf{r}_j - \mathbf{v}| \quad j = 2, 3, ..., 5 \tag{72}$$

$$r_{jk} = |\mathbf{r}_{jk}| = |\mathbf{r}_j - \mathbf{r}_k| \quad j, k = 2, 3, ..., 5, \, and \; j \neq k \tag{73}$$

The vector $\mathbf{v}$ can be evaluated to minimize strain energy density as shown in Appendix. It had been found that the optimization of vector $\mathbf{v}$ is crucial to figure out the stablest configuration of atoms at each deformation [43].

Because overall non-uniform deformation field can be interpreted by a set of bulk elements with piece-wise uniform deformations, we can assume the first order Cauchy-Born rule to represent deformation of the bulk element.

$$\mathbf{r}_i = \mathbf{F}\mathbf{R}_i \tag{74}$$

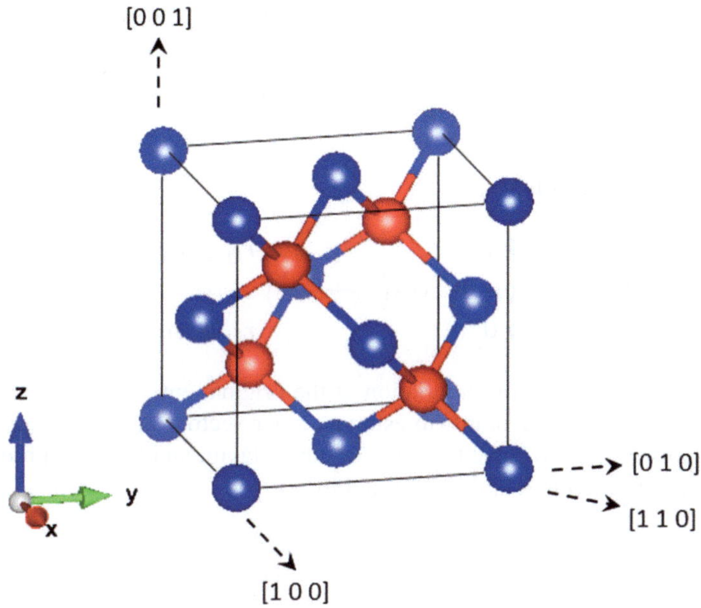

Fig. 6 Unit cell of silicon crystal diamond cubic. Red and blue colors indicate atoms belong to two different FCC lattices

where, $\mathbf{R}_i$ is original position of atom i in the unit cell, and $\mathbf{r}_i$ is the position after deformation. $\mathbf{F}$ is the first order deformation gradient of each element.

$$\mathbf{F} = \frac{\partial \mathbf{x}}{\partial \mathbf{X}} \tag{75}$$

4.2 *Tersoff Potential*

The strain energy density W can be written as follows,

$$W = \frac{U_1}{\Omega_0} = \frac{1}{2\Omega_0} \sum_{j=2}^{5} V_{1j} \tag{76}$$

where, U_1 is the potential energy for the central atom 1, and Ω_0 is the volume occupied by an atom in the initial configuration. V_{1j} is the Tersoff potential energy between the central atom 1 and a surrounding atom j as

$$V_{1j} = f_C(r_{1j})[f_R(r_{1j}) + b_{1j} f_A(r_{1j})] \tag{77}$$

In the Tersoff potential, the functions f_R, f_A are the cut off function,

$$f_R(r_{1j}) = Aexp(-\lambda_{1j} r_{1j}) \tag{78}$$
$$f_A(r_{1j}) = -Bexp(-\mu_{1j} r_{1j}) \tag{79}$$

and f_C are defined as follows;

$$f_C(r_{1j}) = \begin{cases} 1 & r_{1j} \leq R_{1j} \\ \frac{1}{2} + \frac{1}{2}\cos\left(\frac{\pi(r_{1j}-R_{1j})}{S_{1j}-R_{1j}}\right) & R_{1j} < r_{1j} < S_{1j} \\ 0 & r_{1j} \geq S_{1j} \end{cases} \tag{80}$$

Note here that, secondary differentiation of the original cutoff function (80) is discontinuous, it is thus unfavorable to estimate inner vector $\mathbf{v}$ at the transition points $r = R_{1j}$ and $r = S_{1j}$. To solve this inexpedience, Izumi and Suzuki introduced the dumping cutoff function Izumi and Sakai [46] as;

$$f_C(r_{1j}) = \frac{1}{2} - \frac{1}{2}\tanh\left[\frac{\pi}{2}\left(\frac{r_{1j} - R_{1j}}{S_{1j} - R_{1j}}\right)\right] \tag{81}$$

In this study, we employ the modified cutoff function of Eq. (81) instead of Eq. (80) without changing parameters R_{1j} and S_{1j} of the original Tersoff potential.

By taking account three body interaction, the parameter b_{1j} in Eq. (77) will explicitly depend on the location of the third atom k as follows,

$$b_{1j} = (1 + \beta^n \zeta_{1j}^n)^{\frac{-1}{2n}} \tag{82}$$

$$\zeta_{1j} = \sum_{k \neq i,j}^{5} f_c(r_{1k}) g(\theta_{1jk}) \tag{83}$$

$$g(\theta_{1jk}) = 1 + \frac{c^2}{d^2} - \frac{c^2}{d^2 + (h - cos\theta_{1jk})^2} \tag{84}$$

$$cos\theta_{1jk} = \frac{\mathbf{r}_{1j}\mathbf{r}_{1k}}{r_{1j} r_{1k}} = \frac{(r_{1j}^2 + r_{1k}^2 - r_{jk}^2)}{2 r_{1j} r_{1k}} \tag{85}$$

The parameter set of Tersoff potential used in this study is adopted from Tersoff's second paper [47], which can reproduce elastic properties of Silicon more accurately.

According to the strain energy density, the first Piola-Kirchhoff stress tensor is derived as,

$$\mathbf{P} = \frac{\partial W}{\partial \mathbf{F}} = \frac{1}{2\Omega_0}\frac{\partial}{\partial \mathbf{F}}\left(\sum_{j=2}^{5} V_{1j}\right) = \frac{1}{2\Omega_0}\sum_{j=2}^{5}\left(\frac{\partial V_{1j}}{\partial \mathbf{F}} + \frac{\partial V_{1j}}{\partial \mathbf{v}}\frac{\partial \mathbf{v}}{\partial \mathbf{F}}\right) \tag{86}$$

The inner displacement **v** can be evaluated by minimizing of the strain energy density W, i.e.

$$\left(\frac{\partial V_{1j}}{\partial \mathbf{v}}\right)_{\mathbf{F}} = 0 \tag{87}$$

This condition is achieved by using Newton's method to find the minimum of the strain energy density, and as a result, inner displacement vector **v** can be obtained. The procedure is presented and documented in Appendix. Finally, we have

$$\begin{aligned}\mathbf{P} &= \frac{1}{2\Omega_0}\sum_{j=2}^{5}\left(\frac{\partial V_{1j}}{\partial \mathbf{F}}\right)\\ &= \frac{1}{2\Omega_0}\sum_{j=2}^{5}\left[\frac{\partial V_{1j}}{\partial r_{1j}}\frac{\partial r_{1j}}{\partial \mathbf{F}} + \sum_{k=2,k\neq j}^{5}\left(\frac{\partial V_{1j}}{\partial r_{1k}}\frac{\partial r_{1k}}{\partial \mathbf{F}} + \frac{\partial V_{1j}}{\partial cos\theta_{1jk}}\frac{\partial cos\theta_{1jk}}{\partial \mathbf{F}}\right)\right]\end{aligned} \tag{88}$$

where,

$$\frac{\partial r_{ij}}{\partial \mathbf{F}} = \frac{\partial r_{ij}}{\partial \mathbf{r}_{ij}}\frac{\partial \mathbf{r}_{ij}}{\partial \mathbf{F}} = \frac{\mathbf{r}_{ij}\otimes \mathbf{R}_{ij}}{r_{ij}} \tag{89}$$

$$\begin{aligned}\frac{\partial cos\theta_{1jk}}{\partial \mathbf{F}} &= \left(\frac{1}{r_{1k}} - \frac{cos\theta_{1jk}}{r_{1j}}\right)\frac{\partial r_{1j}}{\partial \mathbf{F}} + \left(\frac{1}{r_{1j}} - \frac{cos\theta_{1jk}}{r_{1k}}\right)\frac{\partial r_{1k}}{\partial \mathbf{F}}\\ &\quad - \left(\frac{r_{jk}}{r_{1j}r_{1k}}\right)\frac{\partial r_{jk}}{\partial \mathbf{F}}\end{aligned} \tag{90}$$

According to Eq. (77), the derivative of V_{1j} by distance r is

$$\frac{\partial V_{1j}}{\partial r_{1j}} = \frac{\partial f_C}{\partial r_{1j}}(f_R(r_{1j}) + b_{1j}f_A(r_{1j})) \tag{91}$$

$$+ f_C(r_{1j})\left(\frac{\partial f_R(r_{1j})}{\partial r_{1j}} + b_{1j}\frac{\partial f_A(r_{1j})}{\partial r_{1j}}\right) \tag{92}$$

where,

$$\frac{\partial f_C(r_{1j})}{\partial r_{1j}} = -\frac{\pi}{4(S_{1j}-R_{1j})}\cosh\left[\frac{\pi(r_{1j}-R_{1j})}{2(S_{1j}-R_{1j})}\right] \tag{93}$$

$$\frac{\partial f_R(r_{1j})}{\partial r_{1j}} = -A\lambda_{1j}exp(-\lambda_{1j}r_{1j}) \tag{94}$$

$$\frac{\partial f_A(r_{1j})}{\partial r_{1j}} = B\mu_{1j}exp(-\mu_{1j}r_{1j}) \tag{95}$$

Similarly we can take derivatives of V_{1j} by r_{1k} and θ_{1jk} to estimate Eq. (88), and they are,

$$\frac{\partial V_{1j}}{\partial r_{1k}} = f_C(r_{1j})f_A(r_{1j})\Big[-\frac{1}{2}(1+\beta^n\zeta_{1j}^n)^{\frac{-1}{2n}-1}\Big]\beta^n\zeta_{1j}^{n-1} \cdot \frac{\partial f_C(r_{ik})}{\partial r_{1k}} g(\theta_{1jk}) \tag{96}$$

$$\frac{\partial V_{1j}}{\partial cos\theta_{1jk}} = f_C(r_{1j})f_A(r_{1j})\Big[-\frac{1}{2}(1+\beta^n\zeta_{1j}^n)^{\frac{-1}{2n}-1}\Big]\beta^n\zeta_{1j}^{n-1} \cdot f_C(r_{1k})\frac{2c^2(\cos\theta_{1jk}-h)}{[d^2+(h-\cos\theta_{1jk})^2]^2}\,. \tag{97}$$

4.3 From MMMD to FEM

The coupling between MMMD and FEM is a two-way coupling. We first discuss how to transfer information from MMMD to FEM.

A schematic illustration of the multiscale coupling is shown in Fig. 7 to demonstrate the coupling strategy at the boundary of FEM region and MMMD region.

First, in order to pass the mechanical information from MMMD to FEM, the FEM nodes at the multiscale boundary are chosen as the centers of mass of boundary MMMD cells shown in Fig. 7; and we call these nodes as the intermediate node. The global arrangement of the intermediate nodes is also shown in Fig. 8.

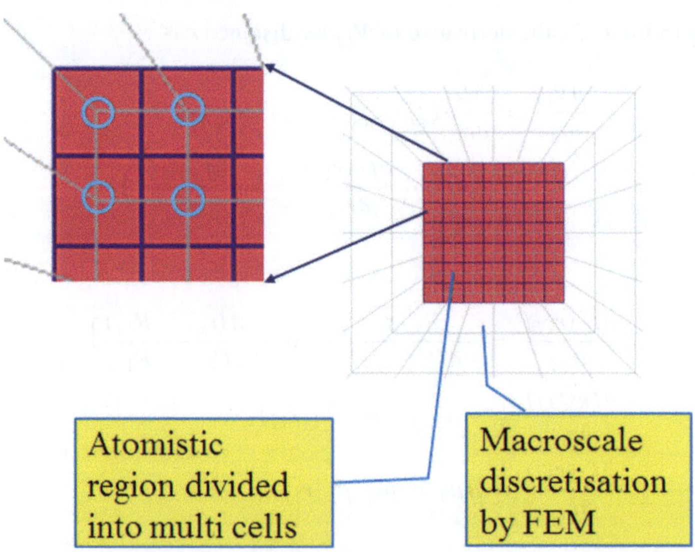

Fig. 7 The multiscale coupling strategy: From MMMD to FEM

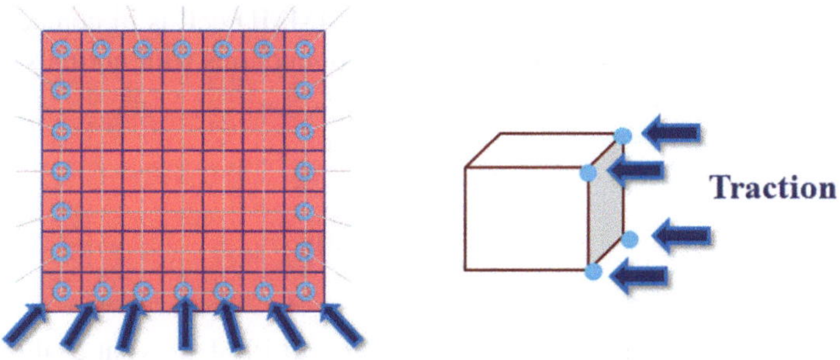

Fig. 8 The multiscale coupling strategy: From FEM to MMMD

Second, at each center of mass of MMMD cells, we can calculated the first Piola-Kirchhoff (PK-I) virial stress $\mathcal{P}_{\alpha}^{Virial}$ based on Eq. (46). We can then use the PK-I virial stress to calculate traction force of the boundary between FEM region and MMMD region. As an example, for a four-node quadrilateral element surface, it contains four centers of mass of MMMD cells, and we assume that the normal of the quadrilateral element is $\mathbf{N}$. The traction vector on the surface of this quadrilateral element will be,

$$\mathbf{T}_m = \sum_{\alpha=1}^{4} \mathbf{N} \cdot \mathcal{P}_{\alpha}^{Virial} \tag{98}$$

where the subscript m denotes multiscale boundary. Thus, the traction boundary Γ_t in Eq. (71) has two parts, i.e. $\Gamma_t = \Gamma_T \bigcup \Gamma_m$, where Γ_T is the macroscale traction boundary and Γ_m is the multiscale traction boundary.

4.4 From FEM to MMMD

Shown in Fig. 7, each center of mass of MMMD boundary cell may be in contact with two elements (for two-dimensional quadrilateral FEM mesh). In three-dimensional space, it may in contact with four or even elements. For each MMMD cell, there is applied external load from it environment, i.e.

$$M_{\alpha} \ddot{\mathbf{r}}_{\alpha} = - \sum_{\beta \neq \alpha} \mathbf{r}_{\alpha\beta} + \mathcal{B}_{\alpha}, \tag{99}$$

in which the external force,

$$\mathcal{B}_{\alpha} = S_{\alpha} \bar{\mathbf{t}}_{\alpha} + \Omega_{\alpha} \bar{\mathbf{b}}_{\alpha} \ . \tag{100}$$

Suppose that a center of mass of boundary MMMD cell is in contact with two elements and the MMMD cell has two exterior boundaries with normals $\mathbf{N}_1$ and $\mathbf{N}_2$ and surface facts S_1 and S_2.

We choose the following expression of $\mathcal{B}_\alpha$,

$$\mathcal{B}_\alpha = \sum_{I=1}^{2} S_I \mathbf{N}_I \cdot \mathbf{P}_I \tag{101}$$

where the PK-I stress is the weighted average of all PK-I stress at evaluated at the each quadrature point of that element, e.g.

$$\mathbf{P}_I = \sum_{i=1}^{4} \frac{\partial W}{\partial \mathbf{F}} \bigg|_{\mathbf{X}_i} w_i, \tag{102}$$

where w_i are the Gaussian quadrature weights. The explicit expression for $\frac{\partial W}{\partial \mathbf{F}}$ for the Tersoff potential is given in Eq. (86) and related equations. A schematic illustration of the multiscale coupling (from FEM to MMMD) is provided in Fig. 8.

5 Numerical Examples

We employ the coupled FEM-MMMD method to simulate nano-indentation of single crystal silicon. The multiscale system contains the MD region and FEM region. In MD region, we partitioned $7 \times 7 \times 5 = 245$ cells, and in each cell there are 64 Silicon atoms so that there are total 15680 atoms in MD system. We used the Tersoff potential to model single crystal Silicon. The FEM region is wrapped around the MD region, it contains total 1860 brick elements as shown in Fig. 9.

Two types of indenters are used in simulations: a 20 nm diameter spherical indenter and Vicker's indenter. In actual simulations, we used analytical expressions to express the indenters' contact surfaces, and we prescribed the motion of the indenters. Since the indentation contact region is only restricted to the MD region, we therefore only apply the indentation contact condition to MD region. We may view that the indentation contact boundary is the coarse scale MMMD boundary condition. Thus we apply this prescribed indentation boundary condition only on the centers of MMMD surface cells. On the other hand, the induced stress and force on the indenter can be calculated by projecting and summing the virial stress of each MMMD cell that is in contact of the indenter.

We have simulated nano-indentations various forms of Silicon. We listed the simulation condition, i.e. the equilibrium bond distance and coordination number in Table 1.

The simulation results for Si-I are displayed in Figs. 10 and 11. From Fig. 10, one may find the the simulated relation between indentation depth and the force exerted by the indenter. Since this is a multiscale simulation, the both MMMD and FEM

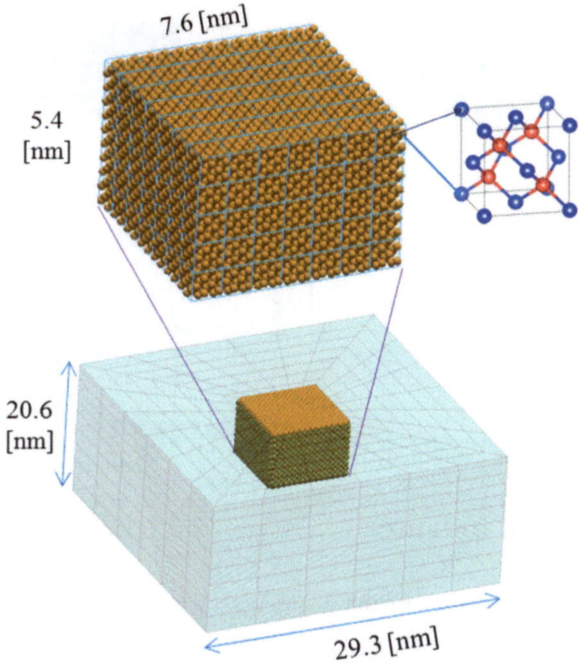

Fig. 9 The multiscale coupling strategy: From FEM to MMMD

Table 1 Equilibrium bond length and coordination number

Crystal	Bond [Å]	Coordination No.
Surface/Amorphous	2.35	<4
Si-I	2.35	4
bct-5	2.31	4
	2.44	1
Si-III, XII	2.39	4
	3.2–3.4	1
Si-II	2.42	4
	2.57	2

parts contribute to silicon substrate deformation or indentation depth. In Fig. 10, we plot the total indentation depth and load curves as well as the the contribution from MD (MMMD) and FEM responses. In Fig. 11, show the molecular contact surface morphologies with two different indenters.

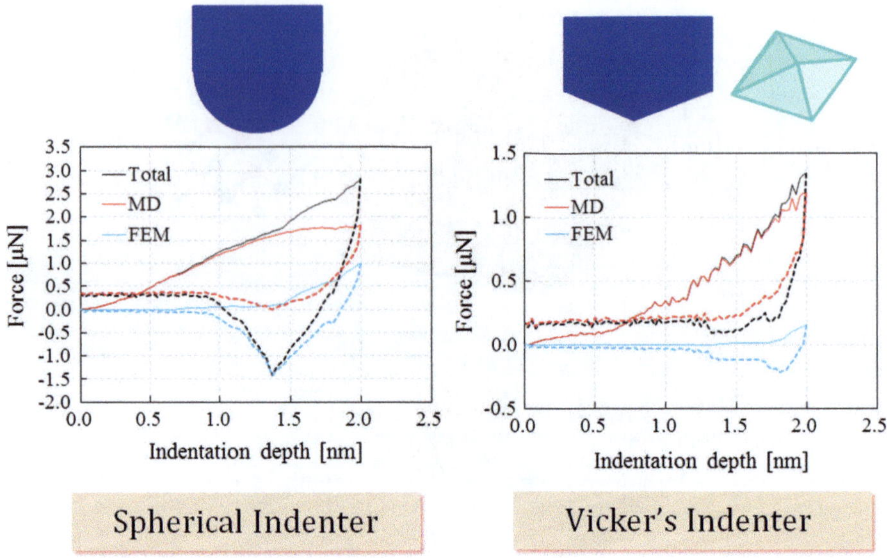

Fig. 10 Nano-indentation simulation result

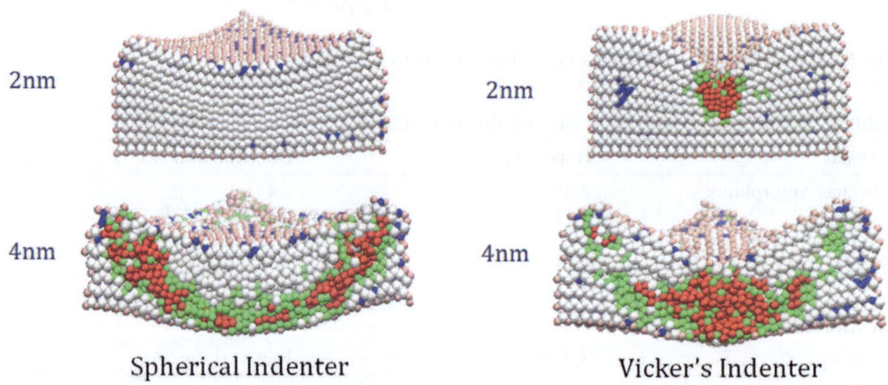

Fig. 11 Indentation contact region morphology

Acknowledgements Dr. S. Urata is sponsored by Asahi Glass Co., Ltd., Japan, and this support is greatly appreciated.

References

1. R.E. Rudd, J.Q. Broughton, Coarse-grained molecular dynamics and atomic limit of finite elements. Phys. Rev. B **58**, R5893–R5896 (1998)
2. G.J. Wagner, W.K. Liu, Coupling of atomistic and continuum simulations using a bridging scale decomposition. J. Comput. Phys. **190**, 249–274 (2003)

3. S.P. Xiao, T. Belytschko, A bridging domain method for coupling continua with molecular dynamics. Comput. Methods Appl. Mech. Eng. **193**, 1645–1669 (2004)
4. A.C. To, S. Li, Perfectly matched multiscale simulations. Phys. Rev. B **72**, 035414 (2005)
5. S. Li, X. Liu, A. Agrawal, A.C. To, Perfectly matched multiscale simulations for discrete systems: extension to multiple dimensions. Phys. Rev. B **74**, 045418 (2006)
6. S. Li, N. Sheng, On multiscale non-equilibrium molecular dynamics simulations. Int. J. Numer. Methods Eng. **83**, 998–1038 (2010)
7. S. Li, N. Sheng, X. Liu, A non-equilibrium multiscale simulation paradigm. Chem. Phys. Lett. **451**, 293–300 (2008)
8. H.S. Park, E.G. Karpov, W.K. Liu, Non-reflecting boundary conditions for atomistic, continuum and coupled atomistic/continuum simulations. Int. J. Numer. Methods Eng. **64**, 237–259 (2005)
9. E.G. Karpov, H. Yu, H.S. Park, W.K. Liu, Q.J. Wang, D. Qian, Multiscale boundary conditions in crystalline solids: theory and application to nanoindentation. Int. J. Solids Struct. **43**, 6359–6379 (2006)
10. S. Namilae, D.M. Nicholson, P.K.V.V. Nukala, C.Y. Gao, Y.N. Osetsky, D.J. Keffer, Absorbing boundary conditions for molecular dynamics and multiscale modeling. Phys. Rev. B **76**, 144111 (2007)
11. R.E. Miller, E.B. Tadmor, A unified framework and performance benchmark of fourteen multiscale atomistic/continuum coupling methods. Model. Simul. Mater. Sci. Eng. **17**, 053001 (2009)
12. R.E. Rudd, J.Q. Broughton, Atomistic simulation of MEMS resonators through the coupling of length scales. J. Model. Simul. Microsyst. **1**, 29–38 (1999)
13. Y. Chen, J.D. Lee, Connecting molecular dynamics to micromorphic theory (I). Instantaneous and average mechanical variables. Phys. A **322**, 359–376 (2003)
14. R.E. Rudd, J.Q. Broughton, Coarse-grained molecular dynamics: nonlinear finite elements and finite temperature. Phys. Rev. B **72**, 144104 (2005)
15. X. Liu, S. Li, Nonequilibrium multiscale computational model. J. Chem. Phys. **126**, 124105 (2007)
16. X. Li, A coarse-grained molecular dynamics model for crystalline solids. Int. J. Numer. Methods Eng. **83**, 986–997 (2010)
17. Y. Chen, J.D. Lee, Connecting molecular dynamics to micromorphic theory (II). Balance laws. Phys. A **322**, 377–392 (2003)
18. Y. Chen, Reformulation of microscopic balance equations for multiscale materials modeling. J. Chem. Phys. **130**, 134706 (2009)
19. L. Xiong, G. Tucker, D.L. McDowell, Y. Chen, Coarse-grained atomistic simulation of dislocations. J. Mech. Phys. Solids **59**(2), 160–177 (2011)
20. A.C. Eringen, *Mechanics of Micromorphic Continua* (Springer, 1968)
21. H.C. Andersen, Molecular dynamics simulations at constant pressure and or temperature. J. Chem. Phys. **72**, 2384–2393 (1980)
22. M. Parrinello, A. Rahman, Crystal structure and pair potentials: a molecular dynamics study. Phys. Rev. Lett. **14**, 1196–1199 (1980)
23. M. Parrinello, A. Rahman, Polymorphic transitions in single crystals: a new molecular dynamics method. J. Appl. Phys. **12**, 7182–7190 (1981)
24. A. Laio, J. VandeVondele, U. Rothlisberger, A Hamiltonian electrostatic coupling scheme for hybrid CarParrinello molecular dynamics simulations. J. Chem. Phys. **116**, 6941–6947 (2002)
25. R. Martonik, A. Laio, M. Parrinello, Predicting crystal structures: the Parrinello-Rahman method revisited. Phys. Rev. Lett. **90**, 075503 (2003)
26. R. Martonik, D. Donadio, A. Oganov, M. Parrinello, Crystal structure transformations in S_iO_2 from classical and *ab initio* metadynamics. Nat. Mater. **5**, 623–626 (2006)
27. P. Podio-Guidugli, On (Andersen)-Parrinello-Rahman molecular dynamics, the related metadynamics, and the use of the Cauchy-Born rule. J. Elast. **100**, 145–153 (2010)

28. M.H. Ulz, Coupling the finite element method and molecular dynamics in the framework of the heterogeneous multiscale method for quasi-static isothermal problems. J. Mech. Phys. Solids **74**, 1–18 (2015)
29. M.H. Ulz, A multiscale molecular dynamics method for isothermal dynamic problems using the seamless heterogeneous multiscale method. Comput. Merthods Appl. Mech. Eng. **295**, 510–524 (2015)
30. S. Li, Q. Tong, A concurrent multiscale micromorphic molecular dynamics. J. Appl. Phys. **117**, 154303 (2015)
31. Q. Tong, S. Li, From molecular systems to continuum solids: A multiscale structure and dynamics. J. Chem. Phys. 143, 064101 (2015), https://doi.org/10.1063/1.4927656
32. S. Li, S. Urata, An atomistic-to-continuum molecular dynamics: theory, algorithm, and applications. Comput. Methods Appl. Mech. Eng. **306**, 452–478 (2016)
33. S. Li, W. Liu, Reproducing kernel hierarchical partition of unity. Part I: formulation and theory. Int. J. Numer. Methods Eng. **45**, 251–288 (1999)
34. S. Silling, M. Epton, O. Weckner, J. Xu, E. Askari, Peridynamic states and constitutive modeling. J. Elast. **88**, 151–184 (2007)
35. E.B. Tadmor, M. Ortiz, R. Phillips, Quasicontinuum analysis of defects in solids. Philos. Mag. A **73**(6), 1529–1563 (1996)
36. J. Marsden, T. Hughes, *Mathematical Foundations of Elasticity* (Prentice-Hall, Inc.,1983)
37. J. Irving, J.G. Kirkwood, The statistical mechanical theory of transport processes. IV. The equations of hydrodynamics. J. Chem. Phys. **18**, 817–829 (1950)
38. D.H. Tsai, The virial theorem and stress calculation in molecular dynamics. J. Chem. Phys. **70**, 1375 (1979)
39. M. Zhou, A new look at the atomic level virial stress: on continuum-molecular system equivalence. Proc. R. Soc. Lond. Ser. A **459**, 2347–2392 (2003)
40. A. Murdoch, A critique of atomistic definition of the stress tensor. J. Elast. **88**, 113–140 (2007)
41. A.K. Subramaniyan, C.T. Sun, Continuum interpretation of virial stress in molecular simulations. Int. J. Solids Struct. **45**, 4340–4346 (2008)
42. Q. Tong, S. Li, Multiscale coupling of molecular dynamics and peridynamics. J. Mech. Phys. Solids **95**, 169–187 (2016)
43. A.R. Khoei, H. DorMohammadi, A. Aramoon, A temperature-related boundary Cauchy-Born method for multi-scale modeling of silicon nano-structures. Phys. Lett. A **378**, 551–560 (2014)
44. H.S. Park, P.A. Klein, A surface Cauchy-Born model for silicon nanostructures. Comput. Methods Appl. Mech. Eng. **197**, 3249–3260 (2008)
45. A.R. Khoei, H. DorMohammadi, Validity and size-dependency of Cauchy-Born hypothesis with Tersoff potential in silicon nano-structures. Comput. Mater. Sci. **63**, 168–177 (2012)
46. S. Izumi, S. Sakai, Internal displacement and elastic properties of the silicon Tersoff model. JSME Int. J. Ser. A Solid Mech. Mater. Eng. **47**, 54–61 (2004)
47. J. Tersoff, Empirical interatomic potential for silicon with improved elastic properties. Phys. Rev. B **38**, 9902 (1988)

Modeling Electronic Properties of Twisted 2D Atomic Heterostructures

Stephen Carr, Daniel Massatt, Shiang Fang, Paul Cazeaux, Mitchell Luskin and Efthimios Kaxiras

Abstract We present a general method for the electronic characterization of aperiodic 2D materials using ab-initio tight binding models. Specifically studied is the subclass of twisted, stacked heterostructures, but the formalism provided can be implemented for any 2D system without long-range interactions. This new method provides a multi-scale approach for dealing with the ab-initio calculation of electronic transport properties in stacked nanomaterials, allowing for fast and efficient simulation of multi-layered stacks in the presence of twist angles, magnetic field, and defects. We calculate the electronic density of states in twisted bilayer systems of graphene and MX_2 transition metal dichalcogenides (TMDCs). We comment on the interesting features of their density of states as a function of twist-angle and local configuration and how these features are experimentally observable. These results support the bilayer twist-angle as a new variable for controlling electronic properties in artificial nanomaterials ("Twistronics").

Keywords Electronic Structure · Two-Dimensional · Twist
Multiscale · Tight-binding · Graphene

S. Carr (✉) · S. Fang · E. Kaxiras
Department of Physics, Harvard University, Cambridge, MA 02138, USA
e-mail: stephencarr@g.harvard.edu

S. Fang
e-mail: sfang@physics.harvard.edu

E. Kaxiras
e-mail: kaxiras@physics.harvard.edu

D. Massatt · P. Cazeaux · M. Luskin
School of Mathematics, University of Minnesota, Minneapolis, MN 55455, USA
e-mail: massa067@umn.edu

P. Cazeaux
e-mail: pcazeaux@umn.edu

M. Luskin
e-mail: luskin@umn.edu

L. L. Bonilla et al. (eds.), *Coupled Mathematical Models for Physical and Biological Nanoscale Systems and Their Applications*, Springer Proceedings in Mathematics & Statistics 232, https://doi.org/10.1007/978-3-319-76599-0_13

1 Introduction

Since the experimental isolation of graphene, many other two-dimensional van der Waals materials have been successfully grown or fabricated from bulk samples in a laboratory setting. Although a single-layer 2-D system has exciting physical properties, there has also been great interest in developing and understanding artificial heterostructure materials composed of multiple 2-D layers [10]. Each layer is only coupled to the other through the weak van der Waals interaction, allowing for stacked materials to still exhibit 2-D phenomenons. These heterostructures are created by taking different monolayer materials and mechanically placing them on top of one another using nano-fabrication techniques. This fabrication method allows for any possible relative angle between the two (or more) layers. Twisting will generally break the translational symmetry of the heterostructure as the layers are no longer aligned. The study of the symmetry of these materials has classified them into two distinct categories: commensurate stacking (in which there exists a periodic supercell structure) and incommensurate stacking (no periodic structure).

Experiments in stacked van der Waals heterostructures have shown interesting results. Bilayer graphene studies have found very clear twist-dependent features in both the electronic density of states and the conductivity [13, 15]. Careful atomic structure studies in bilayer graphene have also shown that at very small twist-angles, a domain-wall phase appears where the relevant phase parameter is the crystal stacking configuration [18]. 2D layers of transition metal dichalcogenides (TMDCs) have also begun to be investigated experimentally [1, 12]. Although commensurate systems have been studied using traditional Density Functional Theory (DFT) when the supercell is not unreasonable large, not much is known about incommensurate systems. Are there distinct physical characteristics that distinguish between the incommensurate and commensurate case? Are there unique phenomenon or phases present only in incommensurate materials? Do the properties of incommensurate systems limit to those of commensurate systems (perhaps as one takes infinitesimal variations in twist-angle)? We investigate these questions by introducing a robust framework for the simulation of incommensurate heterostructures and provide the results of numerical implementation in some bilayer systems.

The starting point for our method is the generation of ab-initio tight-binding models for stacks of 2D materials. A series of small supercell DFT simulations where one layer is translated with respect to the other provides a large sample of tight-binding parameters for specific distances and angles between atomic orbitals. An analytic tight-binding model is then fit to these samples, providing a general formula for computing "hopping parameters" between orbitals. Tight-binding models generated in this manner reproduce band structure with the accuracy, but not the computational cost, of DFT. We use bilayer tight-binding models created from this method for graphene [8] and MX_2 TMDCs [9] where M = (Mo, W) and X = (S, Se).

2 Method

2.1 Formalism

Our methodology is inspired by two previous mathematical works on disordered tight-binding models. First, an algebraic treatment of electronic transport in disordered lattice systems [2, 11] allows for the rigorous definition of quantum-mechanical operators in a disordered material. Second, an argument that local tight-binding models create exponentially localized local observables, i.e. one can controllably remove finite-size and edge effects from tight-binding calculations [7]. Using these previous works, we are able to develop a useful vocabulary for treating the problem of modeling twisted 2D layered materials.

A tight-binding model in d-dimensions is described by orbitals in a d-dimensional lattice ($i \in \mathbb{Z}^d$) and the hopping parameters between orbitals (usually labeled t_{ij}). To describe disorder in this model, we consider the space of all possible defects and apply one specific set of defects in a stochastic manner. This is formulated by defining a configuration space Ω with specific local configurations $\omega \in \Omega$ with a probability distribution of configurations $dP(\omega)$. In short, Ω describes all possible environments that an orbital can see around itself. In this manner, we wish to simulate operators by sampling over the space of disorder configurations Ω.

Viewing the interlayer van der Waals force as a perturbative potential, the twist-angle can be interpreted as an aperiodic disorder to a single-layer system. Although we wish to keep the twist-angle fixed throughout the material, the location of the orbital in the "potential" of another layer does vary. This "shift" between the origin of unit-cells in different layers completely describes the local configuration ω. For each one of these shifts, we can construct a finite-system which contributes a finite-size error to the calculation. However, this error decays exponentially with the radius, so it can be taken controllably to zero. Our results show that this is a computationally feasible strategy.

To describe the difference between an incommensurate and commensurate angle in this model is now easy: a commensurate angle has a finite configuration space (because a periodic super-cell exists) while an incommensurate angle has an infinite configuration space. If two angles, θ commensurate and ϕ incommensurate, are extremely close then a specific "shift" will look effectively identical between them. Therefore, the results of a single ω calculation will not vary significantly between θ and ϕ. Rather, its the sampling of ω that varies. To see the incommensurability/commensurability of a twist angle as a phase transition in a real material, some physical observable must vary strongly enough over Ω and the commensurate twist angle must not sample Ω too finely.

In conclusion, we wish to calculate local operators over the space of all possible configurations Ω. To do this, we choose a local configuration ω and construct a large finite tight-binding system which has the orbital described by ω at its center. Then, the physical property of interest (density of states, conductivity, etc.) is calculated from the model. This method allows for other sources of disorder to be treated exactly as well. Physical defects such as vacancies, ripples, and edges are easy to implement. Simply introduce extra dimensionality to Ω to represent all possible forms the defects can take and place the defects directly into each ω tight-binding model. Although our ab-initio tight-binding model may not be accurate in the presence of large defects or edges, point vacancies should be well captured by just removing a few orbitals without modifying tight-binding parameters.

Magnetic field can be introduced to the tight-binding model with a Peierl's substitution. As we will show in the next section, we have gauge freedom between different configurations ω when computing local density of states. This means we can choose a gauge for each ω calculation and not worry about the vector-potential $\mathbf{A}$ agreeing between each finite-sized sample of our infinite material (Fig. 1).

2.2 *Implementation*

We will now walk-through the details of how this method calculates electronic density of states in twisted heterostructures.

First, create a heterostructure model out of layered disks of a cut-off radius R. They are centered at a point with "zero-shift", which is just one specific ω configuration.

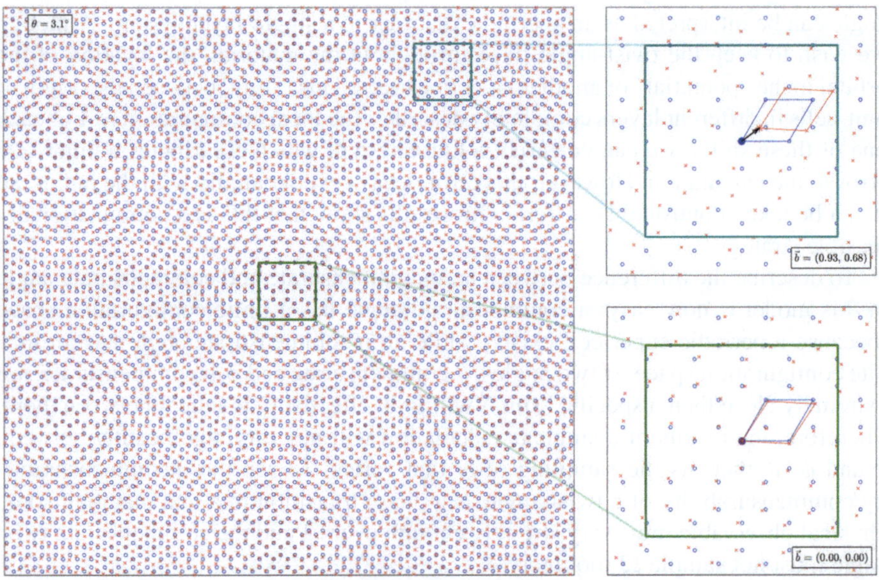

Fig. 1 Large Moiré pattern of bilayer graphene, with the red sheet twisted 3.1° with respect to the blue sheet. The locality method calculates global properties for an infinite material by computing values at the center of many finite-size systems. Two such finite-sized systems are shown above in blue and green outline, with the blue-orbital's local environment described completely by the "**b**-shift" between the blue and red unit-cells

Next, find all relevant hopping indices H_{ij}^{ω} in the sparse Hamiltonian by only looking for pairs that are less than a few nanometers apart. For each desired configuration ω, shift each layer by the proper amount (with respect to a fixed layer, usually the bottom one) and compute H_{ij}^{ω} for each non-zero hopping index. We implement Local EDoS, but can do conductivity with current-current correlation fitting [16]. The Local EDoS can be easily derived from the global EDoS, $g(\varepsilon)$, if we carefully consider all eigenstates (indexed by s) and orbitals (indexed by x):

$$\begin{aligned} g(\varepsilon) &= \frac{1}{N}\sum_{s=1}^{N}\delta(\varepsilon-\varepsilon_s) = \frac{1}{N}\sum_{s=1}^{N}\delta(\varepsilon-\varepsilon_s)\sum_{x}|\psi_s(x)|^2 \\ &= \sum_{x}\frac{1}{N}\sum_{s=1}^{N}\delta(\varepsilon-\varepsilon_s)|\psi_s(x)|^2 = \sum_{x} g_x(\varepsilon) \end{aligned} \tag{1}$$

where we have identified the local density of states $g_x(\varepsilon)$ as:

$$g_x(\varepsilon) = \frac{1}{N}\sum_{s=1}^{N}\delta(\varepsilon-\varepsilon_s)|\psi_s(x)|^2 \tag{2}$$

Finally, we apply the operator of interest to H_{ij}^{ω} with a Kernel Polynomial Method (Chebyshev polynomials) [16]. The Chebyshev polynomials T_i form a complete basis for square integrable functions which take values in the range $[-1, 1]$. A linear combination of them can be chosen to approximate the eigenspectrum of a tight-binding Hamiltonian after a simple rescaling to ensure all eigenvalues lie in $[-1, 1]$. Of interest is the Hamiltonian H_b of a specific orbital at a configuration ω represented by shift b. First, approximate the delta-function operator as a linear combination of the first p Chebyshev polynomials:

$$\delta(\varepsilon - H^{\omega}) \approx \sum_{i=0}^{p} a_i(\varepsilon) T_i(H^{\omega}) \tag{3}$$

$T_i(H^{\omega})$ is as large as the tight-binding Hamiltonian matrix and thus it is very unwieldy to keep track of all of them numerically. So instead we define a projection of this object onto site x as:

$$T_i^{xx}(\varepsilon) = v_x^{\dagger} T_i(H^{\omega}) v_x,\ v_x = (0, 0, \ldots 0, 1, 0 \ldots, 0, 0) \tag{4}$$

This is exactly the definition needed to succinctly write our local density of states operator:

$$
\begin{aligned}
g_x(\varepsilon) &= \frac{1}{N}\sum_{s=1}^{N} \psi_s^\dagger(x)\delta(\varepsilon-\varepsilon_s)\psi_s(x) \\
&\approx \sum_{i=0}^{p} v_x^\dagger(x) a_i(\varepsilon) T_i(H^\omega) v_x = \sum_{i=0}^{p} a_i(\varepsilon) T_i^{xx}(H^\omega)
\end{aligned} \tag{5}
$$

We compute only the center site (far from the boundary of this finite system), so we are interested in:

$$
g_0(\varepsilon) \approx \sum_{i=0}^{p} a_i(\varepsilon) T_i^{00}(H^\omega) \tag{6}
$$

Note that $T_i^{00}(H^\omega)$ is simply a number, and that all the ε dependence occurs in the coefficients $a_i(\varepsilon)$. Thus we can compute the local density of states for any sampling of energy values very cheaply after doing the expensive calculation of $T_i^{00}(H^\omega)$. Fortunately, the error introduced by the KPM adds linearly to the finite-size error and both have the same functional exponential form. This means we can find an optimal ratio of finite cut-off radius R and KPM polynomial order p for a small system and then increase both R and p accordingly for optimized calculations of large systems.

For electric field, only the on-site energies (diagonal elements) of the Hamiltonian change. If the hopping elements are $H_{ij}^\omega = t_{ij}^0$ without the presence of the electric field, the new hopping terms are:

$$
t_{ij} = t_{ij}^0 + (\mathbf{E}(\mathbf{r}_i)\cdot\mathbf{r}_i)\delta_{ij} \tag{7}
$$

where $\mathbf{E}(r)$ is the electric field and $\mathbf{r}_i$ is the position of orbital i.

For magnetic field, we apply the Peierl's substitution for the hopping element $H_{ij}^\omega = t_{ij}$ between orbital i and orbital j

$$
t_{ij} = t_{ij}^0 e^{i\frac{e}{\hbar}\int_{R_i}^{R_j}\mathbf{A}\cdot d\mathbf{l}} \tag{8}
$$

where R_i is the position of orbital i, t_{ij}^0 is the hopping element in the absence of magnetic field, and $\mathbf{A}$ is a vector potential satisfying $\nabla\times\mathbf{A} = \mathbf{B} = (0, 0, B_z)$. The simple gauge $\mathbf{A} = (-yB_z, 0, 0)$ is chosen for our infinite material. A slight complication arises from our local calculation method here. Although we have "gauge freedom" in choosing the initial vector potential for the infinite material, this does not necessarily mean each local configuration can have completely independent vector potentials. In other words, it might be necessary to have the vector potentials of two different H_b systems agree if we were to place them back into the infinite material.

To show that, at least for local density of states, we have gauge freedom between local configurations, we prove that T_i^{00} is independent of the choice of gauge. T_i^{00} is a polynomial in H^ω, so this is equivalent to proving $(H^\omega)_{00}^n$ is gauge independent.

We begin by noting the form of $(H^{\omega})^n_{00}$ is explicitly

$$(H^{\omega})^n_{00} = \sum_{\{a_i\}} t_{0a_1} t_{a_1 a_2} \dots t_{a_{n-2} a_{n-1}} t_{a_{n-1} 0} \tag{9}$$

where $\{a_i\}$ represents the sum over all possible hopping paths of length n from site 0 back to itself.

Since only the phase factor of t_{ij} changes with gauge choice, we only need to show that the phase factor ϕ of the terms in this sum are independent. The phase factor is:

$$\begin{aligned}\frac{\hbar}{e}\phi = \int_{R_0}^{R_{a_1}} \mathbf{A} \cdot d\mathbf{l} + \sum_{i=1}^{n-2} \int_{R_{a_i}}^{R_{a_{i+1}}} \mathbf{A} \cdot d\mathbf{l} \\ + \int_{R_{an}}^{R_0} \mathbf{A} \cdot d\mathbf{l} = \oint_{\{a_i\}} \mathbf{A} \cdot d\mathbf{l}\end{aligned} \tag{10}$$

where the final integral represents the closed path traveled in real-space from the center site R_0 through all the R_{a_i} and back to R_0. Changing our gauge from $\mathbf{A} \rightarrow \mathbf{A} + \Delta\lambda$ for any real-valued function $\lambda(\mathbf{r})$ causes a phase change of:

$$\Delta\phi = \frac{e}{\hbar} \oint_{\{a_i\}} \Delta\lambda \cdot d\mathbf{l} = \frac{e}{\hbar}(\lambda(R_0) - \lambda(R_0)) = 0 \tag{11}$$

Thus our local density of states operator does not change under a gauge transformation. So for computational convenience we can place the center of the finite H^{ω} system at $\mathbf{r} = (0, 0, 0)$ and choose $\mathbf{A} = (-yB_z, 0, 0)$. For any operator whose local projection is a physical observable gauge independence should hold. However, if the operator's local projection is not a physical observable (as in the case of computing topological character, or perhaps conductivity) care must be taken. In these cases, the model needs the local projection to be gauge-independent or to have the gauge "agree" between local configurations.

This method lends itself to excellent parallelization efficiency in comparison to DFT super-cell methods. To compute the global operator we must do a large number of independent computations of the local operator in different configurations (i.e. each H^{ω}). To manage all of these configurations, a master-worker parallelization scheme was implemented using the standard Message Passing Interface (MPI) in C++. Because each H^{ω} 'job' is completed autonomously by each worker, the computation is load distributed (the fastest cores get the most work, the slowest cores get the least). Since each local operator is computed using only sparse matrix-vector operations, we add a second layer of panelization on each worker by using multithreaded implementations of highly optimized matrix-vector operator subroutines (Table 1).

Table 1 Parallelization efficiency on a 750 Å cut-off radius graphene monolayer over 100 configurations

	2 Ranks (s)	4 Ranks (s)	8 Ranks (s)
1 Threads	9638	3373	1758
2 Threads	6082	2139	1076
4 Threads	3717	1521	681

3 Results

3.1 *Twisted Bilayer Graphene (tBLG)*

Twisted bilayer graphene (tBLG) provides an excellent system to test this method on, as it has already been well characterized by many experimental works and low-energy theory [3, 14]. To compute the EDoS of tBLG we used a two-band ab-initio tight-binding model [8]. The bands corresponded to the p_z orbital on the first and second Carbon atom in the Graphene unit cell (i.e. π bonding and anti-bonding interactions only). This tight-binding model already showed very accurate results when used with a periodic super-cell method. We find that the local EDoS modeled here shows no large qualitative difference from the results of periodic systems. However, we are now able to simulate arbitrary twist-angles and also study how the local operator changes in real-space, allowing us to compare to the results of recent STM experiments.

The main feature of twisted bilayer graphene is the presence of Van hove Singularities (VHVS) above and below the fermi energy. The origin of these VHVS can be best understood by considering the low-energy band-structure of tBLG as four dirac cones at the valleys K_l and K_l', where $l = (1, 2)$ labels the layers. At 0° twist, K_1 and K_2 are at the same point in momentum-space. As the layers are twisted, the dirac cones move away from one another in momentum-space, and a partial band-gap opening occurs where the cones now overlap. These hybridizations at the overlap of the dirac-cones cause the VHVS.

First, we compare the spatial dependence of four tBLG angles to experimental results [19] in Fig. 2. This is possible because sampling "shifts" over the diagonal of one graphene's unit-cell is the same as moving linearly from an AA to AB type stacking in the real-space Moiré pattern. We see excellent agreement between simulation and result by comparing the features of the VHVS.

Next, we sample the configuration space Ω for a fixed twist angle of 5.73° (0.1 rad) for 100 configurations along the diagonal of the unit-cell (Fig. 3). The results show that the local DoS varies extremely smoothly as a function of ω, and that the only regions of major configuration dependence are at the VHVS. The AB and BA type stacking have much lower DOS at the VHVS than any other stacking configuration.

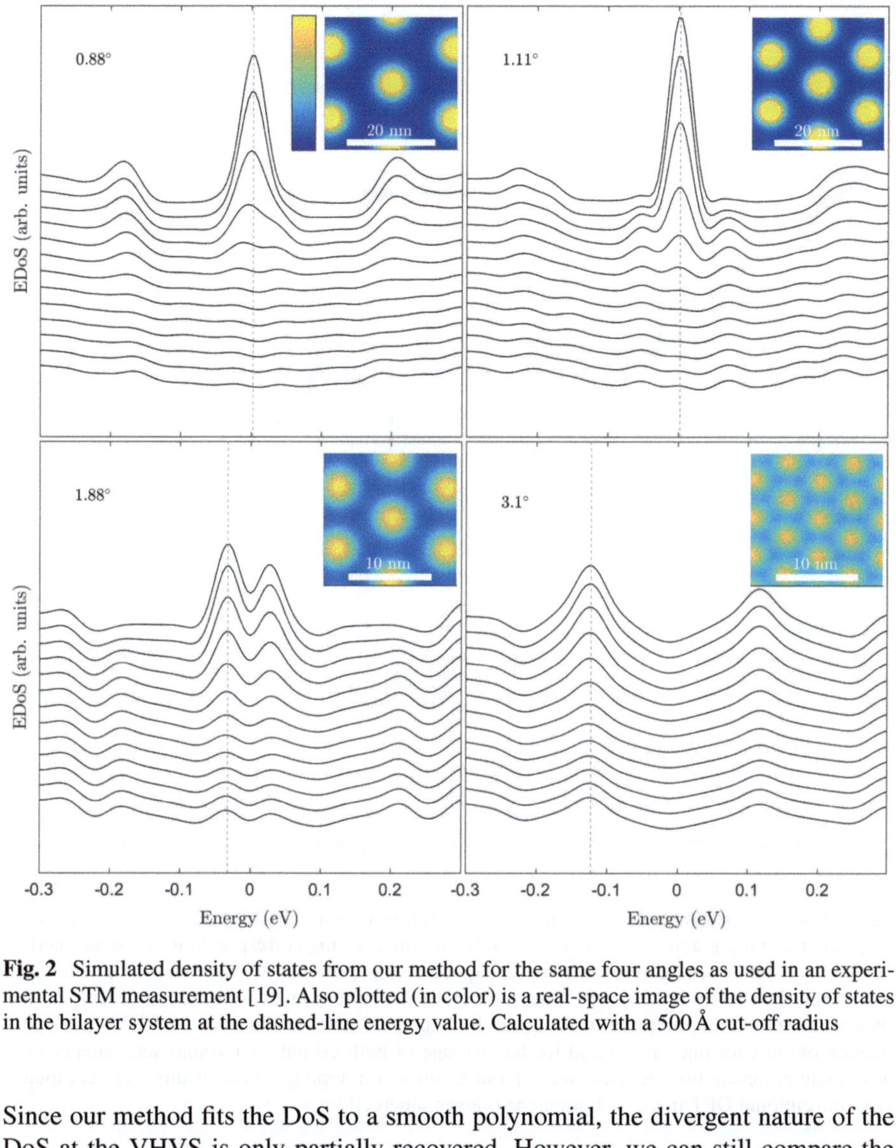

Fig. 2 Simulated density of states from our method for the same four angles as used in an experimental STM measurement [19]. Also plotted (in color) is a real-space image of the density of states in the bilayer system at the dashed-line energy value. Calculated with a 500 Å cut-off radius

Since our method fits the DoS to a smooth polynomial, the divergent nature of the DoS at the VHVS is only partially recovered. However, we can still compare the intensity of the VHVS by looking at the spectral weight near the singularity (i.e. the area under the DoS curve). Also plotted in Fig. 3c are pure DFT ground-state energy (GSE) calculations for non-twisted bilayer graphene over the same diagonal of shifts with and without Li+ intercalation. With Li+ intercalation, the relative stability of the AB and AA stacking inverts. This means Li+ may act as a "lubricant" between the layers, allowing macroscopic samples to change their twist-angle. Interestingly, there

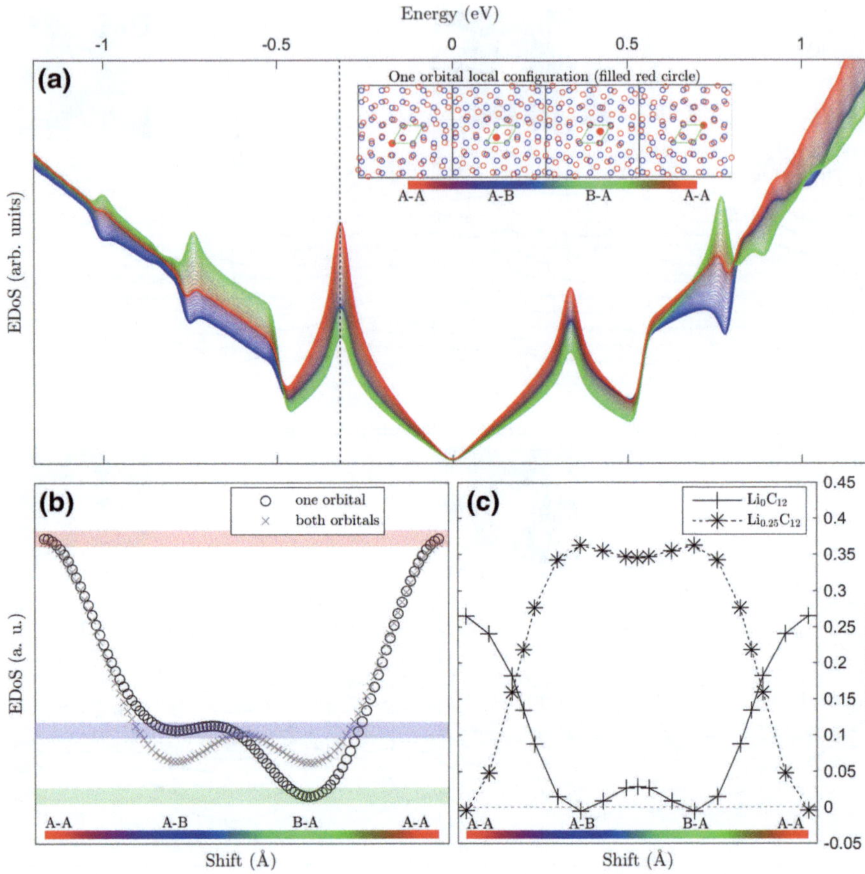

Fig. 3 Local electronic density of states as a function of shift distance acrosss the unit cell's diagonal. **a** The top graph is a scan of the EDoS with the coloring corresponding to distance across the diagonal. The inset figure shows the real-space configuration ω, with the first graphene layer in blue and the second layer in red. The unit-cell of the bottom layer is plotted in green, and the calculated orbital is highlighted in solid red. **b** Density of states at the selected VHVS peak as a function of shift for one orbital and for the average of both orbitals. **c** Ground state energy of 0 twist-angle graphene bilayer with and without Lithium ion doping. These results were computed with conventional DFT in a 2×2 graphene bilayer supercell

are great similarities between the VHVS DoS and the DFT GSE calculation, perhaps meaning we can use DoS spectral weight in our method as a quantity analogous to ground state energy. Sampling across the entire unit-cell provided equally smooth results, which are just a smooth extension of LEDoS values at the three critical stacking configurations.

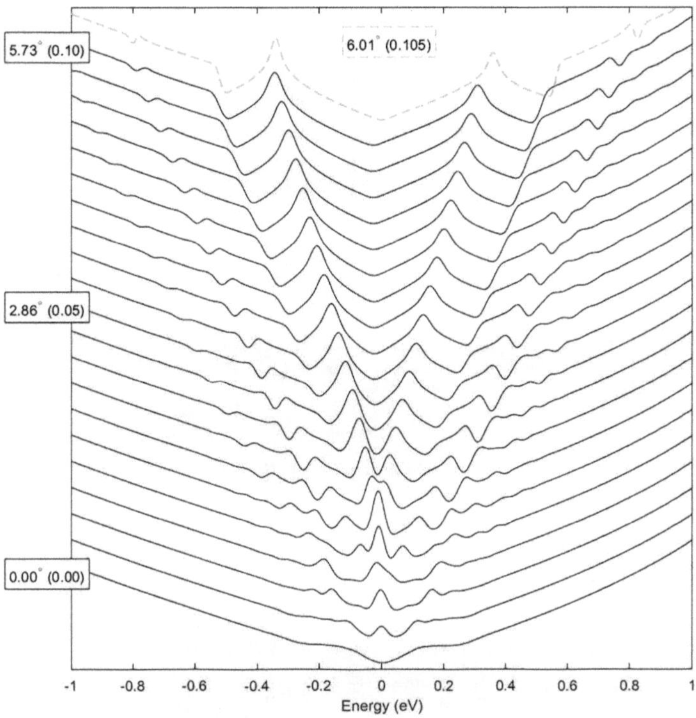

Fig. 4 Electronic density of states computed for twisted bilayer graphene from 0° to 5.73° twist-angle. Some of the twist angles are labeled next to their EDoS line in the format "degrees (radians)". Between 0 and 0.1 rad a simulation was done for every 0.005 rad interval. In dashed green are results from a previous super-cell simulation with this model

We plot the angle-dependent EDoS for bilayer Graphene in Figs. 4 and 5. The first and second Van-Hove Singularities (VHVS) can be seen clearly in the low-angle regime, and they move away from the Fermi Energy linearly with twist angle. This is in agreement with many STM experimental results [4, 17].

Introducing a magnetic field to our model allows us to compute the fermi-velocity, which is defined by the slope of the EDoS at the fermi-level in momentum-space. This traditionally requires a band-structure calculation, which is not performed for our aperiodic model. By calculating how the monolayer's EDoS changes with magnetic field in Fig. 6, and using the low-energy model for the localized states (Landua Levels, or LLs) [6]

$$E(n) = \pm v_F \sqrt{2eBN} \tag{12}$$

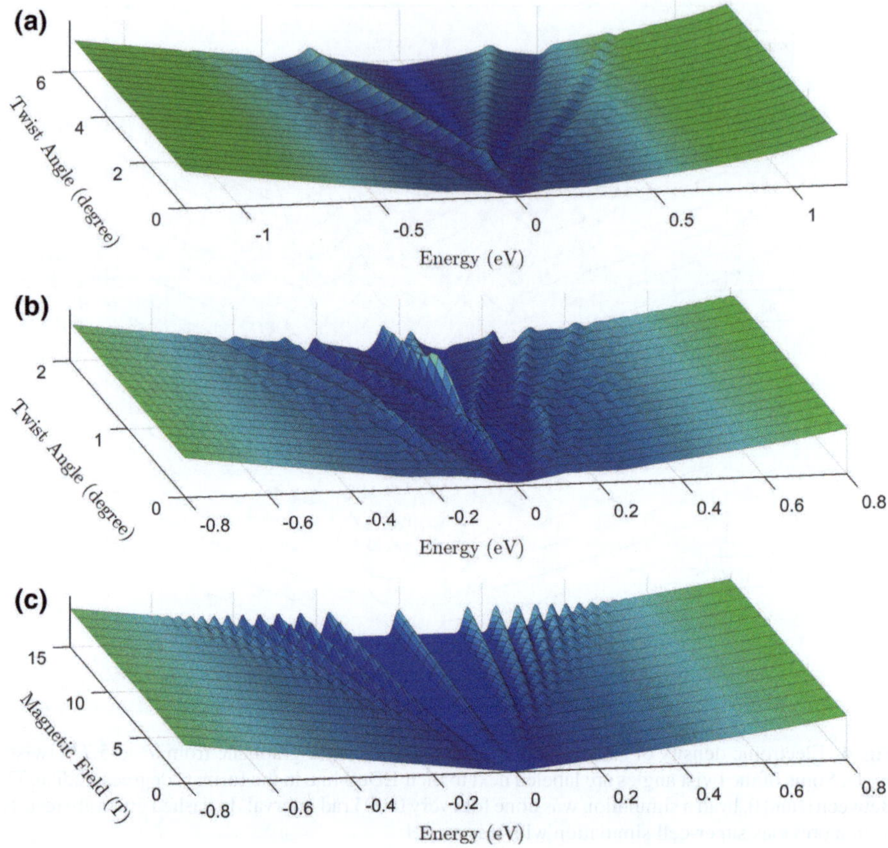

Fig. 5 Average EDoS as a function of **a, b** twist angle in bilayer graphene, or **c** magnetic field in monolayer graphene. Height and color correspond to the value of the EDoS

gives us a fitted Fermi velocity $v_F = 1.2 \times 10^6$ m/s. This is in good agreement with most other theoretical predictions and experimental results, which all find pristine monolayer graphene's Fermi velocity to be on the scale of 10^6 m/s [6].

We also simulate the magnetic-field dependence of the tBLG DOS in Fig. 7 for a 5.73° twist angle. Landau levels appear as the magnetic field increases, with gaps opening in the EDoS for $B > 5$T. This would correspond with the experimentally observered phenomenon of the conductivity in tBLG going to zero in the presence of magnetic field for values of the chemical potential near the fermi level [5]. Note that the effect of magnetic field, and the formations of Landau Levels, is much less straightforward in the twisted system than it was in the monolayer system. This is

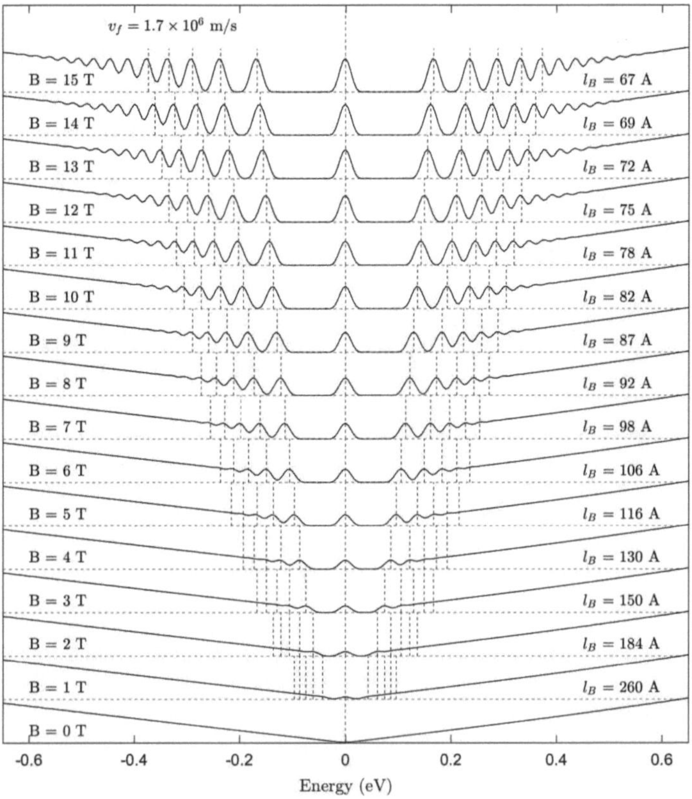

Fig. 6 Single-orbital local EDoS for mono-layer graphene for different values of perpendicular magnetic field. The magnetic length and Landau levels (up to $n = \pm 5$) for the given Fermi-velocity are plotted for each magnetic field strength. Simulation run with a 750 Å cut-off radius

because in the monolayer, only the magnetic field is causing aperiodicity, through a vector gauge potential which is generally incommensurate with the lattice. In the twisted bilayer systems, the single-sheet periodicity is being broken by both the (incommensurate) twist angle and the magnetic field.

With a smaller twist-angle, the "competition" between the twist-disorder and the magnetic disorder becomes even more complicated, as in the 2.0° twist of Fig. 8. At 0 magnetic field, the low-energy EDoS is visible as two distinct van Hove singularities. As the magnetic field increases, the VHVS become less and less distinct, as Landau Levels start to form. But even at $B = 14$T, no clear gap has been opened by the magnetic field, which for the 5.73° twist occurred as early as $B = 10$T. Other methods

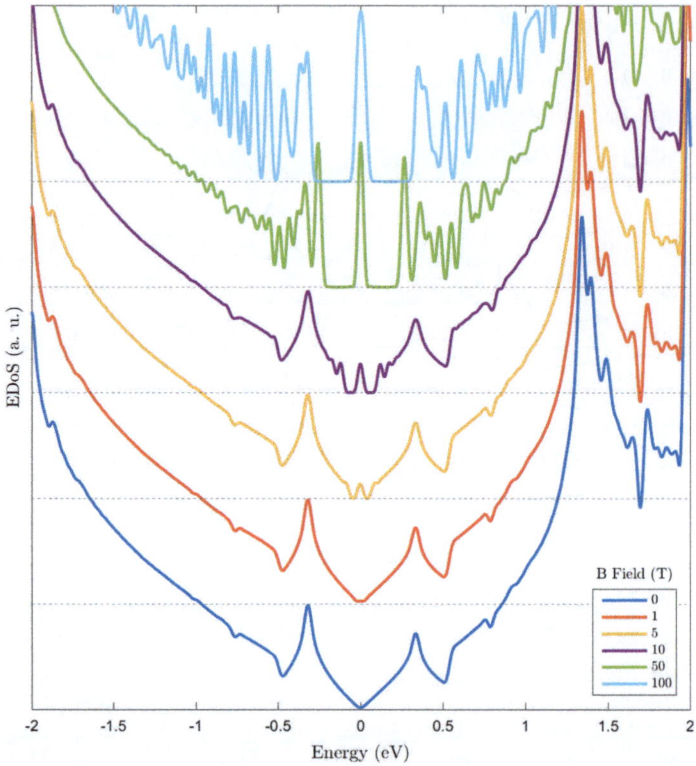

Fig. 7 The local EDoS for tBLG at a 5.73° twist-angle for different values of magnetic field. Simulation run for one configuration (A-A type-stacking) at a 500 Å cut-off radius

of simulating the electronic structure of twisted 2D materials require either very specific choices of the magnetic field (real-space, periodic supercells) or only work when a simple effective Hamiltonian is available ($\mathbf{k} \cdot \mathbf{p}$ theories). Our multi-scale ab-initio method allows for a scalable calculation for arbitrary materials, twist-angle, and magnetic field (among other disorders).

3.2 Twisted MX_2 TMDC Bilayers

Unlike bilayer graphene, heterostructures of transition metal dichalcogenides will not be well-described by low-energy theory due to their large band-gaps (about 1 eV). We use an 11 band model, consisting of 5 d orbitals on the M type atom and 3 p orbitals (p_x, p_y, p_z) on each of the X type atoms. The interlayer interaction is

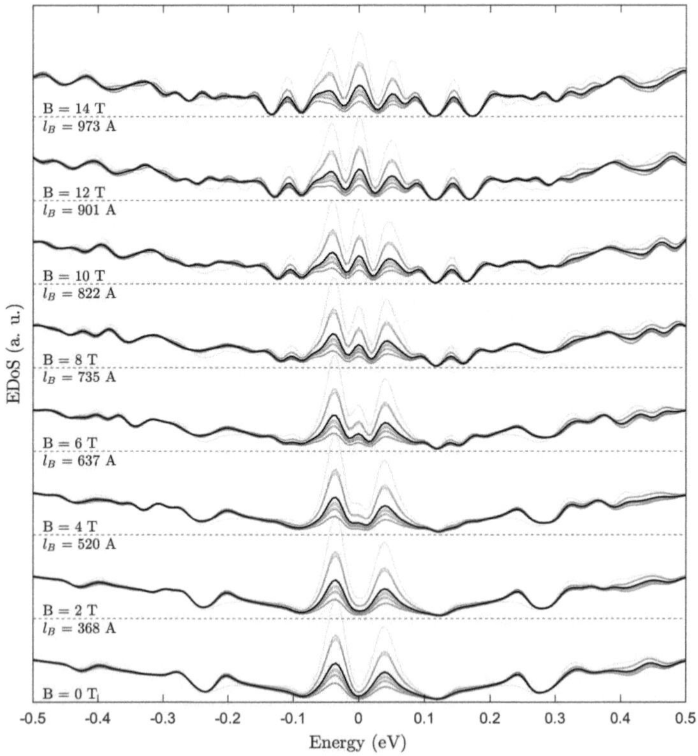

Fig. 8 The local EDoS for tBLG at a 2.0° twist-angle for different values of magnetic field. Each configuration is plotted in light-grey, with the average EDoS for each magnetic field plotted in black. Simulation averaged over 25 configurations at a 750 Å cut-off radius

modeled only between X type atoms directly on the bilayer interface. Thus only the "inner" X atoms interact between layers, with no hopping terms between M type or "outer" X type atoms across the bilayer. We review the results for both a MoS_2 and a WSe_2 ab-initio tight-binding model. The MoS_2 model includes GW corrections to the band structure, allowing for more accurate modeling of band-gap effects.

We find the configuration dependence of bilayer MoS_2, calculated for a 5.73° twist-angle in Fig. 9. We see large amounts of configuration dependence in the p_z orbitals, and interesting configuration dependence near the valence band edge for "inner" Sulfur p_z orbital (bright green in figure). However, as a all the bilayer TMDCS are strongly insulating materials, these sort of effects are difficult to study with the STM methods noted in the bilayer graphene section. We also show the configuration-dependence for WSe_2 bilayer calculated in a similar manner in Fig. 10.

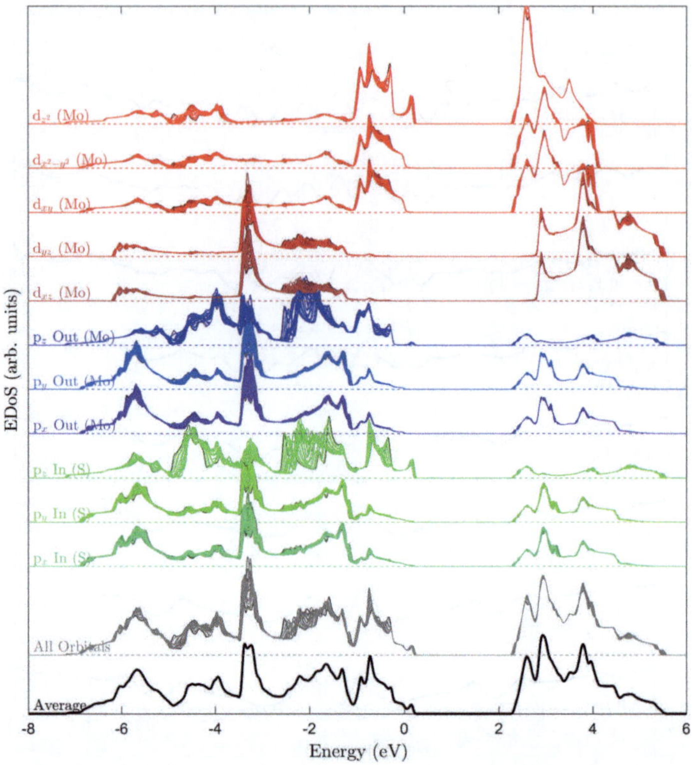

Fig. 9 Electronic density of states of bilayer MoS_2 for a 0.1 rad (5.73°) twist-angle. The average density of states is on the bottom in black, and the projected configuration-dependent EDoS is above in color. The band-gap is dominated by the d_{z^2} orbital (in bright red). The other d orbitals are in faded red and the outer and inner p orbitals are in blue and green. Simulation was run with a 300 Å cut-off radius over 100 configurations of the upper MoS_2 unit-cell

The twist-angle dependence of the density of states for bilayer MoS_2 is shown in Fig. 11. There are significant changes in the EDoS of the valence band, but again it is difficult to probe this region of the material experimentally. Instead, we focus on the valence and conduction band edges, noting that the band-gap is a twist-angle dependent feature. The band gap increases by 64 meV when going from 0° to roughly 30° twist-angle. This represents an approximately 4% change in the band-gap. However, our model does not take into account changes in the distance between

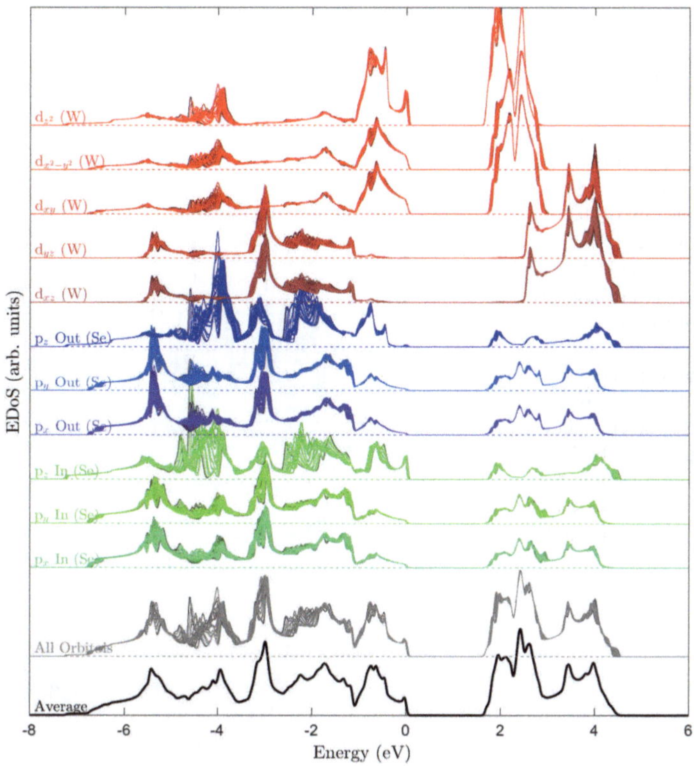

Fig. 10 Electronic density of states of bilayer WSe_2 for a 0.1 rad (5.73°) twist-angle. See Fig. 9 for details on the calculation

the two layers as a function of twist-angle. The band-gap may be more dependent on this distance than the twist-angle, but we still expect a noticeable change from the twist-angle alone.

Our methodology can also give insight into how dopants effect the electronic properties of 2D materials. Here we study the case of a double Sulfur vacancy in MoS_2, which is when both the sulfur atoms of a single unit-cell are missing from an otherwise perfect monolayer. We do not change electronic hoppings between sites in the presence of the defect, we simply remove the two atoms (6 orbitals) from our tight-binding model. This "0-order" approximation may not be accurate, but it can be compared to DFT results of monolayer vacancy systems. If the results agree well between our tight-binding approximation and a full DFT calculation, properties

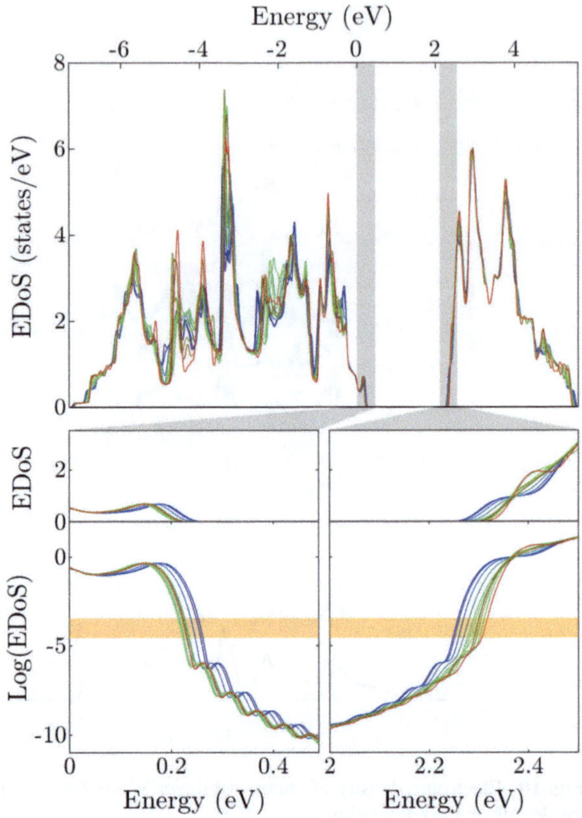

Fig. 11 Simulations of twisted bilayer MoS_2 from 0° (red) to 28.65° (blue) twist angle. The plots above the dashed black line show the EDoS for the various twist angles, the plots below show the logarithm of the EDoS to allow for comparison of band gap widths. There are changes to the density of states in various areas of the conduction and valence bands as well as non-trivial change in the bilayer's band gap. The gap widens by 76 meV when going from a 0° to 28.65° twist angle (a 4% relative increase in the band gap)

of multi-layer materials with vacancies is a reasonable extension. To improve the accuracy of these tight-binding results, a more careful study of how the hopping parameters change in the presence of the defect is necessary.

The results for a double Sulfur vacancy in monolayer MoS_2 are shown in Fig. 12. Three distinct energy values in the band-gap of MoS_2 appear. These are assumed to be localized eigenstates, which is common for vacancy defects. By computing the local EDoS on each orbital near the vacancy, we are able to plot the shape and symmetry of the eigenstates. This is visualized by drawing colored circles (representing Mo and S orbitals) with radii corresponding to the sum of the square of their local density of states at each site.

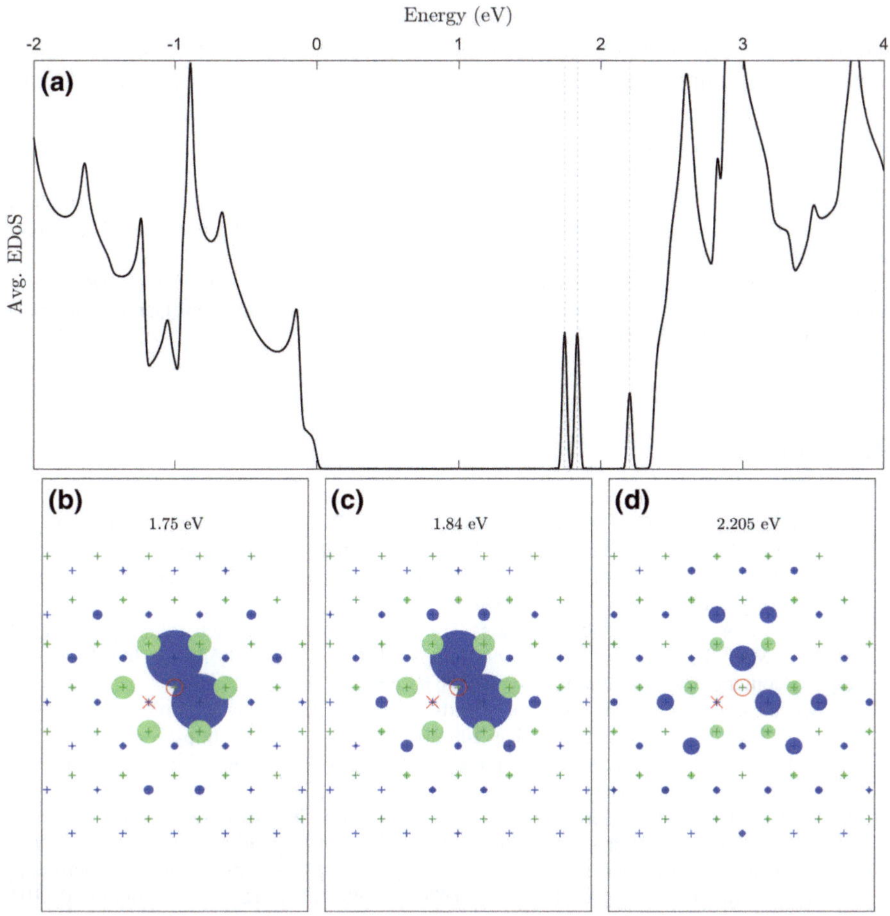

Fig. 12 **a** Local EDoS near a double sulfur vacancy defect of an MoS_2 monolayer. Three defect states appear in the MoS_2 bandgap, at 1.75, 1.84, and 2.205 eV. The real-space wavefunction density for these defect states is plotted in (**b–d**). The Sulfur vacancy defect is plotted as the red circle, and the Mo and S densities are plotted by the area of the blue and green (respectively) circles. One Mo site was not calculated (red 'X') because it shares a unit-cell with the defect. This is omitted due to a small technical limitation in the current method, but is fixable with additional development work. Simulation ran with a 500 Å cut-off radius (~270,000 atoms)

4 Conclusion

We have presented a new method for computing electronic structure by way of a multi-scale ab-initio tight-binding model. This allows us to deal with 2D twisted heterostructures exactly, regardless of twist angle, and include other sources of disorder and fields easily. We present results for graphene and transition metal dichalcogenide in the presence of magnetic field and vacancies. We see many interesting twist-dependent features, an accurate implementation of arbitrary magnetic fields, and a new method of studying the electronic properties of vacancies. We hope to further develop our method and apply it to a large number of 2D systems, allowing for robust and direct simulation of multi-layered meta materials in the presence of arbitrary fields and defects.

Acknowledgements We acknowledge S. Shirodkar for providing the Li-ion itercalated graphene calculations shown in Fig. 3c and B.I. Halperin and D. Huang for helpful discussions. The computations in this paper were run on the Odyssey cluster supported by the FAS Division of Science, Research Computing Group at Harvard University. This work was supported by the ARO MURI Award No. W911NF-14-0247. SF is supported by the STC Center for Integrated Quantum Materials, NSF Grant No. DMR-1231319.

References

1. A. Ayari, E. Cobas, O. Ogundadegbe, M.S. Fuhrer, Realization and electrical characterization of ultrathin crystals of layered transition-metal dichalcogenides. J. Appl. Phys. **101**(1), 014,507 (2007), https://doi.org/10.1063/1.2407388, http://scitation.aip.org/content/aip/journal/jap/101/1/10.1063/1.2407388
2. J. Bellissard, A. van Elst, H. Schulz-Baldes, The noncommutative geometry of the quantum Hall effect. J. Math. Phys. **35**(10), 5373 (1994), https://doi.org/10.1063/1.530758, http://scitation.aip.org/content/aip/journal/jmp/35/10/10.1063/1.530758
3. R. Bistritzer, A.H. MacDonald, Moire bands in twisted double-layer graphene. Proc. Natl. Acad. Sci. United States of America **108**(30), 12,233–7 (2011), https://doi.org/10.1073/pnas.1108174108, http://www.pubmedcentral.nih.gov/articlerender.fcgi?artid=3145708&tool=pmcentrez&rendertype=abstract
4. I. Brihuega, P. Mallet, H. González-Herrero, G. Trambly de Laissardière, M.M. Ugeda, L. Magaud, J.M. Gómez-Rodríguez, F. Ynduráin, J.Y. Veuillen, Unraveling the intrinsic and robust nature of van Hove singularities in twisted bilayer graphene by scanning tunneling microscopy and theoretical analysis. Phys. Rev. Lett. **109**(19), 196,802 (2012), https://doi.org/10.1103/PhysRevLett.109.196802, http://www.ncbi.nlm.nih.gov/pubmed/23215414
5. Y. Cao, J.Y. Luo, V. Fatemi, S. Fang, J.D. Sanchez-Yamagishi, K. Watanabe, T. Taniguchi, E. Kaxiras, P. Jarillo-Herrero, Superlattice-induced insulating states and valley-protected orbits in twisted bilayer graphene. Phys. Rev. Lett. **117**(11), 116,804 (2016), https://doi.org/10.1103/PhysRevLett.117.116804, http://link.aps.org/doi/10.1103/PhysRevLett.117.116804
6. A.H. Castro Neto, F. Guinea, N.M.R. Peres, K.S. Novoselov, A.K. Geim, The electronic properties of graphene. Rev. Mod. Phys. **81**(1), 109–162 (2009), https://doi.org/10.1103/RevModPhys.81.109, http://link.aps.org/doi/10.1103/RevModPhys.81.109
7. H. Chen, C. Ortner, QM/MM methods for crystalline defects. Part 1: locality of the tight binding model. SIAM Multiscale Model. Simul. **14**(1), 232–264 (2016), https://doi.org/10.1137/15M1022628, arXiv:1505.05541
8. S. Fang, E. Kaxiras, Electronic structure theory of weakly interacting bilayers. Phys. Rev. B **93**(23), 235,153 (2016), https://doi.org/10.1103/PhysRevB.93.235153, http://link.aps.org/doi/10.1103/PhysRevB.93.235153
9. S. Fang, R. Kuate Defo, S.N. Shirodkar, S. Lieu, G.A. Tritsaris, E. Kaxiras, Ab initio tight-binding Hamiltonian for transition metal dichalcogenides. Phys. Rev. B **92**(20), 205,108 (2015), https://doi.org/10.1103/PhysRevB.92.205108, http://link.aps.org/doi/10.1103/PhysRevB.92.205108

10. A.K. Geim, I.V. Grigorieva, Van der Waals heterostructures. Nature **499**(7459), 419–425 (2013), https://doi.org/10.1038/nature12385, http://www.ncbi.nlm.nih.gov/pubmed/23887427, http://www.nature.com/doifinder/10.1038/nature12385
11. E. Prodan, Quantum transport in disordered systems under magnetic fields: a study based on operator algebras. Appl. Math. Res. eXpress **2**, 176–265 (2012), https://doi.org/10.1093/amrx/abs017, http://amrx.oxfordjournals.org/content/early/2012/10/23/amrx.abs017.short, http://amrx.oxfordjournals.org/cgi/doi/10.1093/amrx/abs017
12. B. Radisavljevic, A. Radenovic, J. Brivio, V. Giacometti, A. Kis, Single-layer MoS2 transistors. Nat. Nanotech. **6**(3), 147–50 (2011), https://doi.org/10.1038/nnano.2010.279, http://www.ncbi.nlm.nih.gov/pubmed/21278752
13. A. Rozhkov, A. Sboychakov, A. Rakhmanov, F. Nori, Electronic properties of graphene-based bilayer systems. Phys. Rep. **648**, 1–104 (2016), https://doi.org/10.1016/j.physrep.2016.07.003, http://linkinghub.elsevier.com/retrieve/pii/S0370157316301612
14. P. San-Jose, J. González, F. Guinea, Non-Abelian gauge potentials in graphene bilayers. Phys. Rev. Lett. **108**(21), 216,802 (2012), https://doi.org/10.1103/PhysRevLett.108.216802, http://journals.aps.org/prl/abstract/10.1103/PhysRevLett.108.216802, http://www.ncbi.nlm.nih.gov/pubmed/23003289
15. K. Uchida, S. Furuya, J.I. Iwata, A. Oshiyama, Atomic corrugation and electron localization due to Moiré patterns in twisted bilayer graphenes. Phys. Rev. B **90**(15), 155,451 (2014). https://doi.org/10.1103/PhysRevB.90.155451, http://link.aps.org/doi/10.1103/PhysRevB.90.155451
16. A. Weisse, G. Wellein, A. Alvermann, H. Fehske, The kernel polynomial method. Rev. Mod. Phys. **78**(1), 275–306 (2006), https://doi.org/10.1103/RevModPhys.78.275, http://link.aps.org/doi/10.1103/RevModPhys.78.275
17. D. Wong, Y. Wang, J. Jung, S. Pezzini, A.M. DaSilva, H.Z. Tsai, H.S. Jung, R. Khajeh, Y. Kim, J. Lee, S. Kahn, S. Tollabimazraehno, H. Rasool, K. Watanabe, T. Taniguchi, A. Zettl, S. Adam, A.H. MacDonald, M.E. Crommie, Local spectroscopy of moiré-induced electronic structure in gate-tunable twisted bilayer graphene. Phys. Rev. B **92**(15), 155,409 (2015), https://doi.org/10.1103/PhysRevB.92.155409, http://link.aps.org/doi/10.1103/PhysRevB.92.155409
18. C.R. Woods, L. Britnell, A. Eckmann, R.S. Ma, J.C. Lu, H.M. Guo, X. Lin, G.L. Yu, Y. Cao, R.V. Gorbachev, A.V. Kretinin, J. Park, L.A. Ponomarenko, M.I. Katsnelson, Y.N. Gornostyrev, K. Watanabe, T. Taniguchi, C. Casiraghi, H.J. Gao, A.K. Geim, K.S. Novoselov, Commensurateincommensurate transition in graphene on hexagonal boron nitride. Nat. Phys. **10**(6), 451–456 (2014), https://doi.org/10.1038/nphys2954, http://www.nature.com/nphys/journal/v10/n6/abs/nphys2954.html, http://www.nature.com/doifinder/10.1038/nphys2954
19. L.J. Yin, J.B. Qiao, W.J. Zuo, W.T. Li, L. He, Experimental evidence for non-Abelian gauge potentials in twisted graphene bilayers. Phys. Rev. B **92**(8), 081,406 (2015), https://doi.org/10.1103/PhysRevB.92.081406, http://link.aps.org/doi/10.1103/PhysRevB.92.081406

Molecular Dynamics and Related Computational Methods with Applications to Drug Discovery

Jordane Preto, Francesco Gentile, Philip Winter, Cassandra Churchill, Sara Ibrahim Omar and Jack A. Tuszynski

Abstract The main objective of this review chapter is to give the reader a practical toolbox for applications in quantitative biology and computational drug discovery. The computational technique of molecular dynamics is discussed, with special attention to force fields for protein simulations and methods for the calculation of solvation free energies. Additionally, computational methods aimed at characterizing and identifying ligand binding pockets on protein surfaces are discussed. Practical information about available databases and software of use in drug design and discovery is provided.

Keywords Molecular dynamics · Molecular docking · Structure-based drug design · Virtual screening · Scoring functions

---

J. Preto · P. Winter · S. I. Omar · J. A. Tuszynski (✉)
Department of Oncology, University of Alberta, Edmonton, AB T6G 1Z2, Canada
e-mail: jackt@ualberta.ca

J. Preto
e-mail: preto@ualberta.ca

P. Winter
e-mail: pwinter@ualberta.ca

S. I. Omar
e-mail: siomar@ualberta.ca

F. Gentile · C. Churchill · J. A. Tuszynski
Department of Physics, University of Alberta, Edmonton, AB T6G 2E1, Canada
e-mail: fgentile@ualberta.ca

C. Churchill
e-mail: churchil@ualberta.ca

L. L. Bonilla et al. (eds.), *Coupled Mathematical Models for Physical and Biological Nanoscale Systems and Their Applications*, Springer Proceedings in Mathematics & Statistics 232, https://doi.org/10.1007/978-3-319-76599-0_14

1 Introduction

Computational drug discovery is a conceptual approach to finding drug-like molecules by rational design, based on the information regarding their intended biomolecular target. A drug target is an important molecule, usually a protein, involved in a particular metabolic or signaling pathway that is specific to a disease condition. Most approaches attempt to inhibit the functioning of an aberrant or over-expressed pathway in the diseased state by interfering with the normal activity of the target. Medicinal compounds as candidate drugs can have their structures rationally designed at a molecular level in such a way as to optimize their binding to the active region of their target biomolecule in order to inhibit its activity and to simultaneously minimize their effects on other important biomolecules that may cause undesired side effects. Since many challenges are posed by the large chemical and biological spaces involved in designing drugs with high specificity and selectivity, serendipity has traditionally played an additional important role in finding potential new drugs. Conversely, structure-based drug design requires knowledge of the structure of the biomolecular target, and it utilizes 3D information about biomolecules obtained from techniques such as x-ray crystallography and NMR spectroscopy.

The first step in the rational drug design process is usually the identification and characterization of the biomolecular target, such as a protein or a DNA sequence. From here, computational techniques can be used to model a drug within the binding site of the biomolecular target, and this information can be used to design novel drug panels with enhanced activity. Of the computational techniques available, molecular dynamics (MD) is particularly important in the investigation of target characterization and drug-target interactions. In Sect. 2, an overview of the main aspects of MD simulations—including force field descriptions—and related methods intended to characterize drug-target binding is provided. In Sect. 3, other computational drug-discovery strategies, such as binding pocket prediction and molecular docking, are described. Virtual screening (VS) techniques are also discussed.

2 Molecular Dynamics

2.1 General

Like most experimentally-measured properties of molecular systems, the binding affinity of a drug to its target is a thermodynamic quantity, i.e., an ensemble average over a representative statistical ensemble of a system. As a result, the knowledge of a single 3D structure of a given protein complex—obtained, e.g., from x-ray crystallography or cryo-electron microscopy—even if associated with a global energy minimum, is not enough to theoretically predict such macroscopic properties. Instead, it is necessary to generate a representative ensemble of conformations of the same system at a given (typically physiological) temperature. Two popular computational methods may be applied to this end: molecular Monte Carlo simulations (MC) [1] and MD [2]. For the study of dynamic or non-equilibrium properties (e.g., the transport of molecules across biomembranes, chemical reactions, etc.), only the second method may be utilized. Although MC simulations are simpler than MD ones, they usually do not lead to any better statistics in a given amount of time [3]. That is why MD is generally preferred over MC. Popular MD engines include Amber [4], GROMACS [5], LAMMPS [6], NAMD [7].

MD simulations usually involve the numerical integration of Newton's equations of motion for a system of N interacting atoms representing the system of interest, possibly including the molecules of the surrounding solvent:

$$m_i \frac{d^2 \boldsymbol{r}_i}{dt^2} = \boldsymbol{F}_i, \quad i = 1 \ldots N, \tag{1}$$

where $\boldsymbol{r}_i$ is the position of atom i, m_i its mass and $\boldsymbol{F}_i$ is the force acting on it, equal to the negative derivative of the molecular potential U, i.e., $\boldsymbol{F}_i = -\partial U / \partial \boldsymbol{r}_i$.

Using Newton's equations of motion automatically implies the use of classical physics, classical MD having the advantage of being far less computationally demanding than real quantum-dynamical simulations, which require solving the time-dependent Schrödinger equation for the system of interacting particles forming the molecule. However, because of classical approximations, standard MD simulations suffer from several limitations that the reader should be aware of. First, electronic motions are not considered per se. Instead, it is supposed that electrons are always in their ground state adjusting their dynamics instantly when atoms are moved (Born-Oppenheimer approximation). Secondly, most potential energy functions U used to model atomic interactions, commonly referred to as *force fields* in chemistry and biology, are empirical thus approximate. They usually consist of a summation of bonded forces and non-bonded pair-additive forces. Such analytical potentials include free parameters (e.g., coupling constants, equilibrium bond lengths, van der Waals radii, etc.), which are estimated by fitting against detailed electronic calculations or experimental properties (e.g., spectroscopy measurements, elastic constants) in order to reproduce observed experimental equilibrium behaviors [8–10]. Typical classical MD force fields adopt the following functional form:

$$U = \sum_{\text{bonds}} K_b (b - b_0)^2 + \sum_{\text{angles}} K_\theta (\theta - \theta_0)^2 + \sum_{\text{torsions}} K_\phi [\cos(n\phi + \varphi) + 1] + \sum_{i<j}^{N} \left(\frac{A_{ij}}{r_{ij}^{12}} - \frac{B_{ij}}{r_{ij}^{6}} \right) + \sum_{i<j}^{N} \frac{q_i q_j}{4\pi \varepsilon_0 r_{ij}}. \tag{2}$$

The first term in Eq. (2) represents the potential between two chemically-bound atoms, modeled as a simple harmonic potential, b being the distance between the two atoms and b_0 the equilibrium bond length. The proximity between three atoms, which are connected via chemical bonds can be described with an angle. The second term in Eq. (2) stands for this angle-dependence involving three atoms and is also modeled by a harmonic potential, θ being the angle between the three atoms in the structure and θ_0 its equilibrium value. The third term represents the dihedral angle (torsion) potential and depends on four atom coordinates. Such a potential is periodic and is represented by a cosine function with n, the number of maxima and φ, the angular offset. The variable ϕ is obtained from dihedral angles in the structure. Noticeably, an additional term may be included in Eq. (2) to model out-of-plane bending motions, i.e., improper dihedral angles. This is usually done through a cosine or a harmonic function. The last two terms in Eq. (2) account for non-bonded interactions and are calculated pairwise between atoms i and j. The fourth term is the van der Waals potential, which is typically represented by a Lennard-Jones 6-12 potential. The $1/r^6$ term is the attractive component while the $1/r^{12}$ term approximates Pauli repulsion. Parameters A_{ij} and B_{ij} are atom specific while r_{ij} stands for the distance between atoms i and j. The final term corresponds to the electrostatic potential between atoms, and is modeled as a Coulomb potential. Parameters q_i and q_j represent (fixed) charges on atoms i and j, while the constant ε_0 is the vacuum permittivity. Electrostatic interactions dominate over van der Waals forces for long-range intermolecular interactions and they play a significant role in non-chemical binding.

MD simulations result in trajectories, which contain information about the changes of atomic positions over time, which can be analyzed in great detail to extract pertinent information regarding the dynamics of the system. This includes the root-mean-square deviation (RMSD) of ligand and protein atoms, supramolecular (non-covalent) interactions, changes in the potential energy of the system, short-lived reaction intermediates [11], conformational changes, flexibility, and optimum binding modes [12] among many various properties of the biomolecule and its environment. In a computer-aided drug design process, the mobility of crystal water molecules near proteins observed in MD simulations can help identify the amino acid residues that play an important role in ligand binding. MD simulations can also be used for studying ionic conductivity [13, 14], where the simulations provide atomic level insights into ionic mobility. In terms of particular applications, MD has been successfully used to study clinically important proteins such as HIV-1 gp120 [15], binding sites [16], drug resistance mechanisms [17], and protein folding [18, 19] to name but a few.

2.2 Polarizable Force Fields and Quantum Dynamics

Most standard force fields do not incorporate atomic polarizability effects other than adjusting atomic partial charges obtained from quantum chemical computations. However, polarizable force fields include extra degrees of freedom in order to model electronic charges, usually attached to the nucleus by a spring as in the case of the shell model [20]. This allows for a dynamic redistribution of atomic dipoles, which responds to the local chemical environment. More realistic MD simulations, called ab initio molecular dynamics (AIMD), can be applied in order to reproduce electronic dynamics more accurately [21, 22]. Instead of using a prescribed potential for U, AIMD implies solving the time-dependent Schrödinger equation for the many-body wave function of the electrons assuming the atom nuclei fixed (*Born-Oppenheimer* approximation). The Schrödinger equation is generally solved at each MD step using density functional theory (DFT) in order to get the potential energy as a function of the nuclear coordinates. The potential energy is then used to integrate the classical Newton's equations given by Eq. (1). Due to the cost of treating electronic degrees of freedom, the computational cost is far higher than classical MD, implying that AIMD is only applicable to small molecular systems and short time scales (picosecond). A good trade-off between accuracy and speed is achieved by hybrid quantum mechanics/molecular mechanics (QM/MM) force fields [23]. In such simulations, the region of the system in which quantum effects (e.g., bond breaking, quantum resonance ...) take place is treated at an appropriate level of quantum chemistry theory, while the rest is described by a classical molecular mechanics force field [24]. Recently, machine-learning-based algorithms have been suggested as a way to accelerate ab initio methods [25]. A significant advantage of AIMD and QM/MM simulations is the ability to study reactions that involve breakage or formation of covalent bonds, which correspond to multiple electronic states. AIMD has also proved useful for reproducing typical dynamics and spectral features of liquids such as water [26]. Despite the accuracy provided by quantum methods, classical MD remains a reliable method to study biomolecular processes in large systems over long simulation times (up to a few milliseconds) including folding dynamics, conformational molecular changes as well as non-covalent bindings of drugs to their biological target.

2.3 Molecular Dynamics and Drug Discovery

Molecular dynamics can be used together with other methods to solve a host of problems in biomolecular modeling [27, 28]. In the case of VS methods that involve large libraries of chemical compounds in order to identify a high-affinity small molecule that is expected to act as an enzyme inhibitor, or a protein-protein interaction blocker, the calculation of the binding energy of potential hits may help prioritize compounds for experimental testing.

While docking and scoring remain the most widely used computational techniques to predict the binding mode and affinity of a drug to its target due their low computational cost, these methods are not particularly accurate. More precise approaches utilize appropriate sampling of the molecular system generated beforehand with MD simulations as is required when estimating ensemble-averaged quantities like binding free energies. *End-point methods* such as linear interaction energy (LIE) and the molecular mechanics Poisson-Boltzmann Surface Area (MM/PBSA) technique, which rely only on appropriate samplings of the end states, i.e., the complex and possibly the free receptor and ligand, have intermediate efficiencies. The LIE method, originally introduced by Aqvist et al. [29], assumes that the binding free energy can be written as a linear combination of average interaction energies between the ligand

and the rest of the system (protein, water and ions). More explicitly, the binding free energy of the ligand is expressed as [30]:

$$\Delta G_{bind} = \alpha\langle\Delta E_{vdW}^{L-S}\rangle + \beta\langle\Delta E_{el}^{L-S}\rangle, \tag{3}$$

where $\langle\Delta E_{vdW}^{L-S}\rangle = \langle E_{vdW}^{L-S}\rangle_{bound} - \langle E_{vdW}^{L-S}\rangle_{unbound}$ refers to the change in van der Waals interactions between the bound and unbound states of the ligand. The averages stand for ensemble averages obtained from MD simulations whereas the L-S label indicates that the interaction energies are computed only between the ligand and the surroundings. Similarly, $\langle\Delta E_{el}^{L-S}\rangle = \langle E_{el}^{L-S}\rangle_{bound} - \langle E_{el}^{L-S}\rangle_{unbound}$ corresponds to the change in intermolecular electrostatic interactions between the bound and unbound states. Parameters α and β are generally obtained empirically using an appropriate fitting procedure.

Alternatively, MM/PBSA [31], which is arguably the most popular end-point method, turned out to be successful in a number of drug-design case studies [32–34]. The method basically provides an estimate of the binding free energy as:

$$\Delta G_{bind} = \langle\Delta E_{MM}\rangle - T\Delta S + \Delta G_{solv}, \tag{4}$$

where $\langle\Delta E_{MM}\rangle - T\Delta S$ can be regarded as the change in the free energy of the system in vacuum (gas phase); it includes the change in the molecular mechanics energy due to the binding $\langle\Delta E_{MM}\rangle = \langle E_{MM}\rangle_{bound} - \langle E_{MM}\rangle_{unbound}$ and the change in the conformational entropy ΔS, usually estimated from normal mode analysis (NMA) performed on the complex structure and on the free ligand and protein structures. As in the LIE method, every average quantity corresponds to an ensemble average obtained from output MD trajectories. The entropy contribution, which is relatively time-consuming and inaccurate to compute using NMA, can be neglected if a comparison of states of similar entropy is desired such as in the case of comparing two or more ligands binding to the same protein binding site. Finally, ΔG_{solv} stands for the difference of solvation free energies due to the binding, it is given as $\Delta G_{solv} = \Delta G_{solv}^{complex} - \Delta G_{solv}^{lig} - \Delta G_{solv}^{prot}$ where every term on the right-hand side is given as the sum of polar and nonpolar contributions. The polar parts are obtained by solving the Poisson-Boltzmann (PB) equation or by using the Generalized-Born (GB) model (as in the MM/GBSA method) whereas the nonpolar terms are estimated from a linear relation to the solvent accessible surface area (SASA). Despite the fact that MM/PBSA and MM/GBSA are computationally-inexpensive methods, they contain several crude and questionable approximations, e.g., due to the use of implicit solvent models to compute the solvation energies [35]. The capability of the MM/PBSA method to predict the correct binding free energy turns out to be more sensitive to the investigated system compared to the MM/GBSA method, the latter being more useful in multi-target comparisons [36]. Noticeably, the MM/PBSA and GBSA techniques can be used to perform per-residue-free-energy decompositions. The benefit of such decompositions is twofold: providing important information about residues which significantly contribute to the binding energy (hot spots) and giving insights into the changes in binding free energies due to mutations, especially single point mutation.

2.4 Alchemical Free Energy Calculations

Another important category of MD-based methods used to estimate the binding free energy of ligand-protein complexes is called alchemical free energy methods, which include, for example, free energy perturbation (FEP) and thermodynamic integration (TI). Alchemical techniques make use of a parameter-dependent Hamiltonian to smoothly switch between the dynamics of systems with different chemical compositions. Applied to drug design, one may think about connecting the bound and unbound states of a given protein-ligand complex or transforming one ligand to another one within the same binding pocket. In the simplest case, such connections can be achieved by a linear combination of the corresponding Hamiltonians:

$$H(\boldsymbol{x}, \boldsymbol{p}; \lambda) = (1 - \lambda)H_a(\boldsymbol{x}, \boldsymbol{p}) + \lambda H_b(\boldsymbol{x}, \boldsymbol{p}), \tag{5}$$

where λ is a parameter which varies from 0 to 1, and, H_a and H_b are the physical Hamiltonians associated with the two states a and b, e.g., the bound and unbound systems. The FEP and TI approaches differ in the way of estimating the free energy difference $\Delta G_{a \to b}$ between states a and b. Within the FEP framework, $\Delta G_{a \to b}$ is given by the following identity:

$$\Delta G_{a \to b} = -k_B T \ln \left\langle \exp\left(-\frac{(H_b(\boldsymbol{x}, \boldsymbol{p}) - H_a(\boldsymbol{x}, \boldsymbol{p}))}{k_B T} \right) \right\rangle_a, \tag{6}$$

where $\langle \ldots \rangle_a$ stands for an ensemble average over configurations representative of the initial state a. In contrast, the free energy difference in the TI approach is computed from:

$$\Delta G_{a \to b} = \int_0^1 \left\langle \frac{\partial H(\boldsymbol{x}, \boldsymbol{p}; \lambda)}{\partial \lambda} \right\rangle_\lambda d\lambda. \tag{7}$$

Since both FEP and TI identities are based on Boltzmann equilibrium averages, which require extensive sampling of the complex and free ligand in solution, such techniques are generally computationally extensive. Other types of methods involving averages over non-equilibrium trajectories can, however, be used. This is the case of steered molecular dynamics (SMD) [37]. Applied to drug binding studies, SMD introduces a non-conservative force used to pull out the ligand from its binding site at constant speed [38]. The free energy difference during this non-equilibrium process can be obtained using the Jarzynski equality [39]:

$$\Delta G_{a\to b} = -k_B T \, \ln\left\langle \exp\left(-\frac{W_{a\to b}}{k_B T}\right)\right\rangle. \tag{8}$$

$W_{a\to b}$ represents the external work performed on the system during one steered simulation whereas $\langle \ldots \rangle$ corresponds to a non-equilibrium average performed over all the steered trajectories.

2.5 Enhanced Sampling Methods

The above-discussed alchemical methods including SMD can be integrated into a more general class of method called enhanced sampling methods [40]. As mentioned above, MD is a very useful and inexpensive tool to study the behavior of macromolecular systems in silico. However, the use of MD is limited by the computational time required to carry out a reasonable length of simulation. This, in turn, is dependent on the availability of computational hardware and time allocation for high-performance computing. Depending on the time required for a biomolecular system to reach equilibrium, MD can be run long enough to represent the evolution of a system from a few nanoseconds to a few microseconds. Long time-scale dynamical processes, such as slow conformational changes, protein assembly or ligand-protein binding processes, are notoriously difficult to model by MD.

In the past few decades, many computational methods have been developed based on MD simulations to explore relevant transitions between stables states of a system. In some of these techniques, a biased potential is added to the dynamics along one or a few collective variables. Such methods require to carefully choose the collective variables in such a way to facilitate the diffusion towards critical unexplored regions of the configuration space. The resulting trajectories are eventually unbiased to obtain the equilibrium distribution as a function of the collective variables used. Popular techniques performing MD with a biased potential along specific collective variables include metadynamics and umbrella sampling, both of them being related to a number of successful applications to biological systems.

In metadynamics, a positive Gaussian potential is added at regular time intervals to the free energy landscape of a system [41]. In this way, the system is discouraged to come back to the previous point, thus favoring the exploration of yet unvisited values of the collective variable. As more and more Gaussians are added to the true potential, the system is able to diffuse more freely along this variable. Papers on metadynamics research have reported a lot of successful applications to biological systems [42, 43]. As an important application, metadynamics—combined with nudged elastic band—was used to investigate the unbinding process of a ligand away from its specific protein target and the estimate of the associated binding energy [44].

Umbrella sampling is an enhanced sampling technique where a series of biased MD simulations are conducted independently. This works by splitting the reaction coordinate into a series of windows and applying a harmonic potential which acts to force the reaction coordinate to remain close to the center of each window. Assuming the biased trajectories overlap in the space defined by the collective variable, the resulting biased free energy landscapes can be used to compute the true free energy profile. This last step is performed by means of the weighted histogram analysis method (WHAM), which combines the information of the biased simulations so as to minimize the statistical error made on the resulting probability distribution [45]. In the context of drug-binding studies, umbrella sampling was used to estimate the potential of mean force for ion permeation and ligand binding to ion channels [46] and to predict protein–ligand binding structures in kinase systems [47].

3 Other Computational Drug-Discovery Methods

This section is intended to provide an account of other popular useful techniques for drug design and VS studies that can be used instead or in conjunction with MD-based methods discussed in the previous section.

3.1 Binding Pocket Prediction

Identifying and characterizing a suitable binding pocket in a 3D protein structure is a central aspect of any drug discovery study. This step is also relevant to shed light on biomolecular functions as many proteins are biologically functional only after interacting with cofactors or other biological molecules.

A common way to define a binding pocket, if a ligand is already bound to it, is to introduce a distance cut-off. Binding pocket atoms are typically defined whenever their distance to the ligand is below 4–8 Å. Following is a list of physicochemical key properties of binding pockets:

- The solvent-accessible surface area (SASA) which is usually computed as the atoms of the pocket reachable by a solvent probe sphere rolling over the protein surface.
- The volume of the pocket and its depth which corresponds to the average distance of the pocket atoms to their nearest water molecules from bulk solvent [48].
- The pocket hydrophobicity which depends on polar and non-polar residues involved in the binding site.
- The number of hydrogen bond donors and acceptors on the pocket surface.
- The conservation of residues over similar binding pockets of other proteins, which is particularly relevant for functional sites.

In addition to experimental binding site detection techniques such as NMR-based methods [49], a number of computational methods can be found which may be helpful in the process of binding pocket identification of a molecular target. Pocket finder algorithms are usually tested and validated on protein and ligand datasets. Such tests are intended to estimate the reliability of identifying the correct binding pocket within the first one to three hits provided by an algorithm. Two popular publicly-available databases are the Protein Data Bank (PDB) [50], which provides 3D protein structures for input into the pocket finding algorithms, and the PDBbind database [51], which contains bound protein structures filtered from the PDB database. The current version of PDBbind has around 3100 protein-ligand complexes, 1300 of these having been manually selected to form the refined set with the focus on the quality of structures and binding data. Due to its large size and manual curation, the refined set of the PDBbind database provides a suitable benchmark for most case studies. Further reduced from this, is the core set of 210 complexes. Optimal databases for pocket prediction testing should include high-resolution, diverse and non-redundant protein-ligand complexes. Pocket finder algorithms are generally split into two classes, namely, geometric-based and energetic-based approaches.

Geometry-based algorithms have the advantage of a low computational cost. The underlying assumption behind such methods is that the ligand binding pocket corresponds to the larger cleft within the protein structure [52, 53]. Therefore, geometrical criteria may be sufficient to identify the correct binding location on a protein. One such example, SURFNET [54] is an early-developed program which fits spheres between pairs of atoms so that they do not contain more than one atom. The binding pocket is defined as the volume containing the largest number of adjacent spheres. An improvement of the program, called SURFNET-ConSurf [55], refines the binding pocket prediction also considering the residue conservation within the binding site. The SURFNET-ConSurf algorithm was tested on a set of 244 non-redundant, diverse and representative ligand-protein complexes, obtained by a filtered version of the PDB database. A 75% rate of successfully recognized native ligand pockets is reported in the original paper about this method [55].

Another algorithm called VisGrid [56] is based on geometrical hashing and identifies cavities by considering the visibility of each point in a 3D grid, that is, the fraction of directions that are not blocked by protein atoms. In this way, a cluster of closely-located grid points with limited visibility indicates a pocket. VisGrid was compared with other pocket prediction methods, including SURFNET and LIGSITE, and the observed success rates on a set of bound and unbound structures were comparable with existing methods.

LIGSITE [57] uses a grid-based method in which points are either assigned to the solvent or protein category, and cavities are defined as groups of points in which solvent points are surrounded by protein points. Although the LIGSITE original validation identified the correct binding pocket in all the testing cases, a big limitation of this study was the reduced size of the dataset, with only ten ligand-receptor complexes. Its extension, LIGSITE^{csc} [58], improves the original algorithm by calculating more accurately the contact between protein surface and solvent using the Connolly surface, and by re-ranking the identified pockets by their degree of residue conservation in homolog proteins. The LIGSITE^{csc} testing process is more significant than the LIGSITE one, and a comparison with other geometry-based methods is also provided. The success rates calculated on a set of 210 non-redundant bound structures were 75% for LIGSITE^{csc}, 65% for LIGSITE and 42% for SURFNET.

The algorithm also showed good performances in recognizing the correct binding pocket in unbound structures.

Another class of methods dedicated at identifying binding pockets are *energy-based methods* which rely on the energetic properties of a binding site. A common approach of these methods is to use molecular probes to search for favorable interaction sites on a protein, and cluster them together to identify putative pockets. An early effort resulted in the GRID Fortran code [59]. The probes employed by this algorithm include water, methyl group, the hydroxyl, amine nitrogen and carboxy oxygen. Energetic contours are calculated with a function considering a 12-6 Lennard-Jones term, an electrostatic term and a hydrogen bond term, and negative energy levels indicate promising interaction sites for each probe.

Laurie and Jackson's Q-SiteFinder method [60] calculates the interaction energy of a methyl probe and the grid points generated on the protein structure. A clustering analysis step links favorable interaction sites to rank putative binding pockets based on their total interaction energy. Q-SiteFinder was tested on a diverse set of bound and unbound protein conformations, resulting in success rates of 74% and 71%, respectively.

EasyMIFs and SITEHOUND [61] are two complementary energy-based tools developed at the Sanchez lab. The first algorithm calculates the interaction energy between grid points and molecular probes using the GROMOS force field, while SITEHOUND recognizes putative binding sites by filtering and clustering the spatial variation of the interaction energy fields calculated by EasyMIFs or any other grid-based program. Multiple probes are used, as well as different site clustering algorithms. SITEHOUND's success rate was evaluated on a set of 77 complexes and it was reported as 95% (bound structures) and 79% (unbound structures) considering the binding pocket identified when present in the top three ranked sites.

Another similar energy-based algorithm, AutoLigand [62], was created by the developers of the popular molecular docking software Autodock. AutoLigand was reported to have a success rate of 73% when tested on a set of 187 bound structures and 80% when tested on 96 unbound structures.

3.2 Ligand-Receptor Docking

Molecular docking methods have been developed to predict how a given compound naturally binds to its biomolecular target, i.e., its binding mode, and to provide an estimate of its binding affinity. Docking software usually rely on optimization algorithms which include both a search algorithm and a scoring function. Such methods require at least one ligand structure and one target structure as inputs. The location of the targeted site should be provided although blind docking approaches [63, 64] can help deal with unknown binding locations in addition to the pocket prediction methods discussed in the previous section.

The search algorithms are dedicated to exhaustively explore the conformational space of the ligand within the targeted pocket. Three groups of such functions have been described, namely matching, systematic and stochastic methods. *Matching algorithms* are based on shape complementarity between the ligand and the receptor site, and, possibly, chemical complementarity. *Systematic search algorithms* explore the degrees of freedom of the ligand in a progressive way. This class of methods can be divided in three subgroups:

- Exhaustive algorithms systematically rotate all ligand dihedral angles until an optimal solution is reached.
- Fragment-based methods break down the ligand into different fragments which are separately placed within the binding site and re-connected in the last step of the process.
- Ensemble-based methods pre-generate a large number of ligand conformations, which are then rigidly placed within the binding site.

The last class of search algorithms includes *stochastic methods* such as MC and evolutionary algorithms, which introduce random changes of the degrees of freedom of the ligands to rapidly reach an optimal solution.

In molecular docking, the binding free energy is calculated using a position-dependent scoring function. This is required not only to identify the correct binding pose corresponding to the lowest binding energy, but also to rank a set of tested compounds according to their affinity to a target. *Force-field-based scoring functions* used for docking are similar to MD force fields discussed in Sect. 2.1. In *empirical scoring functions*, the different contributions in the binding energies are weighted with coefficients, set beforehand to reproduce experimental dissociation constants of known ligand-receptor complexes. The Autodock4 scoring function [65] is an example of an empirical scoring function where a non-bonded interaction potential is calculated as

$$V = W_{vdW} \sum_{ij} \left(\frac{A_{ij}}{r_{ij}^{12}} - \frac{B_{ij}}{r_{ij}^{6}} \right) + W_{hb} \sum_{ij} E(\theta) \left(\frac{C_{ij}}{r_{ij}^{12}} - \frac{D_{ij}}{r_{ij}^{10}} \right)$$
$$+ W_{elec} \sum_{ij} \frac{q_i q_j}{\varepsilon(r_{ij}) r_{ij}} + W_{sol} \sum_{i,j} (S_i V_j + S_j V_i) e^{\frac{-r_{ij}^2}{2\sigma^2}}. \quad (9)$$

The first term represents van der Waals contributions, the second is a hydrogen bond term, the third is a Coulombic term for electrostatic interactions, and the last is a desolvation potential. W represents the empirical coefficients, obtained from the training over 188 bound complexes from a PDB calibration subset. The coefficients A and B derived from the AMBER force field, while C and D are Autodock internal parameters. $E(\theta)$ depends on the angle of deviation θ from the ideal hydrogen bond geometry. S and V are a solvation parameter and the volume of the atoms surrounding one atom, respectively. σ is distance-weighting factor of Autodock. The total Autodock score of a binding pose is calculated by summing the difference of intra-molecular energies between the bound and unbound forms of the ligand and the protein, then subtracting the difference of inter-molecular energies. A simple entropic term is also included in the final score to model the variation of the system entropy upon binding. *Knowledge-based scoring functions* use the potential of mean force (PMF), derived from protein-ligand structures and calculated for each ij ligand-receptor atom pair type as:

$$w_{ij}(r) = -k_B T ln(\rho(r_{ij})/\rho^*(r_{ij})), \quad (10)$$

where k_B is the Boltzmann constant, T is the temperature in Kelvin degrees, $\rho(r_{ij})$ is the number density of the ij atom pair derived from the structural training set and $\rho^*(r_{ij})$ is the number density in a reference state. Although knowledge-based scoring functions do not provide a precise interaction energy potential due to difficulties arising from the reference state calculation, they directly connect the atomic interactions between ligand and protein to structural data instead of to kinetics as is the case for empirical methods. Knowledge-based scoring functions turn out also to be more computationally efficient than force-field-based methods [66, 67]. Finally, *consensus scoring methodologies* combine different scoring function outputs of the same ligand to obtain a single, consensual score. Different combination strategies can be employed, such as weighting and summing up the ranks or performing a regression analysis [68].

Accounting for the flexibility of the binding site is an important issue in molecular docking. As a result, different approaches have been proposed to address this problem. Soft docking algorithms use modified short-range repulsion parameters for the binding site atoms, which allow the ligand to slightly penetrate through the surface of the pocket to mimic the induced fit of the binding. Many algorithms also include the possibility to treat the side chains of pocket residues as flexible, although such methods still ignore backbone dynamics, while increasing noticeably the computational cost. Using multiple receptor conformations when performing molecular docking is a popular way to take into account the backbone flexibility [69]. This approach, called Relaxed Complex Scheme (RCS), relies on NMR or MD-derived conformational ensembles, which are used as molecular docking targets.

Testing of docking software is usually performed by evaluating the percentage of docking poses with small enough RMSD (typically 2 Å) compared to the co-crystallized poses extracted from a high-resolution structural database.

The original implementation of DOCK [70] is an example of geometry-matching algorithm, where the binding site and the ligand atoms are represented as spheres that are systematically matched using a shape-based routine. DOCK 6 [71], the latest version of the program, applies a fragment-based algorithm and a set of different force-field-based scoring functions which can be selected, such as PBSA, GBSA and Amber scoring methods. In addition, a minimization step is performed for the ligand in order to remove minor protein-ligand clashes and relax its internal geometry. The success rate of the latest DOCK release was estimated around 73% in reproducing crystallographic poses. The authors tested the algorithm on 1043 structures obtained from a ligand-receptor database designed as a benchmark for assessing docking software performances [72].

Autodock [73], probably the most popular docking software, uses a Lamarckian genetic algorithm to independently generate a large number of binding poses, scored with the empirical scoring function described in Eq. (9). A clustering algorithm can be optionally used to identify the most populated portion of the conformational space of the ligand, from which the lowest energy pose should match the native one. The success rate of the latest version, Autodock4 [74], was around 53% when the software was tested on the calibration structural set.

Autodock Vina [75] utilizes an iterated local search global optimizer searching method. The Vina scoring function combines aspects from knowledge-based and empirical potentials. Tested on the same set used for Autodock4, Vina was able to identify the correct binding pose in 78% of the cases. Noteworthy, Vina scoring function was trained with the PDBbind refined set, much bigger than the training set used for the Autodock scoring function.

Glide [76] is a docking software included within the Schrödinger molecular modeling package. It is based on an exhaustive systematic search algorithm used to sample the ligand conformational space, followed by a minimization step. An optimal choice for the scoring method [76] is given as a combination of a force-field-based function, an empirical function (GlideScore) and the strain energy of the ligand conformation. The pose success rate was reported around 66% when tested on 282 ligand-receptor complexes selected from the PDB database.

GOLD [77] is another popular docking program. The software maps together complementary chemical features of the ligand and the receptor within the binding site. A genetic algorithm is then used to explore different binding modes. Three main scoring functions are available, namely, Goldscore (force-field-based), Chemscore (empirical) and Astex Statistical Potential [78] (knowledge-based). Testing of GOLD on the CCDC/Astex database [79], a PDB subset designed to test docking software, resulted in success rates up to 87% depending on the scoring function used. The correlation coefficients (R^2) between experimentally-measured and GOLD binding affinities were reported between 0.51 and 0.55.

3.3 Virtual Screening

The discovery of a new drug is an expensive and long process. It is estimated that up to two and half billion dollars and twenty years are required to bring a new product from the bench to the clinic [80, 81]. Consequently, efforts are made to shorten the process and reduce the cost. Some of the time and funding savings are expected to result from a wider use of computational techniques applied to drug discovery. One of such techniques is called virtual screening (VS). VS refers to an in silico active compound search against biomolecular targets [82]. It has the advantage of being fast and inexpensive compared with traditional high throughput screening. Nowadays, libraries including billions of compounds can be virtually screened depending on the available computational resources.

A typical VS workflow consists of sequential series of filtering and scoring steps aimed at providing a set of promising compounds for experimental validation. VS methods can be divided in ligand-based VS (LBVS) and structure-based VS (SBVS). LBVS approaches are computationally faster but they do not provide any estimate of the binding energy of the ligand. On the other hand, SBVS methods are more computationally expensive but they enable to rank potential hits based on their predicted binding affinity. Regardless of the chosen approach, a compound database is always required as starting point for VS. Examples of extensive small molecule repositories are PubChem [83, 84], ZINC [85] and the National Cancer Institute databases [86]. Such collections usually include millions of compounds that can be downloaded for screening purposes [87].

LBVS techniques rely solely on the 2D or 3D structure of ligands, ignoring the biological target. The main assumption behind these methods is that structurally related compounds share similar activities [88]. Therefore, the structure of at least one known active compound should be available as template for the computational search, and a measure of the distance between structures needs to be computed. Simple ways to represent the chemical structure of a compound in a computer-readable format are chemical fingerprints or pharmacophore representations. Chemical fingerprints [89] are binary strings in which each bit codes for the presence or the absence of particular chemical groups. A widely-used way to compare two fingerprints is to use the Tanimoto index, given by

$$T_{A,B} = \frac{c}{a + b - c}, \tag{11}$$

where A and B are the two fingerprints, c is the number of bits set to 1 at the same position in both the fingerprints, and, a and b are the total number of bits set to 1 in A and B, respectively. A publicly-available package which can be used for fingerprint-based VS is chemfp [90].

Another way to perform LBVS is to use a pharmacophore model of the active compounds [91], which provides a representation of the ligand from its spatially-distributed chemical features (hydrogen bond acceptor, hydrogen bond donor, hydrophobic moiety, ring structure, polar or charged residue) including the distances between centers forming a chemical structure. In pharmacophore-based VS, the distance between two ligand structures is usually calculated as the RMSD between the superposed pharmacophore points. The main benefit of this approach is to identify molecules with different chemical groups but similar generic features, providing novel scaffolds to medicinal chemistry. Contrary to most chemical fingerprints, pharmacophore models also include 3D properties of the ligand.

Data mining and machine learning methods including support vector machines, neural networks, Bayesian networks and decision trees, are also utilized for LBVS [92]. LBVS methods are useful in case a 3D structure of the target is not available, but they can also be used to clean up large databases in order to generate focused libraries [93]. Indeed, these structurally-related subsets are designed to interact with a specific target and they are built by screening larger and diverse databases. Focused libraries have a limited size compared with the parent databases. Therefore, they can be rapidly and efficiently screened with SBVS or experimental techniques. LBVS methods have led to the discovery of novel and promising compounds with low-range potency [94]. Examples of such successes are the discovery of anti-cancer tubulin dimerization inhibitors [95], inhibitors of the 17β-HSD2 enzyme for osteoporosis treatment [96] and novel scaffolds for the inhibition of the HIV-1 integrase [97].

SBVS methods provides a ranking of the screened compounds based on their computed binding affinities. Therefore, one or multiple structures of the target are required. SBVS always relies on molecular docking methods, which are used to place the compounds within the targeted pocket, and to estimate the binding affinity from the resulting binding poses. We have already discussed all the docking-related aspects in Sect. 3.2. The docking scoring functions are designed to quickly estimate the binding energy from a ligand pose, therefore they often do not lead to very accurate results. Several strategies have been developed to deal with this, including the already mentioned consensus scoring, MD simulations of the complex structures and/or more accurate scoring functions (e.g., MM/PBSA or GBSA) [98, 99]. Recently, SBVS techniques were successfully applied to the discovery of DNA repair inhibitors [98, 100–103], anti-malarian compounds [104], kinase inhibitors [105] and HIV-1 inhibitors [106, 107].

4 Conclusions

This review chapter provides introductory information regarding the computational tools currently used in the drug design and discovery process. We have given an overview of molecular dynamics methods that are very useful in biomolecular target characterization for drug action. We have also given practical information regarding the identification of binding pockets for putative inhibitors of proteins, as well as an overview of molecular docking techniques that are based on the protein-ligand interactions. These interactions and their ranking involving the binding free energy of the ligand-target pair are used in massive searches for specific and selective inhibitors of particular protein, a methodology referred to as virtual screening. The latter methodology relies on large and diverse databases of pharmacologically-acceptable compounds. Lists of databases and software packages used in all stages of computational drug design have been presented in this chapter to assist in practical aspects of research in this area.

Acknowledgements The authors are grateful to the Natural Sciences and Engineering Research Council of Canada, the Li Ka Shing Institute of Applied Virology and the Alberta Cancer Foundation for funding support.

References

1. A. Vitalis, R.V. Pappu, Annu. Rep. Comput. Chem. **5**, 49 (2009)
2. M. Karplus, Acc. Chem. Res. **35**, 321 (2002)
3. W.L. Jorgensen, J. Tirado-Rives, J. Phys. Chem. **100**, 14508 (1996)
4. D.A. Pearlman, D.A. Case, J.W. Caldwell, W.S. Ross, T.E. Cheatham, S. DeBolt, D. Ferguson, G. Seibel, P. Kollman, Comput. Phys. Commun. **91**, 1 (1995)
5. S. Pronk, S. Pall, R. Schulz, P. Larsson, P. Bjelkmar, R. Apostolov, M.R. Shirts, J.C. Smith, P.M. Kasson, D. van der Spoel, B. Hess, E. Lindahl, Bioinformatics **29**, 845 (2013)
6. S. Plimpton, J. Comput. Phys. **117**, 1 (1995)
7. J.C. Phillips, R. Braun, W. Wang, J. Gumbart, E. Tajkhorshid, E. Villa, C. Chipot, R.D. Skeel, L. Kalé, K. Schulten, J. Comput. Chem. **26**, 1781 (2005)
8. W.D. Cornell, P. Cieplak, C.I. Bayly, I.R. Gould, K.M. Merz, D.M. Ferguson, D.C. Spellmeyer, T. Fox, J.W. Caldwell, P.A. Kollman, J. Am. Chem. Soc. **117**, 5179 (1995)
9. V. Hornak, R. Abel, A. Okur, B. Strockbine, A. Roitberg, C. Simmerling, Proteins **65**, 712 (2006)
10. A.D. MacKerell, D. Bashford, R.L. Dunbrack, J.D. Evanseck, M.J. Field, S. Fischer, J. Gao, H. Guo, S. Ha, D. Joseph-McCarthy, L. Kuchnir, K. Kuczera, F.T.K. Lau, C. Mattos, S. Michnick, T. Ngo, D.T. Nguyen, B. Prodhom, W.E. Reiher, B. Roux, M. Schlenkrich, J.C. Smith, R. Stote, J. Straub, M. Watanabe, J. Wiórkiewicz-Kuczera, D. Yin, M. Karplus, J. Phys. Chem. B **102**, 3586 (1998)
11. M. Fujihashi, T. Ishida, S. Kuroda, L.P. Kotra, E.F. Pai, K. Miki, J. Am. Chem. Soc. **135**, 17432 (2013)
12. G. Tiwari, D. Mohanty, PLoS One **8** (2013)
13. J.A. Tuszynski, C. Wenger, D.E. Friesen, J. Preto, Int. J. Environ. Res. Public Health **13** (2016)
14. D. Zahn, J. Mol. Model. **17**, 1531 (2011)
15. N.T. Wood, E. Fadda, R. Davis, O.C. Grant, J.C. Martin, R.J. Woods, S.A. Travers, PLoS One **8** (2013)
16. A. Chaudhuri, I. Sarkar, S. Chakraborty, J. Biomol. Struct. Dyn. **32**, 1969 (2014)
17. G. Leonis, T. Steinbrecher, M.G. Papadopoulos, J. Chem. Inf. Model. **53**, 2141 (2013)
18. T. Yoda, Y. Sugita, Y. Okamoto, Biophys. J. **99**, 1637 (2010)
19. T. Yoda, Y. Sugita, Y. Okamoto, Proteins-Struct. Funct. Bioinform. **82**, 933 (2014)
20. B.G. Dick, A.W. Overhauser, Curr. Contents/Phys. Chem. Earth Sci. **24** (1985)
21. B. Kirchner, P.J. di Dio, J. Hutter, Multiscale Mol. Methods Appl. Chem. **307**, 109 (2012)
22. D. Marx, J. Hutter, Mod. Methods Algorithms Quantum Chem. **1**, 141 (2000)
23. A. Warshel, M. Levitt, J. Mol. Biol. **103**, 227 (1976)
24. E. Brunk, U. Rothlisberger, Chem. Rev. **115**, 6217 (2015)
25. V. Botu, R. Ramprasad, Int. J. Quantum Chem. **115**, 1074 (2015)
26. A.A. Hassanali, J. Cuny, V. Verdolino, M. Parrinello, Philos. Trans. A Math. Phys. Eng. Sci. **372**, 20120482 (2014)
27. S.Y. Hu, H.J. Yu, Y.J. Liu, T. Xue, H.B. Zhang, J. Mol. Model. **19**, 3087 (2013)
28. W. Hu, S.W. Deng, J.Y. Huang, Y.M. Lu, X.Y. Le, W.X. Zheng, J. Inorg. Biochem. **127**, 90 (2013)
29. J. Aqvist, C. Medina, J.E. Samuelsson, Protein Eng. **7**, 385 (1994)
30. H. Guitierrez-de-Teran, J. Aqvist, Comput. Drug Discov. Des. **819**, 305 (2012)
31. P.A. Kollman, I. Massova, C. Reyes, B. Kuhn, S. Huo, L. Chong, M. Lee, T. Lee, Y. Duan, W. Wang, O. Donini, P. Cieplak, J. Srinivasan, D.A. Case, T.E. Cheatham, Acc. Chem. Res. **33**, 889 (2000)
32. F. Chen, H. Liu, H.Y. Sun, P.C. Pan, Y.Y. Li, D. Li, T.J. Hou, Phys. Chem. Chem. Phys. **18**, 22129 (2016)
33. J.M. Sanders, M.E. Wampole, M.L. Thakur, E. Wickstrom, PLoS One **8** (2013)
34. T. Zhu, H. Lee, H. Lei, C. Jones, K. Patel, M.E. Johnson, K.E. Hevener, J. Chem. Inf. Model. **53**, 560 (2013)

35. S. Genheden, U. Ryde, Expert Opin. Drug Discov. **10**, 449 (2015)
36. H.Y. Sun, Y.Y. Li, S. Tian, L. Xu, T.J. Hou, Phys. Chem. Chem. Phys. **16**, 16719 (2014)
37. B. Isralewitz, M. Gao, K. Schulten, Curr. Opin. Struct. Biol. **11**, 224 (2001)
38. J.S. Patel, A. Berteotti, S. Ronsisvalle, W. Rocchia, A. Cavalli, J. Chem. Inf. Model. **54**, 470 (2014)
39. S. Park, F. Khalili-Araghi, E. Tajkhorshid, K. Schulten, J. Chem. Phys. **119**, 3559 (2003)
40. R.C. Bernardi, M.C.R. Melo, K. Schulten, Biochim. Biophys. Acta Gen. Subj. **1850**, 872 (2015)
41. A. Laio, F.L. Gervasio, Reports. Prog. Phys. **71**, 126601 (2008)
42. A. Barducci, R. Chelli, P. Procacci, V. Schettino, F.L. Gervasio, M. Parrinello, J. Am. Chem. Soc. **128**, 2705 (2006)
43. F.L. Gervasio, A. Laio, M. Parrinello, J. Am. Chem. Soc. **127**, 2600 (2005)
44. D. Branduardi, F.L. Gervasio, M. Parrinello, J. Chem. Phys. **126**, 54103 (2007)
45. M. Souaille, B. Roux, Comput. Phys. Commun. **135**, 40 (2001)
46. T. Baştuğ, P.-C. Chen, S.M. Patra, S. Kuyucak, J. Chem. Phys. **128**, 155104 (2008)
47. H. Kokubo, T. Tanaka, Y. Okamoto, J. Chem. Theory Comput. **9**, 4660 (2013)
48. S. Chakravarty, R. Varadarajan, Structure **7**, 723 (1999)
49. P.J. Hajduk, J.R. Huth, S.W. Fesik, J. Med. Chem. **48**, 2518 (2005)
50. H.M. Berman, Nucleic Acids Res. **28**, 235 (2000)
51. R. Wang, X. Fang, Y. Lu, C.Y. Yang, S. Wang, J. Med. Chem. **48**, 4111 (2005)
52. R.A. Laskowski, N.M. Luscombe, M.B. Swindells, J.M. Thornton, Protein Sci. **5**, 2438 (1996)
53. X. Zheng, L. Gan, E. Wang, J. Wang, AAPS J **15**, 228 (2013)
54. R.A. Laskowski, J. Mol. Graph. **13**, 323 (1995)
55. F. Glaser, R.J. Morris, R.J. Najmanovich, R.A. Laskowski, J.M. Thornton, Proteins Struct. Funct. Bioinform. **62**, 479 (2005)
56. B. Li, S. Turuvekere, M. Agrawal, D. La, K. Ramani, D. Kihara, Proteins Struct. Funct. Bioinform. **71**, 670 (2008)
57. M. Hendlich, F. Rippmann, G. Barnickel, J. Mol. Graph. Model. **15**, 359 (1997)
58. B. Huang, M. Schroeder, BMC Struct. Biol. **6**, 19 (2006)
59. P.J. Goodford, J. Med. Chem. **28**, 849 (1985)
60. A.T.R. Laurie, R.M. Jackson, Bioinformatics **21**, 1908 (2005)
61. D. Ghersi, R. Sanchez, Bioinformatics **25**, 3185 (2009)
62. R. Harris, A.J. Olson, D.S. Goodsell, Proteins Struct. Funct. Bioinform. **70**, 1506 (2007)
63. C. Hetényi, D. van der Spoel, Protein Sci. **11**, 1729 (2002)
64. C. Hetényi, D. van der Spoel, FEBS Lett. **580**, 1447 (2006)
65. R. Huey, G.M. Morris, A.J. Olson, D.S. Goodsell, J. Comput. Chem. **28**, 1145 (2007)
66. P.D. Thomas, K.A. Dill, J. Mol. Biol. **257**, 457 (1996)
67. Z. Zheng, K.M. Merz Jr., J. Chem. Inf. Model. **53**, 1073 (2013b)
68. M. Feher, Drug Discov. Today **11**, 421 (2006)
69. R.E. Amaro, R. Baron, J.A. McCammon, J. Comput. Aided Mol. Des. **22**, 693 (2008)
70. I.D. Kuntz, J.M. Blaney, S.J. Oatley, R. Langridge, T.E. Ferrin, J. Mol. Biol. **161**, 269 (1982)
71. W.J. Allen, T.E. Balius, S. Mukherjee, S.R. Brozell, D.T. Moustakas, P.T. Lang, D.A. Case, I.D. Kuntz, R.C. Rizzo, J. Comput. Chem. **36**, 1132 (2015)
72. S. Mukherjee, T.E. Balius, R.C. Rizzo, J. Chem. Inf. Model. **50**, 1986 (2010)
73. G. Morris, D. Goodsell, J. Comput. Chem. 1639 (1998)
74. G.M. Morris, R. Huey, W. Lindstrom, M.F. Sanner, R.K. Belew, D.S. Goodsell, A.J. Olson, J. Comput. Chem. **30**, 2785 (2009)
75. O. Trott, A.J. Olson, J. Comput. Chem. **31**, 455 (2010)
76. R.A. Friesner, J.L. Banks, R.B. Murphy, T.A. Halgren, J.J. Klicic, D.T. Mainz, M.P. Repasky, E.H. Knoll, M. Shelley, J.K. Perry, D.E. Shaw, P. Francis, P.S. Shenkin, J. Med. Chem. **47**, 1739 (2004)
77. M.L. Verdonk, J.C. Cole, M.J. Hartshorn, C.W. Murray, R.D. Taylor, Proteins **52**, 609 (2003)
78. W.T.M. Mooij, M.L. Verdonk, Proteins Struct. Funct. Bioinform. **61**, 272 (2005)

79. J.W.M. Nissink, C. Murray, M. Hartshorn, M.L. Verdonk, J.C. Cole, R. Taylor, Proteins Struct. Funct. Bioinform. **49**, 457 (2002)
80. J. Avorn, N. Engl, J. Med. **372**, 1877 (2015)
81. M. Dickson, J.P. Gagnon, Discov. Med. **4**, 172 (2009)
82. A. Lavecchia, C. Di Giovanni, Curr. Med. Chem. **20**, 2839 (2013)
83. E.E. Bolton, Y. Wang, P.A. Thiessen, S.H. Bryant, Annu. Rep. Comput. Chem. **4**, 217 (2008)
84. X.-Q. Xie, Expert Opin. Drug Discov. **5**, 1205 (2010)
85. T. Sterling, J.J. Irwin, J. Chem. Inf. Model. **55**, 2324 (2015)
86. NCI Compound Sets (The National Cancer Institute (NCI), Bethesda, MD, United States of America, 2017), https://wiki.nci.nih.gov/display/NCIDTPdata/Compound+Sets. Accessed 20 July 2015
87. S. Cosconati, S. Forli, A.L. Perryman, R. Harris, D.S. Goodsell, A.J. Olson, Expert Opin. Drug Discov. **5**, 597 (2010)
88. M.A. Johnson, G.M. Maggiora, *Concepts and Applications of Molecular Similarity* (Wiley, 1990)
89. P. Willett, Drug Discov. Today **11**, 1046 (2006)
90. A. Dalke, J. Cheminform. **5**, 36 (2013)
91. D. Horvath, Methods Mol. Biol. **672**, 261 (2011)
92. J. Melville, E. Burke, J. Hirst, Comb. Chem. High Throughput Screen. **12**, 332 (2009)
93. C.J. Harris, R.D. Hill, D.W. Sheppard, M.J. Slater, P.F.W. Stouten, Comb. Chem. High Throughput Screen. **14**, 521 (2011)
94. P. Ripphausen, B. Nisius, J. Bajorath, Drug Discov. Today **16**, 372 (2011)
95. Y.-K. Chiang, C.-C. Kuo, Y.-S. Wu, C.-T. Chen, M.S. Coumar, J.-S. Wu, H.-P. Hsieh, C.-Y. Chang, H.-Y. Jseng, M.-H. Wu, J.-S. Leou, J.-S. Song, J.-Y. Chang, P.-C. Lyu, Y.-S. Chao, S.-Y. Wu, J. Med. Chem. **52**, 4221 (2009)
96. A. Vuorinen, R. Engeli, A. Meyer, F. Bachmann, U.J. Griesser, D. Schuster, A. Odermatt, J. Med. Chem. **57**, 5995 (2014)
97. A. Kurczyk, D. Warszycki, R. Musiol, R. Kafel, A.J. Bojarski, J. Polanski, J. Chem. Inf. Model. **55**, 2168 (2015)
98. K.H. Barakat, L.P. Jordheim, R. Perez-Pineiro, D. Wishart, C. Dumontet, J.A. Tuszynski, PLoS ONE **7**, e51329 (2012)
99. D.C. Thompson, C. Humblet, D. Joseph-McCarthy, J. Chem. Inf. Model. **48**, 1081 (2008)
100. F. Gentile, J.A. Tuszynski, K.H. Barakat, J. Mol. Graph. Model. **65**, 71 (2016)
101. F. Gentile, J.A. Tuszynski, K.H. Barakat, Curr. Pharm. Des. **22**, 3527 (2016)
102. L.P. Jordheim, K.H. Barakat, L. Heinrich-Balard, E.-L. Matera, E. Cros-Perrial, K. Bouledrak, R. El Sabeh, R. Perez-Pineiro, D.S. Wishart, R. Cohen, J. Tuszynski, C. Dumontet, Mol. Pharmacol. **84**, 12 (2013)
103. E.M. McNeil, K.R. Astell, A.-M. Ritchie, S. Shave, D.R. Houston, P. Bakrania, H.M. Jones, P. Khurana, C. Wallace, T. Chapman, M.A. Wear, M.D. Walkinshaw, B. Saxty, D.W. Melton, DNA Repair (Amst). **31**, 19 (2015)
104. R.R. Nunes, M.D.S. Costa, B.D.R. Santos, A.L. da Fonseca, L.S. Ferreira, R.C.R. Chagas, A.M. da Silva, F.P. de Varotti, A.G. Taranto, Mem. Inst. Oswaldo Cruz **111**, 721 (2016)
105. D. Bajusz, G. Ferenczy, G. Keserű, Curr. Top. Med. Chem. **17**, 2235 (2017)
106. W.-G. Gu, X. Zhang, J.-F. Yuan, AAPS J **16**, 674 (2014)
107. N. Li, R.I. Ainsworth, B. Ding, T. Hou, W. Wang, J. Chem. Inf. Model. **55**, 1400 (2015)

Macroscopic Models for the Bioelectronic Interface of Engineered Artificial Membranes

William Hoiles and Vikram Krishnamurthy

Abstract In this chapter, we present the formation and modeling techniques of two important sensing devices built out of engineered artificial membranes: the Ion Channel Switch (ICS) biosensor, and the Electroporation Measurement Platform (EMP). The ICS biosensor can be used to detect femto-molar concentrations of analyte species in an electrolyte solution, and the EMP is used to study the dynamics of electroporation in engineered membranes. The engineered membrane in the ICS and EMP are design to mimic the electrophysiological properties of real cell membranes. Common to both platforms is the bioelectronic interface for performing electrical measurements. Experimental measurements of the two platforms are performed by estimating the current response of the engineered membrane which depends on the charging dynamics at the bioelectronic interface and membrane, as well as dynamics of aqueous pores and conducting ion-channels in the membrane.

Keywords Tethered membranes · Biosensors · Electroporation · Dynamical models · Ion channel switch

1 Introduction

Biological membranes are ubiquitous from the cell membrane, which encompasses a cell, to membranes for organelles within the cell. In this chapter we focus on engineered artificial membranes and their construction and modeling. We use the term "engineered artificial membranes" to reflect the fact that we will precisely engineer various components of the engineered artificial membrane and model the dynamics of the resulting membrane.

W. Hoiles (✉)
The University of British Columbia, Kaiser Building, 2332 Main Mall,
Vancouver, BC V6T 1Z4, Canada
e-mail: whoiles@ece.ubc.ca

V. Krishnamurthy (✉)
Cornell University and Cornell Tech, 2 West Loop Road, New York, NY 10044, USA
e-mail: vikramk@cornell.edu

L. L. Bonilla et al. (eds.), *Coupled Mathematical Models for Physical and Biological Nanoscale Systems and Their Applications*, Springer Proceedings in Mathematics & Statistics 232, https://doi.org/10.1007/978-3-319-76599-0_15

The cell membrane provides an interface between the cell and its environment as illustrated in Fig. 1. The extracellular fluid contains all components that are not contained in the cell, and the cytosol is composed of all components which are contained in the cell. Cell membranes are dynamic and respond to their environment as a result of an external stimulus. The membrane is composed of three primary components:

1. Lipids. Lipids are composed of a hydrophilic head (attracted to water) and hydrophobic tails (water repulsive) and are the primary components of the cell membrane. Lipids separate the extracellular fluid from the cytosol of the cell.
2. Macromolecules. Macromolecules provide the primary method for cellular communication both in cells and between cells. Given the low permeability of biological membranes, the primary task of macromolecules is to transport molecules from the extracellular fluid to the interior of the cell. Additionally there exist other macromolecules which act as molecular receptors, detect vesicles carrying cellular cargo, and facilitate the transport of other macromolecules across the membrane from the extracellular fluid to the cytosol, or from the cytosol to the extracellular fluid.
3. Cytoskeleton. Cytoskeletal filaments provide structural support for the membrane and determine/modulate the shape of the cell.

The development of novel drugs and therapeutic protocols for effective drug delivery require knowledge of the dynamic processes present in cell membranes which depend on the properties of the lipids, macromolecules, and cytoskeletal filaments.

Why Artificial Membranes? Biological cells can be grown in the laboratory, so why do we need to construct artificial membranes to study the dynamics of cell membranes? The primary reason is that cell membranes contain thousands of protein macromolecules, peptides, lipids, and cytoskeletal filaments which contribute to the dynamics of the cell membrane. If we are interested in studying the affect of a specific protein macromolecule on the dynamics of the cell membrane then it is difficult as we can not isolate if any changes in the membrane were a result of this macromolecule, or the reactions of other proteins and peptides present in the membrane. The aim of an engineered artificial membranes is to create a biomimetic[1] system that mimics the cell membrane and also allows us to study specific biological processes in an experimentally controllable environment.

In this chapter, we discuss two important classes of engineered artificial membrane architectures, namely the Ion Channel Switch (ICS) biosensor, and the Electroporation Measurement Platform (EMP) which are illustrated in Fig. 2. The ICS biosensor is used to detect specific analyte molecules in the electrolyte solution, and the EMP is used to study the process of aqueous pore formation and destruction in engineered membranes. Additionally, we construct fractional order macroscopic models of these membrane devices. Using the fractional order macroscopic models with experimental measurements from these devices allows important biological parameters to be

[1] A biomimetic system is a physical device that contains the features of a biological system. This is similar to an in vivo system in which biological molecules are used outside their normal biological environment.

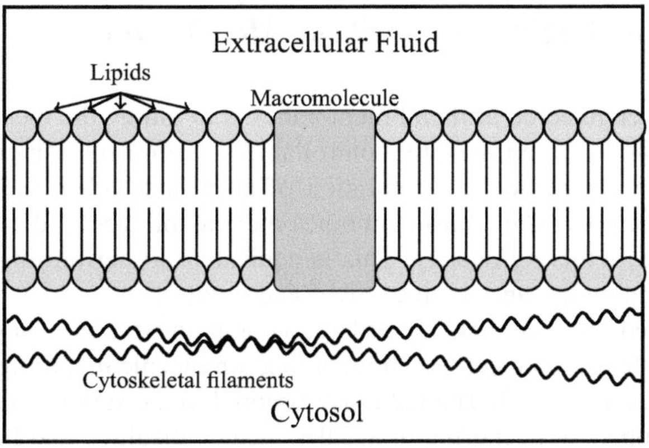

Fig. 1 Schematic of a cell membrane. The extracellular fluid represents the content outside the cell, and the cytosol is the interior of the cell with the membrane separating the two domains. Note that cell membranes have several different types of lipids, macromolecules, and cytoskeletal filaments components. Additionally the cytosol contains a host of other organelles (specialized structures within a living cell) which are not illustrated. It is important to note that the cell membrane is a dynamic system, the individual lipid layers slide on each other

estimated such as the concentration of specific analyte species, and the conductance of the membrane.

2 Engineered Artificial Membranes

In this section we introduce engineered artificial membranes and how they can be used to construct biosensors and sensing platforms. Recall that in this chapter the specific engineered artificial membrane platforms we consider are provided in Fig. 2.

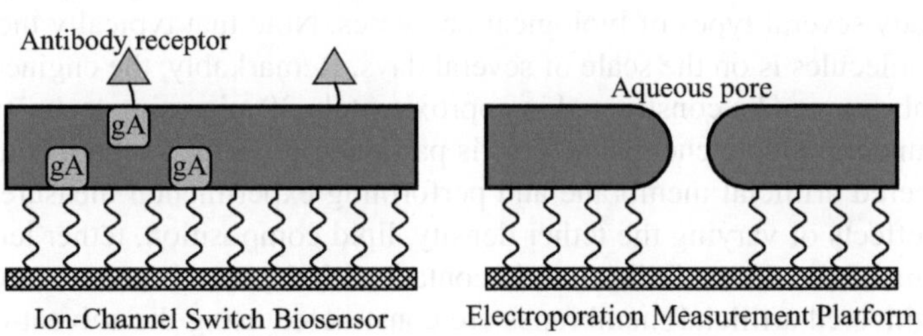

Fig. 2 A schematic diagram of the ICS and EMP engineered artificial membrane devices. The tethered membrane is depicted in gray, and the gold interface by the crosshatch pattern. The unifying theme of the two devices is the use of an inert gold bioelectronic interface and an engineered tethered membrane

2.1 Primer of Engineered Artificial Membranes

The purpose of engineered artificial membranes is to mimic the membranes of living organisms in an experimentally controllable environment. Parameters that are desirable to control include: concentration of electrolyte, chemical molecules in the electrolyte, temperature, transmembrane potential, bilayer lipid membrane composition, and the tethering density. This is equivalent to performing experimental measurements on membranes both ex vivo[2] and in vitro.

All engineered artificial membranes have a common structure: they use a surface tethered molecule that connects to a lipid or macromolecule in the engineered membrane as illustrated in Fig. 3. The tether is anchored to the surface via an anchoring molecule. Examples of anchoring molecules include disulfide and lipoic acid. The surface can be composed of various components including gold, mercury, platinum. Note that silver is never used as it ablates off, and platinum is not typically used as it is not easy to anchor molecules to. The only requirement is that the anchor molecule can bind to the surface. The tethers are designed to provide sufficient space between the engineered membrane and surface support to eliminate the effects of frictional coupling between the surface and lipids. That is, the tethers ensure a tethering reservoir exists that contains a sufficient abundance of water to allow ions to freely cross aqueous pores in the membrane, and ensure the lipids do not interact directly with the surface.

The use of tethers is particularly useful as the tethered lipids or macromolecules mimic the response of the cytoskeletal filaments found in cellular membranes. Therefore the dynamics of the lipids in the engineered membrane do not suffer from the effects of frictional coupling. The key advantage of using tethers is they allow the incorporation of a relatively large amount of water per lipid between the membrane and the supporting surface, and promote the isolation of the membrane from the surface minimizing any membrane distortion. Additionally only tethering some lipids or macromolecules in the engineered membrane can allow a wide range of macromolecule to be inserted into the engineered membrane. The tethers also allow the engineered membrane to have a life-time of several months. This provides sufficient time to study several types of biological processes. Note that typically the lifetime of macromolecules is on the scale of several days. Remarkably, the engineered artificial membrane can be constructed in approximately 20 min using techniques from first-year undergraduate chemistry. This is particularly useful for both constructing the engineered artificial membrane and performing experimental measurements to study the effects of varying the tether density, lipid composition, tether length, and composition of the extracellular fluid in contact with the membrane surface.

All engineered artificial membranes are constructed using the solvent-exchange technique provided in [1].

---

[2]Ex vivo means which take place outside the normal membrane environment in a cell, but with minimal alterations from the natural conditions of the membrane.

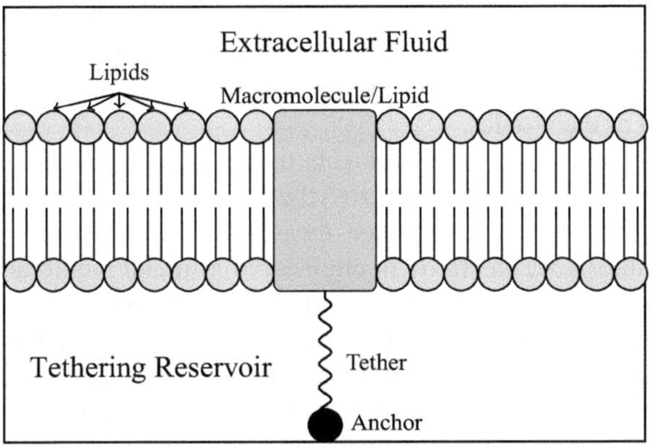

Fig. 3 Schematic diagram illustrating an engineered artificial membrane. The engineered artificial membrane contains a membrane composed of lipids and other molecules (not shown), and a tethered macromolecule or lipid. The tethered molecule contains three components: an anchor which binds the tether to the surface, a tether, and either a macromolecule or lipid that connects the engineered membrane to the tether. The extracellular fluid is any fluid outside the tethering reservoir. Note that the tether can be attached molecules that both span half the lipid bilayer, or the full lipid bilayer as illustrated

2.2 *Experimental Measurements with Engineered Artificial Membranes*

To build a measurement device using engineered artificial membranes requires a *bioelectronic interface* which connects the biological system with electronic instrumentation. The bioelectronic interfaces can be composed of silicon, gold, platinum, or other metals that do not impede the function of the tethered membrane. Figure 2 illustrates the bioelectronic interface of the ICS biosensor and EMP. The bioelectronic interface we consider in this chapter is composed of a gold electrode coated with polyethylene spacers and tethers. Gold electrodes are used as they are biologically inert–that is, the gold electrodes do not react with peptides, proteins, or the lipid bilayer. Another benefit of using gold electrodes is that their oxidation potential is approximately 0.9 V. In biological systems, signal communication is performed using ionic charge carriers, whereas in electronic systems, information is conveyed via electrons. Therefore constructing a biolelectronic interface has several challenges including:

- the diffuse three dimensional distribution of ions in solution relative to the delocalized property of electrons in metals;
- the slow kinetics of ions in solution compared to the relatively instantaneous kinetics of electrons in metals.

These effects have been recruited by nature to be the controlling phenomenon in the time dependent currents in cellular action potentials, in the propagation of action potentials along nerve fibers, and in the transduction mechanisms of the many sensory functions found in sentient life. The bioelectronic interface is the physical barrier that must be understood and controlled to permit communication between the complex ephemeral systems of biology and the far faster, better defined, and more robust inorganic world of electronics.

To perform experimental measurements using engineered artificial membranes involves applying a voltage potential across the gold electrode bioelectronic interface, and a counter gold electrode, and then measuring the current response from the device. The counter gold electrode in engineered artificial membranes is hundreds of micro-meters away from the surface of the membrane. Note that the counter electrode does not contain any spacers, tethers, or lipid membrane. The current response contains information about the charging of the electrical double layer at the gold electrodes, and the conductance of the tethered membrane. To estimate important biological parameter of the tethered membrane using the current response measurements requires a dynamic model of the engineered artificial membrane. For both the ICS biosensor and EMP we construct dynamic models that can be used to estimate important biological parameters.

3 Ion Channel Switch Biosensor

The Ion Channel Switch (ICS) biosensor [1, 2] is a fully functioning nano-machine device constructed out of an engineered membrane with moving parts comprising of gramicidin (gA) monomers and conducting gA dimer channels. Gold electrodes constitute the bioelectronic interface between the electrical instrumentation and the electrolyte solution. The ICS biosensor can detect femto-molar (10×10^{-15} m) concentrations of target species including proteins, hormones, polypeptides, microorganisms, oligonucleotides, DNA segments, and polymers in cluttered electrolyte environments, and has a lifetime of several months. The ICS biosensor has also been used in clinical trials for the detection of Influenza A.

A schematic of the ICS is given in Fig. 4. Useful for experimentalists is the fact the ICS can be designed with specific binding sites with the membrane having a lifetime of several months [1–7]. The engineered membrane is composed of a self-assembled monolayer of mobile lipids and gA monomers, and a self-assembled monolayer of mobile lipids and gA monomers. The tethered components are anchored to the gold electrode *via* polyethylene glycol chains. Spacer molecules are used to ensure the tethers are evenly spread over the gold electrode. The intrinsic spacing between tethers and spacers is maintained by the benzyl disulphide moieties which bond the spacers and tethers to the electrode surface. A time-dependent voltage potential is applied between the electrodes to induce a transmembrane potential of electrophysiological interest; this results in a current $I(t)$ related to the charging of the electrical double-layers and the conductance of the engineered membrane.

Using mathematical models of the ICS biosensor with experimental measurements, it is possible to estimate important biological parameters from the ICS biosensor such as the concentration of analyte present.

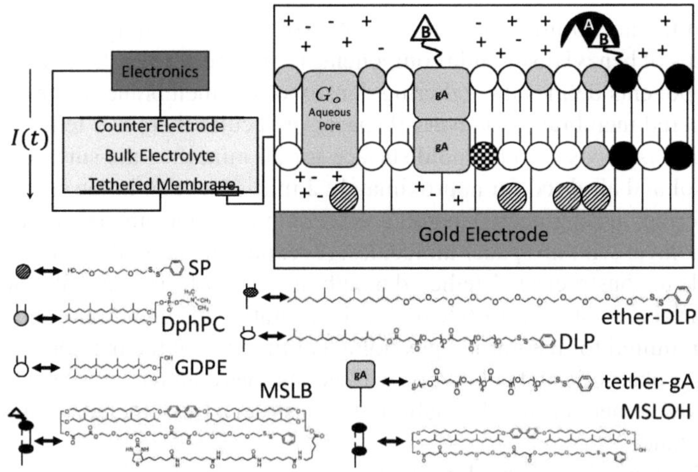

Fig. 4 Overview of the Ion Channel Switch (ICS) biosensor and molecular components. The "Electronics" block represents the electronic instrumentation that produces the drive potential between the gold electrode and gold counter electrode, and records the current response $I(t)$. G_o is a transient aqueous pore. The conducting gramicidin (gA) dimer is shown and is composed of two gA monomers. *A* represents the analyte species, and *B* the analyte receptor. MSLOH denotes synthetic archaebacterial membrane-spanning lipids, DLP half-membrane-spanning tethered lipids, DphPC and GDPE mobile half-membrane-spanning lipids, MSL membrane spanning lipid, and SP a spacer

3.1 Formation of Ion Channel Switch Biosensor

The engineered membrane of the ICS is supported by a 25 × 75 × 1 mm polycarbonate slide. Six 100 nm thick sputtered gold electrodes, each with dimensions 0.7 × 3 mm, rest on the polycarbonate slide. Each electrode is in an isolated flow cell with a common gold return electrode. The formation of the tethered membrane on the gold electrode is performed in two stages using a solvent-exchange technique [1, 8, 9].

Stage 1: The first stage of formation involves anchoring of the tethers and spacers to the gold surface. The tethers provide structural integrity to the membrane and mimic the physiological response of the cytoskeletal supports of real cell membranes. The spacers laterally separate the tethers allowing patches of mobile lipids to diffuse in the membrane. The tethers and spacers both contain benzyl disulphide components (i.e. MSLOH, DLP, ether-DLP, tether-gA, MSLB, and SP in Fig. 4). The benzyl disuphide bonds to the gold surface with the disulphide bond maintained [1, 10]. This bonding structure has been detected experimentally from X-ray photoelectron spectra. The use of the disulphide has the advantage that the thiols do not oxidize on storage allowing the membrane to have a lifetime of several months. From experimental measurements, the electrodesorption of the thiol to gold bond is negligible for electrode potentials below 800 mV [11].

To form the anchoring layer, an ethanolic solution containing 370 μM of engineered ratios of benzyl disulphide components is prepared. The ratio of benzyl disulphide components defines the tethering density of the membrane. For example, for a 10% tethered membrane, for every 9 spacer molecules there is 1 tether molecule. This solution is exposed to the gold surface for 30 min, then the surface is flushed with ethanol and air dried for approximately 2 min. Note that in the special case of 100% tethering, the engineered membrane is composed of a tethered archaebacterial based monolayer with no spacer molecules. As experimentally illustrated in [3], it is not possible to construct a 0% tethered membrane as any formed membrane binds to the gold surface. As the electrolyte reservoir separating the membrane and electrode surface is required for the normal physiological function of the membrane, and noting that all prokaryotic and eukaryotic cell membranes contain cytoskeletal supports with a 1–10% tether density, the inability to construct a 0% tethered membrane is of little importance.

Stage 2: The second stage involves the formation of the tethered membrane. A solution containing a mixture of mobile lipids is brought into contact with the gold bonded components from Stage 1. Several lipid solvents can be used [1, 4]; however, in most cases the lipids selected to form the bilayer are soluble in ethanol. As an example, let us consider the formation of a 70% DphPC (zwittrionic C20 diphytanyl-ether-glycero-phosphatidylcholine) and 30% GDPE (C20 diphytanyl-diglyceride ether). mixed tethered membrane. 8 μL of 3 mM of the 70% DphPC and 30% GDPE ethanolic solution is added to the flow chamber. The ethanol solution also contains biotinylated gA monomers (tether-gA in Fig. 4) with the biotin linked to the gA monomer via a 5-aminocaproyl linker. The solution is incubated for 2 min at 20 °C in which the tethered membrane forms. After the 2 min incubation period, 300 μL of phosphate buffered saline is flushed through each flow chamber. The tethered membrane is equilibrated for 30 min prior to performing any experimental measurements. For the detection of streptavidin a biotin receptor is used. The associated antibody fragments used to detect ferritin, TSH, and hCG are the anti-ferritin F_{ab}, thyrotropin binding inhibitory immunoglobulin, and immunoglobulin G respectively. Details on how the antibodies are connected to the MSLB and mobile gA monomers, as illustrated schematically in Fig. 4, is provided in [1].

3.2 *Fractional Order Model to Estimate Membrane Conductance*

To estimate important biological parameters from the ICS biosensor and EMP requires a method to estimate the conductance of the tethered membrane from current response measurements. In this section, a fractional order macroscopic model is constructed which can be used predict the current response of the engineered membrane given the tethered membrane conductance. If the predicted current and experimentally measured current are in agreement, this indicates that the given membrane

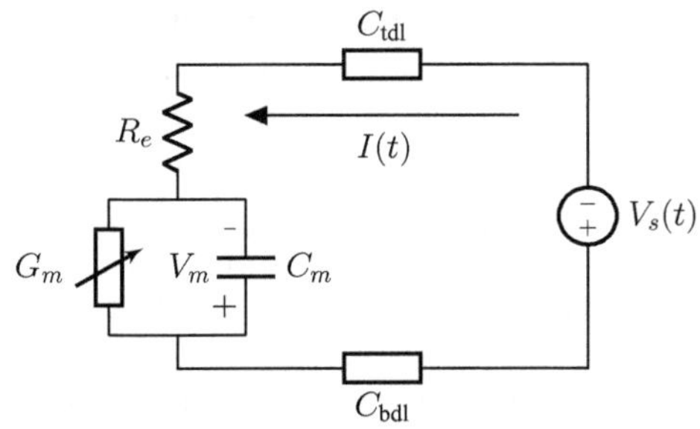

Fig. 5 Lumped circuit macroscopic model of tethered membrane platform. The parameter R_e is the electrolyte resistance, C_m is the membrane capacitance, G_m is the membrane conductance, C_{bdl} is the electrode double-layer capacitance, C_{tdl} is the counter electrode double-layer capacitance, and V_m is the tethered transmembrane potential. Note that C_{tdl} and C_{bdl} are modeled using fractional order elements to account for the charge accumulation, diffusion-limited charge transfer, reaction limited charge transfer, and ionic adsorption dynamics at the electrode and counter electrode surfaces

conductance is equivalent to the conductance of the engineered membrane. Fractional order operators are utilized in the model as the gold surface bioelectronic interface of the tethered membrane may contain diffusion-limited charge transfer, reaction limited charge transfer, and ionic adsorption dynamics. These double-layer charging effects can be modeled using fractional order operators.

The fractional order macroscopic model of the engineered artificial membrane is illustrated schematically in Fig. 5, and is given by the following system of fractional order differential equations

$$\begin{aligned}
\frac{dV_m}{dt} &= -\left(\frac{1}{C_m R_e} + \frac{G_m}{C_m}\right)V_m - \frac{1}{C_m R_e}V_{dl} + \frac{1}{C_m R_e}V_s,\\
\frac{d^p V_{dl}}{dt^p} &= -\frac{1}{C_{dl} R_e}V_m - \frac{1}{C_{dl} R_e}V_{dl} + \frac{1}{C_{dl} R_e}V_s \qquad 0 < p \leq 1,\\
I(t) &= \frac{1}{R_e}\big(V_s - V_m - V_{dl}\big),
\end{aligned} \tag{1}$$

where C_{dl} is the total electrode capacitance of C_{tdl} and C_{bdl} in series with p denoting the order of the fractional order operator, V_m the transmembrane potential, and V_{dl} the double-layer potential. Note that if $p < 1$ then the SI units of C_{dl} are $\frac{\mathrm{s}^{p+3}\mathrm{A}^2}{\mathrm{m}^2\mathrm{kg}}$.

The fractional order macroscopic model of the engineered artificial membrane accounts for both the charging dynamics of the membrane and the restricted diffusion dynamics at the electrode surface.

Typical Lumped Circuit Parameter Values: A high quality membrane with negligible defects will have a low conductance G_m in the range of 0.04–1.00 μS/mm^2, and an associated membrane capacitance C_m in the range of 4–10 nF/mm^2. The gold electrodes typically have a double-layer capacitance C_{dl} between 47 and 85 nF/mm^2. Note that C_{dl} is the series capacitance of the electrode capacitance C_{bdl}, and the counter electrode C_{tdl}. The fractional order parameter p for the tethered membrane is in the range of 0.85 to 1.00 which illustrates that fractional order dynamics are present at the electrode surface of the engineered artificial membrane. Finally, the electrolyte resistance R_e is in the range of 47–380 Ω/mm^2. If the membrane contains significant defects then the values of G_m and C_m will be outside these values.

The two primary domains that the fractional order macroscopic model accounts for are the membrane domain (G_m and C_m), and the electrode double-layer domain (C_{dl} and p). Here we discuss each of these important domains.

Membrane Domain G_m, C_m: The membrane is assumed to be polarizable and to also contain aqueous pores as a result of random thermal fluctuations—that is, random thermal fluctuations allow the energy barrier to be crossed for the conformational change of lipids allowing the formation of transient aqueous pores. This allows the tethered membrane to be modeled by an effective permittivity with capacitance C_m in parallel with the tethered membrane conductance $G_m(t)$ [3, 12, 13]. G_m depends on the population of and size of aqueous pores present as well as the population and conductance dynamics of embedded ion channels. For example, in the ICS biosensor the conductance of the membrane G_m depends on the population of conducting gramicidin A (gA) dimers present.

Electrode Double-Layer Domain C_{dl}, p: The electrode double-layer domain models the charge build-up that results at the electrode to electrolyte interface that results from an applied excitation potential $V_s(t)$. This double-layer domain is typically modeled using Warburg impedance elements [14–16]. For engineered artificial membranes however we use a generalized Warburg impedance as the classic Warburg impedance does not account for all the chemical processes present at the electrode to electrolyte interface. This includes diffusion-limited charge transfer, reaction limited charge transfer, and ionic adsorption dynamics [17]. Note that the generalized Warburg impedance is also known as a constant-phase-element and is composed of a capacitance C_{dl} and a fractional order operator p. The current to voltage relationship at the electrode surface is

$$C_{dl}\frac{d^p V_{dl}}{dt^p} = I(t) \tag{2}$$

where $I(t)$ is the current passing through the double-layer capacitance C_{dl}. For $p = 1$, (2) this is the classical current-voltage relation for an ideal capacitor which accounts for charge accumulation. If $p = 0.5$ then (2) represents the Warburg impedance. In general, for $0 < p < 1$ then (2) can be used to account for diffusion-limited charge transfer, reaction limited charge transfer, and ionic adsorption dynamics. Insight into the charging behaviour of the double-layer capacitance can be obtained if we consider the impedance representation of (2) in the complex domain. The impedance $Z_{dl}(\omega)$ of (2) is

$$Z_{dl}(\omega) = \frac{1}{C_{dl}(j\omega)^p} = \left(\frac{\cos(p\pi/2)}{C_{dl}\omega^p}\right) - j\left(\frac{\sin(p\pi/2)}{C_{dl}\omega^p}\right) \tag{3}$$

where $j = \sqrt{-1}$ is the complex number, and $\omega = 2\pi f$ is the angular frequency with f the frequency in hertz. The magnitude $|Z_{dl}(\omega)|$ and angle $\angle Z_{dl}(\omega)$ of the double-layer impedance (3) is

$$|Z_{dl}(\omega)| = \frac{1}{C_{dl}\omega^p}, \qquad \angle Z_{dl}(\omega) = -p\frac{\pi}{2}. \tag{4}$$

If $p = 1$, then (3) is the impedance of an ideal capacitor which has an angle of $\angle Z_{dl}(\omega) = -\pi/2$. However, for $0 < p < 1$ the associated angle increases such that $\angle Z_{dl}(\omega) > -\pi/2$. Using impedance measurements of engineered artificial membranes, the typical values of p are in the range of 0.85 to 1 suggesting that charge accumulation, diffusion-limited charge transfer, reaction limited charge transfer, and ionic adsorption dynamics are present at the electrode-to-electrolyte interface. Therefore, the generalized Warburg impedance, given by (3) in the complex domain or (2) in the time-domain, is used to account for these chemical processes.

Using the tools of fractional calculus, the fractional derivative in (2) can be evaluated using either the Riemann-Liouville, Caputo, or Grnwald-Letnikov definition of the fractional order derivative.[3] The Riemann-Liouville definition of the fractional order operator in (2) for $p \in (0, 1]$ is

$$\frac{d^p V_{dl}}{dt^p} = \frac{1}{\Gamma(1-p)}\frac{d}{dt}\int_0^t V_{dl}(u)(t-u)^{-p}du \tag{5}$$

where $\Gamma(\cdot)$ is the Gamma function. As seen in (5), the fractional order operator is not local. This means that the value of $d^p V_{dl}/dt^p$ depends on all the values of V_{dl} in the time interval from $[0, t]$. This dependency on time allows the fractional order operator to model path-dependent phenomena such as sub-diffusion processes. In engineered artificial membranes the electrode surface is composed of tethers, spacers, and may contain irregularities that impede the flow of ions in the electrolyte. This impedance of flow causes the electrolyte dynamics to violate Brownian motion (e.g. the mean squared displacement of particles is not proportional to the diffusion times time, or the relative change in position of an ion between time increments are not independent). These type of long-range memory, or path-dependent phenomena can be accounted for using the fractional order operator. For further details on fractional calculus and its importance for modeling nanobiotechnological processes, the reader is referred to [21].

Estimation of Model Parameters: The model parameters in the fractional order macroscopic model (1) can be estimated by minimizing the mean-squared error

[3]The evaluation of the fractional order derivative can be performed using the Adomian decomposition method [18] which is a popular semi-analytical method for solving ordinary and non-linear partial differential equations. Another method is the Variational iteration method [19] typically used for numerically solving non-linear partial differential equations. Linear multistep methods, such as the Adams-Bashforth-Moulton [20], can also be used to evaluate the fractional order derivative.

between the measured current response $I(t)$ and predicted current response by fitting the parameters $C_{tdl}, C_{bdl}, p, C_m, G_m$, and R_e. Typically a low-voltage sinusoidal excitation $V_s(t)$ potential is used allowing (1) to be represented by an algebraic expression. That is, impedance measurements can be used to estimate $C_{tdl}, C_{bdl}, p, C_m, G_m$, and R_e as illustrated in [22]. The parameters C_{tdl}, C_{bdl}, p, C_m, R_e, and the equilibrium membrane conductance $G_m(0)$ (the conductance at a membrane potential $V_m = 0$) will remain constant throughout all experiments. The only variation in the membrane conductance G_m can result from the dynamics of gramicidin dimer association/dissociation, or if the excitation potential V_s is sufficiently larger then from reversible electroporation. If the values of C_{tdl}, C_{bdl}, p, C_m, $G_m(0)$, and R_e change dramatically from their equilibrium value this suggests that the membrane contains significant defects.

3.3 *Model of the Conductance of the Ion Channel Switch Biosensor*

In the previous section we illustrated how the conductance $G_m(t)$ of the ICS biosensor can be estimated from the current response measurements. In this section we construct a reaction-rate limited model for the ICS biosensor of use for estimating the analyte concentration from current response measurements from the biosensor. The model is constructed by combining the fractional order macroscopic model (1) with a system of non-linear differential equations which model the surface binding and surface reaction dynamics in the ICS biosensor. A two-time scale model is then constructed using singular perturbation theory to approximate the conductance of the tethered membrane as a function of the conducting dimer concentration and population of aqueous pores. We illustrate how the model and experimental measurements from the ICS biosensor can be used to estimate the concentration of the Human Chorionic Gonadotropin (hCG) pregnancy hormone.

The tethered membrane conductance $G_m(t)$ in the fractional order model (1) is dependent on the concentration of gA dimers present in the biosensor. As the number of gA dimers decreases, the conductance $G_m(t)$ increases. This results in the overall conductance of the biosensor decreasing as the concentration of gA dimers decreases. To link the membrane conductance G_m to the target analyte concentration requires a model that accounts for the electro-diffusive effects of the analyte in the electrolyte, and the surface reactions present on the tethered membrane surface of the ICS biosensor.

The chemical reactions taking place at the ICS biosensor involve analyte molecules binding to the tethered antibody sites followed by a cross-linking of the mobile gA monomers to the captured analytes. The primary species involved in this process include the analytes a, binding sites b, mobile gA monomers c, tethered gA monomers s, and the dimers d, with respective concentrations $\{A, B, C, S, D\}$. Other chemical complexes present include w, x, y, and z with concentrations $\{W, X, Y, Z\}$.

The chemical reactions that relate these chemical species are described by the following set of reactions [23]:

$$a+b \underset{r_1}{\overset{f_1}{\rightleftharpoons}} w \quad a+c \underset{r_2}{\overset{f_2}{\rightleftharpoons}} x \quad w+c \underset{r_3}{\overset{f_3}{\rightleftharpoons}} y \quad x+b \underset{r_4}{\overset{f_4}{\rightleftharpoons}} y$$
$$c+s \underset{r_5}{\overset{f_5}{\rightleftharpoons}} d \quad a+d \underset{r_6}{\overset{f_6}{\rightleftharpoons}} z \quad x+s \underset{r_7}{\overset{f_7}{\rightleftharpoons}} z. \tag{6}$$

In (6), r_i and f_i, for $i \in \{1, 2, 3, 4, 5, 6, 7\}$, denote the reverse and forward reaction rates for the chemical species $\{a, b, c, d, s, w, x, y, z\}$. The chemical reactions in (6) give a complete symbolic description of the operation of the ICS biosensor that was qualitatively described in Sect. 2 of the companion paper. The forward part of the first equation reports on an analyte molecule a being captured by a binding site b and the resulting complex is denoted by w. The third equation says that a free moving gramicidin monomer c in the outer leaflet of the tethered membrane, can bind to the complex w, thus producing another complex, denoted by y. An analyte molecule can also be captured by the binding site linked to the freely diffusing monomer, c. The second equation says that this results in the production of the complex, x. The complex x can still diffuse on the outer leaflet of the tethered membrane, and so can move towards a tethered binding site, b, and bind to it, resulting in the complex y (fourth equation). On the other hand the complex x can diffuse on top of the tethered ion channel monomer, s, which results in the production of complex z (seventh equation). The event that determines the biosensor conductance (and thus the current flowing through the biosensor) is the binding of the free moving ion channel monomer, c, and the tethered ion channel monomer, s. This results in the formation of a dimer, d (fifth equation). Indeed, the biosensor conductance is proportional to the dimer concentration, i.e., $G(t) = \text{constant} \times D(t)$. Finally, an analyte molecule can also bind to an already formed dimer, which again produces the complex z (sixth equation).

It is possible to convert the chemical reactions in (6) to a set of non-linear ordinary differential equations with reaction rates given by:

$$\begin{aligned} R_1 &= f_1AB - r_1W & R_5 &= f_4CS - r_5D \\ R_2 &= f_2AC - r_2X & R_6 &= f_6AD - r_6Z \\ R_3 &= f_3WC - r_3Y & R_7 &= f_7XS - r_7Z \\ R_4 &= f_4XB - r_4Y. \end{aligned} \tag{7}$$

Using (7), the time-evolution of the chemical species in the ICS biosensor can be determined from the non-linear ordinary differential equations given by:

$$\frac{du}{dt} = Mr(u(t)), \tag{8}$$

$$\text{where } M = \begin{pmatrix} -1 & 0 & 0 & -1 & 0 & 0 & 0 \\ 0 & -1 & -1 & 0 & -1 & 0 & 0 \\ 0 & 0 & 0 & 0 & 1 & -1 & 0 \\ 0 & 0 & 0 & 0 & -1 & 0 & -1 \\ 1 & 0 & -1 & 0 & 0 & 0 & 0 \\ 0 & 1 & 0 & -1 & 0 & 0 & -1 \\ 0 & 0 & 1 & 1 & 0 & 0 & 0 \\ 0 & 0 & 0 & 0 & 0 & 1 & 1 \end{pmatrix}$$

where the vector $u(t) = [B, C, D, S, W, X, Y, Z]^{\mathrm{T}}$ in (8) denotes the concentration of the chemical species, and $r(u) = [R_1, R_2, \ldots, R_7]^{\mathrm{T}}$ denotes the rate of reaction of the ith species in (6). The matrix M in (8) is the stoichiometry matrix relating u and $r(u)$. Note that (8) and (6) does not model the dynamics of the analyte concentration A. Note that in the reaction-rate limited regime the analyte concentration A are assumed negligible. If this were not the case then continuum theories for electrodiffusive flow would be required to model the dynamics of A.

The parameter $\psi(t)$ is the rate at which the analyte molecules are fixed by immobilized species at the electrode surface, which can be found using (6) and (8), and is given by:

$$\psi(t) = -A(f_1 B + f_2 C + f_6 D) + r_1 W + r_2 x + r_6 Z. \tag{9}$$

The total membrane conductance $G_m(t)$ is related to the gA ion channel concentration $D(t)$ by a constant, (i.e. $G(t) \propto D(t)$). Therefore, the time evolution of $G_m(t)$ is computed from the solution of Eqs. (1) and (9) given the initial concentration of the chemical species, reaction rates, diffusivity constant, flow velocity, and assuming we are in the reaction-rate limited regime. The reaction-rate limited regime of the ICS biosensor is satisfied if large analyte flow rates of mL/min, micro-molar analyte concentration, or low binding site densities less than $10^8/\mathrm{cm}^2$ are present.

3.4 Singular Perturbation Analysis of Dimer Concentration

If large analyte flow rates of mL/min, micro-molar analyte concentration, or low binding site densities less than $10^8/\mathrm{cm}^2$ are present, then the assumption that the analyte concentration is approximately constant over space and time is reasonable. In this case, the singular perturbation theory can be used to approximate the time evolution of the dimer concentration $D(t)$ which is related to the ICS biosensor conductance $G_m(t)$ [23]. For the parameter values in Table 2, the decay rate of species Y and Z is much faster than the decay rate of the other species by analysis of the eigenvalues of the linearized version of (8). Let us denote the parameters $\gamma = \{Y, Z\}$ for the fast species, and $\beta = \{B, C, D, S, W, X\}$ for the slow species. We can then represent (7) and (8) as a two-time scale system:

$$\frac{d\beta}{dt} = f(\beta, \gamma), \quad \varepsilon\frac{d\gamma}{dt} = g(\beta, \gamma) \tag{10}$$

where g and f denote vector fields of the fast variables and the slow variables, and $\varepsilon \approx \frac{1}{|\lambda_{max}|}$ with λ_{max} the largest eigenvalue of the linearized version of (8). Tikhonov's theorem combined with the approximation $S \approx S(0)$ allows for the simplification of the above two-time scale system of chemical dynamics equations (10) [24, Sect. 11.1]. The following theorem based on singular perturbation analysis provides an equation to evaluate the evolution of the biosensor conductance versus analyte concentration:

Theorem 1 ([23]) *Consider the two time scale system* (10) *for the dynamics of the chemical species. As* $\varepsilon \to 0$*, the dimer concentration* $D(t)$ *converges to* $\overline{D}(t)$ *given by the following system:*

$$\frac{d}{dt}\overline{D}(t) = -\overline{D}(t)(r_5 + f_6 A^*) + \big(f_5 C + \frac{r_6 f_7 X}{r_6 + r_7}\big)S(0) \tag{11}$$

with constants $r_5, r_6, r_7, f_5, f_6, f_7$ *defined in* (6), A^* *the analyte concentration, and S(0) the initial number of tethered gA monomers. Specifically, if the initial dimer concentration* $D(0)$ *at time* $t = 0$ *is within an* $O(\varepsilon)$ *neighbourhood of* $\gamma = h(\beta)$*, where* $h(\beta)$ *is the solution of the algebraic equation* $g(\beta, \gamma) = 0$ *for g given in* (10)*; then for all time* $t \in [0, T]$*,* $|D(t) - \overline{D}(t)| = O(\varepsilon)$ *where* $T > 0$ *denotes a finite time horizon.* □

The proof of Theorem 1 is given in [24]. It can be shown that as $\varepsilon \to 0$, the fast dynamics approach the quasi steady state, $h(\beta)$ defined in the theorem. This quasi steady state $h(\beta)$ of the fast variables β is then substituted in the slow dynamics in (10), which results in the following approximate dynamics for the slow species: $d\bar{\beta}/dt = f(\bar{\beta}, h(\bar{\beta}))$. We are interested in a specific component of $\bar{\alpha}$, namely, the approximate dimer concentration $\bar{D}$, which can be shown to evolve according to (11).

Using Theorem 1, the conductance of the ICS biosensor for different analyte concentrations A^* can be computed using (11) when the operation of the biosensor is in the reaction-rate-limited regime.

3.5 Detection of Human Chorionic Gonadotropin

To illustrate the application of the ICS and predictive model (11) with (1) for analyte concentration estimation in this section we compare the predicted membrane conductance with the experimentally measured membrane conductance for the analyte species human chorionic gonadotropin (hCG). The concentration of hCG is an

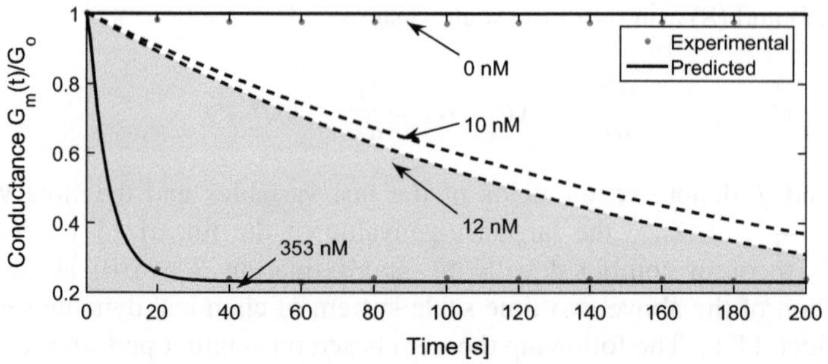

Fig. 6 Experimentally measured (gray dots) and numerically predicted concentration (black dots) of hCG present in an analyte solution using the ICS biosensor. The black lines for the 10 and 12 nM concentrations are associated with the a blastocyst or mammalian embryogenesis being present. Above 10 nM a mammalian embryogenesis is present; below 12 nM a mammalian embryogenesis is not present. Intermediate values can be associated with either a mammalian embryogenesis being present or not. All predictions are performed using the reaction-rate limited model (1) and (11) with the parameters defined in Table 2

excellent indicator of the presence of blastocyst or mammalian embryogenesis (i.e. pregnancy).

Figure 6 presents the measured and numerically predicted conductance of the ICS biosensor for the concentration estimation of hCG. For concentrations of hCG above 10 nM in blood or urine, this suggests a blastocyst or mammalian embryogenesis is present. The experimentally measured concentrations of hCG used are 0 and 353 nM. As seen from Fig. 6, these produce very different membrane conductance dynamics. From Fig. 6, at $t = 40$ s, the change in the number of conducting gA dimers is negligible. Therefore the membrane conductance is at its equilibrium value and results because only aqueous pores are present for $t > 40$ s. As seen from Fig. 6, using the reaction-rate limited dynamic model (11) in combination with the fractional order macroscopic model (1) it is possible to estimate the concentration of hCG present.

4 Electroporation Measurement Platform

In this section we introduce the Electroporation Measurement Platform (EMP), modeling techniques, and experimental measurements from the EMP for membranes containing cholesterol.

Experimental platforms to study electroporation include synthetic bilayer lipid membranes and in vitro cells. However, synthetic non-tethered bilayer lipid membranes do not provide a good representation of physiological systems since the effects caused by the cytoskeletal network are not present. All cells contain a cytoskeletal network that provides structural support for the biological membrane. Using cells provides a physiological system for validation; however, it is impossible to fully define the physiological environment which affects properties associated with electroporation. This motivates the need for an engineered artificial membrane platform which gives the experimentalist control over tethering density, membrane composition, and physiological environment unlike the synthetic lipid membrane and cell based platforms. The EMP is a engineered artificial membrane platform that provides a fully controllable physiological environment with a tethered membrane. The tethers in the EMP mimic the response of the cytoskeletal network in cells. Therefore the EMP can be used to model the dynamics of electroporation; however, a dynamic model is required to relate the experimental measurements from the EMP to important biological parameters such as aqueous pore density and size.

Models of the electroporation process employ the Smoluchowski-Einstein equation, derived from statistical mechanics, and the energy of a pore. The pore energy models are typically constructed by assuming the membrane is a dielectric and elastic continuum. However, the pore energy models can also be constructed from the results of coarse-grained molecular dynamics and all-atom molecular dynamics simulations. A key property that must be estimated is not only the pore energy, but also the conductance of an aqueous pore. In [13] the authors use the Generalized Poisson-Nernst-Planck (GPNP) continuum model to estimate the conductance of aqueous pores. The GPNP continuum is equivalent to the Poisson-Nernst-Planck system of equations, classically used to estimate electroylte dynamics, if steric effects are neglected. Note that near a pore entrance significant nonlinear potential gradients are present which restrict the current flowing through the pore, this effect is denoted as the "spreading conductance" and is dominant for pore radii significantly larger then the membrane thickness causing the pore conductance to scale proportionally to the pore radius. The pore conductance scales proportionally to the pore radius for all pore radii suggesting that the "spreading conductance" is dominant as a result of the electrode in proximity to the membrane surface and the non-linear potential gradients present. This effect only becomes pronounced for large pores in free floating membranes.

4.1 Modeling the Process of Electroporation

The Smoluchowski-Einstein equation [25–27] provides a useful model for estimating the number and size of aqueous pores in an engineered membrane, thereby allowing us to model electroporation. In a membrane, the creation and destruction of thousands of aqueous pores occurs as a result of thermal fluctuations. The Smoluchowski-Einstein encodes the mechanical properties of the membrane and how the membrane reacts to changes in the transmembrane potential. The transmembrane potential is the potential difference across the membrane that results from ionic gradients.

The Smoluchowski-Einstein equation is an advection-diffusion partial differential equation that governs the probability distribution of the number of aqueous pores as a function of their radius r and time t [25–27]. Denoting $n(r, t)$ as the pore density distribution function, then the Smoluchowski-Einstein equation asserts:

$$\frac{\partial n}{\partial t} = D\frac{\partial}{\partial r}\Big[\frac{n}{k_B T}\frac{\partial W(r,n)}{\partial r} + \frac{\partial n}{\partial r}\Big] + S(r,n), \tag{12}$$

where D is the diffusion coefficient of pores, k_B is the Boltzmann constant, T is the temperature, $W(r,n)$ is the change in energy of the membrane to create a pore of radius r, and $S(r,n)$ models the creation and destruction rate of pores. Equation 12 describes the time-evolution of the probability density function of the number and size of pores as a function of temperature and the energy required to form the pore. Intuition tells us that if we increase the temperature of the membrane then this should promote the formation of pores. Interestingly from (12) we see that an increase in temperature T causes a decrease in the contribution from the energy required to form a pore; however, the source term $S(r,n)$ for the creation of pores must increase as T increases. Typical initial and boundary conditions for (12) can be found in [25, 28–30]. A typical assumption is that initially for a membrane with zero transmembrane potential $V_m = 0$, all pores have radius r_*, and that any pores $r < r_*$ close rapidly. Therefore, the solution to (15) can be estimated by solving (12) to (14) with initial condition $n(r,0) = G_m(0)/G_p(r_*)$, and absorbing boundary condition $n(r,t) = 0$ if $r < r_*$.

All pores that are formed by electroporation are aqueous pores (water filled pores that cross the membrane). The creation of these aqueous pores follows a two step process. First, a hydrophobic pore is formed in which the the lipid tails are brought into contact with the water. The hydrophobic pore either spontaneously converts into a stable hydrophilic pore or spontaneously collapses [29, 31, 32]. A hydrophilic pore is an aqueous pore in which the perimeter of the pore is composed of the head-groups of the lipids.

The creation and destruction of aqueous pores $S(r,n)$ can be modeled using:

$$S(r,n) = v_c h_m \frac{\partial U}{\partial r} e^{U/k_B T} - v_d n H(r_* - r) \tag{13}$$

where U is the energy of a non-conducting hydrophilic generally given in terms of modified Bessel's functions, v_c is the attempt-rate density, v_d is the destruction rate density (e.g. frequency of lipid fluctuations), and r_* denotes the minimum radius for the transition from a hydrophilic pore to a hydrophobic pore. The Heaviside step function $H(r_* - r)$ in (13) accounts for the assumption that non-conducting pores with $r < r_*$ have a lifetime on the order of the lipid fluctuations.

The classical free energy model for a hydrophobic aqueous pore $W(r,n)$, in (12), in the membrane consists of four energy terms: the pore edge energy γ, the membrane surface tension $\sigma(n)$, the electrostatic interaction between lipid heads, and the transmembrane potential energy contribution $W_{\text{es}}(r, V_m)$. The pore energy $W(r,n)$ in (12) can be modeled using:

$$W(r,n) = 2\pi\gamma r - \pi\sigma(n)r^2 + (\frac{C}{r})^4 + W_{\text{es}}(r, V_m) + W_{\text{m}}, \tag{14}$$

with the energy contribution from the mechanobiological properties of the tethers included as W_{m}. The surface tension $\sigma(n)$ is dependent on n as the creation of pores will decrease the area occupied by the membrane. If we denote A_m as the area of the membrane, then the effective area of the membrane is given by $A'_m = A_m - A_p$ where A_p is the area of the membrane occupied by the pores and is dependent on the pored density distribution function n. As A_p increases the surface tension of the membrane is expected to decrease. However, if $A_p \ll 1$ then the surface tension dependence on n can be neglected with $\sigma(n) = \sigma$. The linkages of the tethers to the membrane are analogous to "springs" and act to restrain the enlargement of aqueous pores. This is similar to the experimentally measured results in [33] which suggests that irreversible electroporation cannot create pores that are larger than the cytoskeletal network anchors. Modelling the mechanical properties of the membrane as an elastic continuum and assuming a permanent tethered network anchorage, the effect of the tethers is accounted for via the energy required to deform the Hookean springs–formally, the energy contribution can be modelled using $W_m = 0.5K_{\mathrm{t}}r^2$ with K_t denoting the spring constant of the tethers [34]. Note, the energy model for tethers is identical to that of the cytoskeletal network presented in [35–38].

4.2 Equivalent Circuit Model of Electroporation

From the Smoluchowski-Einstein equation (the non-linear partial differential equations defined by (12) to (14)), it is possible to construct an equation for the membrane conductance $G_m(t)$ that accounts for the creation/destruction of aqueous pores. The Smoluchowski-Einstein equation provides the pore distribution function $n(r, t)$ over all possible pore radii as a function of time and transmembrane potential V_m. To compute the total membrane conductance $G_m(t)$ we require an estimate of each of the aqueous pore conductances. Denoting $G_p(r)$ as the conductance of an aqueous pore of size r, the associated membrane conductance can be computed by taking the expected value of the number of pores of radius r multiplied by the associated conductance $G_p(r)$. Formally, with E denoting mathematical expectation,

$$G_m(t) = E[G_p(r)] = \int_0^\infty G_p(r)n(r, t)dr. \tag{15}$$

Numerical estimates of the pore conductance $G_p(r)$ can be found in [13].

The current response of the EMP can be computed by plugging (15) into the fractional order macroscopic model (1). Notice that this requires the simultaneous solution of a coupled system composed of a non-linear partial differential equation and fractional order differential equations. Using singular perturbation approximations it is possible to convert this coupled system of equations into a set of fractional order differential equations. The key is to use the approximations provided in [28] to convert the equations defined by (12) to (14) into the following system of equations:

$$
\begin{aligned}
G_m(t) &= \sum_{i=1}^{\lfloor N(t) \rfloor} G_p(r_i), \\
\frac{dr_i}{dt} &= -\frac{D}{k_B T}\frac{\partial W(r, V_m)}{\partial r_i} \quad \text{for} \quad r_i \in \{1, 2, \ldots, \lfloor N(t) \rfloor\}, \\
\frac{dN}{dt} &= \alpha e^{(\frac{V_m}{V_{ep}})^2}\left(1 - \frac{N}{N_o} e^{-q(\frac{V_m}{V_{ep}})^2}\right),
\end{aligned}
\tag{16}
$$

where $G_p(r)$ is the conductance of a pore of radius r. Note that with the formulation in (16) it is assumed that the initial radius of all created pores is r_*. The complete dynamics of the EMP are given by simultaneously solving the system of ordinary differential equations (16) and the fractional order macroscopic model (1).

4.3 Membranes with Sterol Inclusions

Using the fractional order macroscopic model (1) and (16) with experimental measurements from the EMP allows the estimation of several important parameters related to electroporation. This includes the estimation of the membrane conductance G_m, double-layer capacitance C_{dl}, membrane capacitance C_m, and characteristic voltage of electroporation V_{ep}. In this section we present how variations in % m concentration of cholesterol (molar ratio of cholesterol to lipids) in the engineered artificial membrane effects the electroporation dynamics and the lumped circuit parameters G_m, C_m, and V_{ep}.

Parameter Estimation: Note that prior to all experimental measurements from the EMP, the membrane integrity and static parameters in the fractional order macroscopic model are evaluated using impedance measurements. That is, prior to performing any electroporation measurements, the equilibrium membrane conductance G_m at $V_m = 0$, membrane capacitance C_m, double-layer capacitance C_{dl}, fractional order parameter p, and electrolyte resistance R_e are all estimated from impedance measurements. Additionally, the electroporation parameters α, q, C, D, r_m are obtained from [27, 32, 39, 40], γ, σ from [41], and $W_{es}(r)$ and $G_p(r)$ from [13]. Therefore, the only parameter that is estimated from the electroporation measurements is the characteristic voltage of electroporation V_{ep}.

Background on Sterol Inclusions in Membranes: Cholesterol is an important lipid molecule that can be used to control the properties of engineered artificial membranes. Over the past decade several insights into how cholesterol impacts cell membrane properties have been achieved. For example, in [42], chronopotentiometry measurements of POPC (1-palmitoyl-2-oleoyl-sn-glycero-3-phosphocholine) membranes containing cholesterol illustrated that as the % of cholesterol increases from 0 to 73%, there is an increase in the membrane capacitance C_m and reduction in equilibrium membrane conductance $G_m(0)$. In [43], molecular dynamics simulations of DPPC (1, 2-dipalmitoyl-sn-glycero-3-phosphocholine) containing different % of cholesterol were used to estimate the minimum areal strain (pressure applied to stretch the membrane) required to cause the formation of an aqueous pore. It was found that for % cholesterol from 0 to 40%, the equilibrium conductance $G_m(0)$ decreases; however, for larger % of cholesterol $G_m(0)$ increases. This suggests that DPPC membranes with 40% cholesterol have the highest resistance to electroporation compared with DPPC membranes containing different molar concentrations of cholesterol. In [44], bilayer lipid membranes composed of DphPC (1, 2-diphytanoyl-sn-glycero-3-phosphocholine) were constructed with different % molar concentrations of cholesterol. In [44] it was illustrated that the characteristic voltage of electroporation V_{ep} increases as the % of cholesterol is increased from 0 to 10%. However, as the concentration of cholesterol increased from 10 to 40% there was a decrease in the characteristic voltage of electroporation V_{ep}. Note that the engineered artificial membrane, namley, the EMP is composed of 70% DphPC (zwittrionic C20 diphytanyl-ether-glycero-phosphatidylcholine) and 30% GDPE (C20 diphytanyl-diglyceride ether). Therefore, the EMP may have different V_{ep} characteristics compared to the DphPC membranes in [44].

The above gives insight into how the percent of cholesterol in POPC, DPPC, and DphPC membranes impacts the membrane capacitance C_m, equilibrium membrane conductance $G_m(0)$, and characteristic voltage of electroporation V_{ep}. However, the aim here is to study how varying the % of cholesterol in engineered artificial membranes impacts the membrane properties and electroporation dynamics.

Impedance Measurements: How does the molar % of cholesterol (molar ratio of cholesterol to lipids) effect the membrane capacitance C_m, characteristic voltage of electroporation V_{ep}, and the equilibrium membrane conductance $G_m(0)$ (the conductance at a transmembrane potential $V_m = 0$) in the tethered archaebacterial membrane of the EMP? Table 1 provides the estimated C_m, $G_m(0)$, and V_{ep} for different % concentrations of cholesterol in the EMP. Note that the parameter V_{ep} was estimated using electroporation measurements. From Table 1, we see that as the % cholesterol increases the associated membrane capacitance C_m tends to increase. This is in agreement with the results in [42] for POPC membranes containing cholesterol. Interestingly, the equilibrium membrane conductance $G_m(0)$ decreases for % cholesterol increasing from 0 to 30%; however, beyond 30% $G_m(0)$ increases. These results are in agreement with the results from molecular dynamics simulations in [43] for DPPC membranes. Notice however that the EMP does not have the same equilibrium membrane conductance characteristics as the POPC membrane in [42] where $G_m(0)$ decreased for increasing % of cholesterol from 0 to 73%. This is not surprising given the difference in the atomic structure of the negativley charged POPC lipids compared with the archaebacterial DphPC lipids in the EMP. In fact, it was illustrated in [45] that a 25% cholesterol POPC membrane has similar $G_m(0)$ to that of a 0% cholesterol DphPC membrane. At an atomistic level, a possible explanation for the decrease in

Table 1 Macroscopic model parameters for the EMP

% Cholesterol	C_m (nF)	$G_m(0)$ (μS)	V_{ep} (mV)
0	34.4	0.48	270
10	32.4	0.26	290
20	31.4	0.25	300
30	31.0	0.24	330
40	35.0	1.00	360
50	38.0	1.10	350

the equilibrium membrane conductance $G_m(0)$ for the archaebacterial membrane in the EMP beyond 30% cholesterol is the packing of the hydrocarbon chains in the DphPC and GDPE lipids. For archaebacterial membranes, molecular dynamics simulations suggest that the phytanyl tails form a tightly packed network with neighbouring hydrocarbon chains being interdigitated [46–48]. As such, the introduction of a low percentage of cholesterol (below 30%) likely causes the DphPC and GDPE molecules to condense the phospholipid network resulting in a decrease in $G_m(0)$. However, for larger % cholesterol beyond 30% the interdigitation of the phytanyl tails decreases causing an increase in the permeability of the EMP membrane. Note that ab-initio molecular dynamic simulations are required to perform analysis on the atomistic dynamics of how cholesterol impacts the properties of membranes. Linking molecular dynamics to the fractional order macroscopic model parameters is an active area of research that requires the construction of multi-physics models.

The characteristic voltage of electroporation V_{ep} for the EMP membrane increases for increasing % cholesterol in the membrane from 0 to 40%. However, at 50% we see a slight decrease in V_{ep}. These results are in agreement with the molecular dynamics results of DPPC membranes presented in [43]. It is interesting to note that the V_{ep} results for the EMP membrane are quite different from that of the DphPC membrane in [44] where V_{ep} increases for % cholesterol in the range of 0–10%, but then decreases for higher molar % of cholesterol. This illustrates the unique effects cholesterol can have on different archaebacterial membranes.

Electroporation Measurements: Given the fractional order macroscopic model with the parameters in Tables 1 and 2, Fig. 7 provides the experimentally measured current from the EMP and the numerically predicted current $I(t)$. Additionally, Fig. 7 provides the predicted double-layer charge $Q_{dl}(t)$, membrane charge $Q_m(t)$, and membrane conductance $G_m(t)$ for different molar % concentrations of cholesterol. As seen from Fig. 7b, the computed current $I(t)$ response is in excellent agreement with the experimentally measured current response of the EMP. As expected since $C_{dl} > C_m$, Fig. 7c illustrates that the charge accumulation at the electrodes is larger then at the surface of the membrane. Comparing Fig. 7c, d, we can estimate the

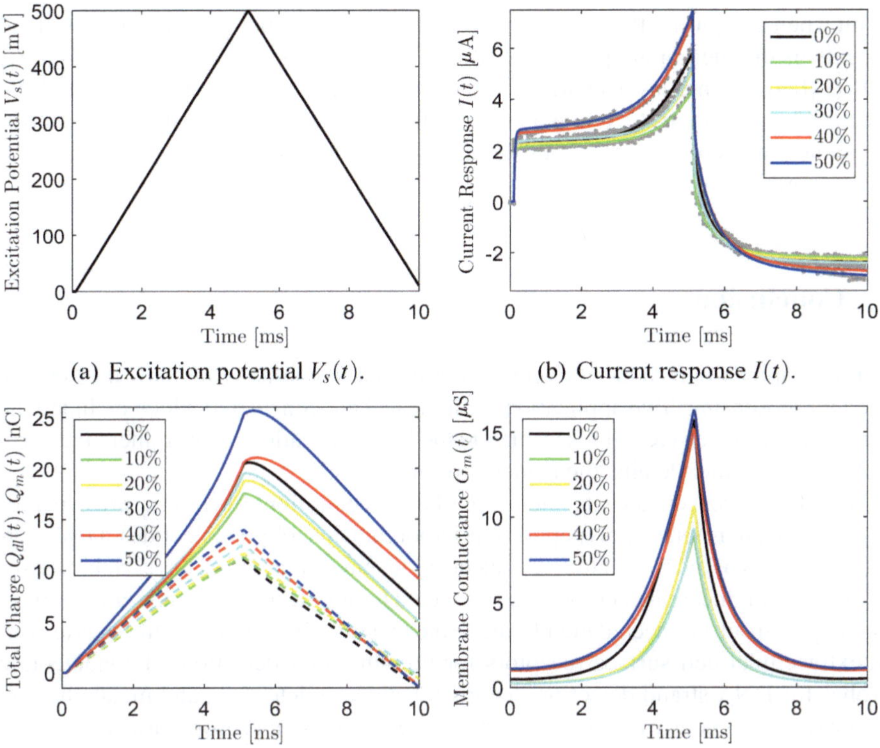

(a) Excitation potential $V_s(t)$.

(b) Current response $I(t)$.

(c) Charge accumulation Q_m (dashed lines) and Q_{dl} (solid lines) at the membrane and electrodes respectivley.

(d) Membrane conductance $G_m(t)$.

Fig. 7 Experimentally measured and numerically predicted current response for the EMP containing different mol % concentrations of cholesterol in the range of 0–50%. The experimentally measured current response is represented by the gray dots, and the predicted current response from (1) and (16) by the colored lines. The parameters for the numerical predictions are given in Table 1. The excitation potential $V_s(t)$ is identical for all experiments and is given by a linearly increasing voltage with rate of 100 V/s for 5 ms, then a linearly decreasing voltage with a rate of -100 V/s for 5 ms

amount of charge Q_m on the engineered membrane required to allow the formation of aqueous pores in the membrane. If no aqueous pores are formed then the membrane conductance would equal the equilibrium membrane conductance $G_m(0)$ in Table 1. For the EMP with % cholesterol in the range of 0–20%, the charge required to form an aqueous pore is $Q_m \approx 2.5$ nC, for 30% the charge is $Q_m \approx 2.8$ nC, and for 40–50% the charge is $Q_m \approx 3.5$ nC. This suggests that the stability of the EMP membrane increases with increasing % of cholesterol. However, if we look at the dynamics of the membrane conductance $G_m(t)$ in Fig. 7d, we see that the EMP membrane most resistant to the process of electroporation is the 10% cholesterol EMP which has a maximum conductance of 8.9 μS at $t = 5$ ms. This illustrates an important point when comparing the membrane resistance to electroporation, it depends on all aspects of the membrane properties (C_m, $G_m(0)$, γ, σ, α, q, C, D, r_m, V_{ep}, $W_{es}(r)$, and $G_p(r)$), double-layer charging effects (C_{dl} and p) of the electrodes, and the electrolyte resistance R_e. Using a single value to compare the resistance to electroporation is not sufficient to establish which membranes have a higher resistance to electroporation.

5 Conclusion

In this chapter we introduced engineered artificial membranes and how they can be used for biosensing. Two application examples where considered which included the Ion Channel Switch (ICS) biosensor, and the Electroporation Measurement Platform (EMP). For further details, the reader is referred to [2].

The ICS biosensor can detect femto-molar concentrations of target species including proteins, hormones, polypeptides, microorganisms, oligonucleotides, DNA segments, and polymers in cluttered electrolyte environments. The ICS biosensor employees engineered receptor sites connected to mobile gA monomers and biotinylated lipids in an engineered membrane. Note that over 50 different antibodies as well as extracellular cell surface receptors for growth factor detection, oligonucleotide probes for DNA strand detection, lectins for glucose detection, and metal chelates for heavy metal detection have been utilized in the ICS biosensor. Further receptors can be designed by exploiting the receptor association dynamics discussed in [49]. The major requirement is an attachment chemistry that will not inactivate the receptor for a particular analyte.

The EMP is a tethered membrane platform that provides a fully controllable physiological environment with a tethered membrane. The tethers in the EMP mimic the response of the cytoskeletal network in cells. Recall that all cells contain a cytoskeletal network that provides structural support for the biological membrane. Using cells provides a physiological system for validation; however, it is impossible to fully define the physiological environment which effects properties associated with electroporation. This motivates the need for the EMP which gives the experimentalist control over tethering density, membrane composition, and physiological environment.

The fractional order macroscopic models presented in this chapter provide a basis to study several interesting phenomenon that occur in engineered artificial membranes. These include, measuring the dynamics of embedded ion channels, estimating analyte concentrations from experimental measurements from the ICS biosensor, and studying the dynamics of electroporation in membranes. Future research in the area of engineered artificial membrane modeling includes constructing multiphysics models which combine the fractional order macroscopic model with continuum and molecular dynamics models. The continuum and molecular dynamics models can be used to estimate parameters of the fractional order macroscopic models from first-principles. Abstractly, the interaction of the macroscopic, mesoscopic, microscopic, and quantum mechanics models is given by

$$\begin{aligned} I(t) &= \text{macroscopic}\,(V(t), G_m, C_m, C_{dl}, p, R_e) \\ G_m, \ldots, R_e &= \text{mesoscopic}\,(D, \gamma, \sigma, \ldots, h_m, h_r) \\ D, \gamma, \sigma, \ldots, h_m, h_r &= \text{microscopic}\,(U, T, P) \\ U &= \text{quantum mechanics} \end{aligned}$$

where U is the interaction potential between atoms, T is the temperature, P is the temperature, h_m is the membrane thickness, and h_r is the tethering reservoir thickness. The interaction potentials between atoms can be estimated from quantum mechanics which has no free parameters. This potential is then used in the microscopic model to estimate important parameters for the mesoscopic models, which are then used to compute the parameters of the macroscopic model. In this chapter we have focused on the macroscopic model of engineered tethered membranes. For interested readers, detailed discussion of continuum and molecular dynamics models of engineered artificial membranes is provided in [2].

The construction of multi-physics models and novel engineered artificial membrane chemistries provide a rational approach to the in vitro study of biological membranes, and the design of biosensors. This is an active area of research in the fields of metabolic engineering, food safety, medicine, and security. The ultimate goal is the design of low-cost biosensors that allow for rapid molecular detection of specific analyte species, and artificial membrane platforms that can be used for rational drug design.

Appendix

All experimental measurements using the ICS and EMP, unless otherwise stated, were conducted at 20 °C in a phosphate buffered solution with a pH of 7.2, and a 0.15 M saline solution composed of Na^+, K^+, and Cl^-. At this temperature the tethered membrane is in the liquid phase. A pH of 7.2 was selected to match that typically found in the cellular cytosol of real cells. The forward and reverse reaction rates in Table 2 are obtained from [10, 23, 50]. The electroporation parameters α, q, C, D, r_m are obtained from [27, 32, 39, 40], γ, σ from [41], and $W_{es}(r)$ and $G_p(r)$ from [13]. Using impedance measurements it was determined that the fractional order operator p is in the range of 0.95 to 0.98 suggesting that a diffusion-limited process is present at the bioelectronic gold interface of the tethered archaebacterial membrane. The associated capacitance C_{dl} is in the range of 120–180 nF.

Table 2 Model parameters for ICS biosensor and EMP

Symbol	Definition	Value
L_w	Width of flow chamber	3.0 mm
L	Length of flow chamber	0.7 mm
h	Height of flow chamber	100 μm
$c^w(0), c^x(0), c^y(0), c^z(0)$	Initial concentration	0 mol/m^2
A	Inlet analyte concentration of hCG	10–353 nM
$c(0)$	Mobile gA monomers	1.6 pmol/m^2
$s(0)$	Tethered gA monomers	16 pmol/m^2
$b(0)$	Tethered binding sites	3 pmol/m^2
$d(0)$	gA dimers	33 pmol/m^2
$f_1 = f_2 = f_6$	Forward reaction rate	5×10^2 m^3/smol
$f_3 = f_4$	Forward reaction rate	3×10^9 m^2/smol
$f_5 = f_7$	Forward reaction rate	3×10^{11} m^2/smol
$r_1 = r_2 = r_6$	Reverse reaction rate	10^{-4} s^{-1}
$r_3 = r_4$	Reverse reaction rate	10^{-4} s^{-1}
$r_5 = r_7$	Reverse reaction rate	1.0×10^{-2} s^{-1}
Q	Flow rate	100 μL/min
γ	Edge energy	1.2×10^{-11} J/m
σ	Surface tension	15×10^{-3} J/m^2
α	Creation rate coefficient	1×10^9 s^{-1}
q	$q = (r_m/r_*)^2$	2.46
C	Steric repulsion constant	9.67×10^{-15} J$^{1/4}$ m
D	Radial diffusion coefficient	1×10^{-14} m^2/s
r_m	Equilibrium pore radius	0.8 nm
$G_p(r_m)$	Equilibrium pore conductance	1.56 nS
R_e	Electrolyte resistance	200 Ω
$W_{es}(r)$	Electrostatic Energy	Refer to [13]
$G_p(r)$	Pore conductance	Refer to [13]

References

1. B. Cornell, V. Braach-Maksvytis, L. King, P. Osman, B. Raguse, L. Wieczorek, R. Pace, Nature **387**, 581 (1997)
2. W. Hoiles, K. Krishnamurthy, B. Cornell, *The Dynamics of Engineered Artificial Membranes and Biosensors* (Cambridge University Press, 2017)
3. F. Heinrich, T. Ng, D. Vanderah, P. Shekhar, M. Mihailescu, H. Nanda, M. Losche, Langmuir **25**(7), 4219 (2009)
4. D. McGillivray, G. Valincius, D. Vanderah, W. Febo-Ayala, J. Woodward, F. Heinrich, J. Kasianowicz, M. Lösche, Biointerphases **2**(1), 21 (2007)
5. C. Cranfield, B. Cornell, S. Grage, P. Duckworth, S. Carne, A. Ulrich, B. Martinac, Biophys. J. **106**(1), 182 (2014)

6. B. Raguse, V. Braach-Maksvytis, B. Cornell, L. King, P. Osman, R. Pace, L. Wieczorek, Langmuir **14**(3), 648 (1998)
7. J. Prashar, P. Sharp, M. Scarffe, B. Cornell, J. Mater. Res. **22**(08), 2189 (2007)
8. S. Oh, B. Cornell, D. Smith, G. Higgins, C. Burrell, T. Kok, Biosens. Bioelectron. **23**(7), 1161 (2008)
9. G. Woodhouse, L. King, L. Wieczorek, P. Osman, B. Cornell, J. Mol. Recognit. **12**(5), 328 (1999)
10. V. Krishanmurthy, S. Monfared, B. Cornell, IEEE Trans. Nanotechnol. (Special Issue on Nanoelectronic Interface to Biomolecules and Cells) **9**(3), 303 (2010)
11. M. Vela, H. Martin, C. Vericat, G. Andreasen, Hern, J. Phys. Chem. B **104**(50), 11878 (2000)
12. W. Hoiles, V. Krishnamurthy, B. Cornell, IEEE Trans. Biomed. Circuits Syst. **PP**(99), 1 (2014)
13. W. Hoiles, V. Krishnamurthy, C. Cranfield, B. Cornell, Biophys. J. **107**(6), 1339 (2014)
14. S. Taylor, E. Gileadi, Corrosion **51**(9), 664 (1995)
15. A. Allagui, T. Freeborn, A. Elwakil, B. Maundy, Sci. Rep. **6** (2016)
16. I. Jesus, T. Machado, Nonlinear Dyn. **56**(1–2), 45 (2009)
17. Z. Hughes, T. Walsh, J. Colloid Interface Sci. **436**, 99 (2014)
18. H. Jafari, V. Daftardar-Gejji, J. Comput. Appl. Math. **196**(2), 644 (2006)
19. S. Momani, Z. Odibat, J. Comput. Appl. Math. **207**(1), 96 (2007)
20. K. Diethelm, A. Freed, Forschung und wissenschaftliches Rechnen **1999**, 57 (1998)
21. S. Das, *Functional Fractional Calculus* (Springer Science & Business Media, 2011)
22. W. Hoiles, V. Krishnamurthy, IEEE Trans. Mol. Biol. Multi-scale Commun. **1**(3), 265 (2015)
23. V. Krishanmurthy, S. Monfared, B. Cornell, IEEE Trans. Nanotechnol. (Special Issue on Nanoelectronic Interface to Biomolecules and Cells) **9**(3), 313 (2010)
24. H. Khalil, *Nonlinear Systems* (Prentice Hall, 2002)
25. V. Pastushenko, Y. Chizmadzhev, V. Arakelyan, J. Electroanal. Chem. Interfacial Electrochem. **104**, 53 (1979)
26. A. Barnett, J. Weaver, Bioelectrochem. Bioenerg. **25**(2), 163 (1991)
27. S. Freeman, M. Wang, J. Weaver, Biophys. J. **67**(1), 42 (1994)
28. J. Neu, W. Krassowska, Phys. Rev. E **59**, 3471 (1999)
29. J. Neu, W. Krassowska, Phys. Rev. E **67**, 021915 (2003)
30. J. Neu, W. Krassowska, Phys. Rev. E **74**, 031917 (2006)
31. I. Abidor, V. Arakelyan, L. Chernomordik, Y. Chizmadzhev, V. Pastushenko, M. Tarasevich, J. Electroanal. Chem. Interfacial Electrochem. **104**, 37 (1979)
32. R. Glaser, S. Leikin, L. Chernomordik, V. Pastushenko, A. Sokirko, Biochimica et Biophysica Acta (BBA)-Biomembranes **940**(2), 275 (1988)
33. D. Chang, T. Reese, Biophys. J. **58**(1), 1 (1990)
34. W. Sung, P. Park, Biophys. J. **73**, 1797 (1997)
35. C. Kanthou, S. Kranjc, G. Sersa, G. Tozer, A. Zupanic, M. Cemazar, Mol. Cancer Therapeutics **5**(12), 3145 (2006)
36. J. Teissie, M. Rols, Ann. New York Acad. Sci. **720**(1), 98 (1994)
37. M. Rols, J. Teissié, Biochimica et Biophysica Acta (BBA)-Biomembranes **1111**(1), 45 (1992)
38. C. Rosazza, J. Escoffre, A. Zumbusch, J. Rols, M, Mol Ther **19**(5), 913 (2011)
39. W. Krassowska, P. Filev, Biophys. J. **92**(2), 404 (2007)
40. K. Smith, J. Neu, W. Krassowska, Biophys. J. **86**(5), 2813 (2004)
41. W. Hoiles, R. Gupta, B. Cornell, C. Cranfield, V. Krishnamurthy, PloS One **11**(10), e0162790 (2016)
42. M. Naumowicz, A. Figaszewski, Cell Biochem. Biophys. **66**(1), 109 (2013)
43. T. Shigematsu, K. Koshiyama, S. Wada, Chem. Phys Lipids **183**, 43 (2014)
44. I. Uitert, S. Gac, A. Berg, Biochimica et Biophysica Acta (BBA)-Biomembranes **1798**(1), 21 (2010)
45. W. Knoll, *Handbook of Biofunctional Surfaces* (Pan Stanford Publishing, 2013)
46. T. Husslein, D. Newns, P. Pattnaik, Q. Zhong, P. Moore, M. Klein, J. Chem. Phys. **109**(7), 2826 (1998)
47. W. Shinoda, M. Mikami, T. Baba, M. Hato, J. Phys. Chem. B **107**(50), 14030 (2003)

48. K. Shinoda, W. Shinoda, T. Baba, M. Mikami, J. Chem. Phys. **121**(19), 9648 (2004)
49. D. Lauffenburger, J. Linderman, *Receptors: Models for Binding, Trafficking, and Signaling* (Oxford University Press, 1993)
50. S. Monfared, V. Krishnamurthy, B. Cornell, Biosens. Bioelectron. **34**, 261 (2012)

MIX
Papier aus verantwortungsvollen Quellen
Paper from responsible sources
FSC® C105338

If you have any concerns about our products,
you can contact us on
ProductSafety@springernature.com

In case Publisher is established outside the EU,
the EU authorized representative is:
Springer Nature Customer Service Center GmbH
Europaplatz 3, 69115 Heidelberg, Germany

Printed by Libri Plureos GmbH
in Hamburg, Germany